An Introduction to Plasmonics

Advanced Textbooks in Physics

Print ISSN: 2059-7711
Online ISSN: 2059-772X

The *Advanced Textbooks in Physics* series explores key topics in physics for MSc or PhD students.

Written by senior academics and lecturers recognised for their teaching skills, they offer concise, theoretical overviews of modern concepts in the physical sciences. Each textbook comprises of 200–300 pages, meaning content is specialised, focussed and relevant.

Their lively style, focused scope and pedagogical material make them ideal learning tools at a very affordable price.

Published

An Introduction to Plasmonics
by Olivier Pluchery & Jean-François Bryche

Principles of Astrophotonics
by Simon Ellis, Joss Bland-Hawthorn & Sergio Leon-Saval

Topics in Statistical Mechanics (Second Edition)
by Brian Cowan

Physics of Electrons in Solids: Design and Applications
by Jean-Claude Tolédano

Astronomical Spectroscopy: An Introduction to the Atomic and Molecular Physics of Astronomical Spectroscopy (Third Edition)
by Jonathan Tennyson

A Guide to Mathematical Methods for Physicists: Advanced Topics and Applications
by Michela Petrini, Gianfranco Pradisi & Alberto Zaffaroni

Quantum States and Scattering in Semiconductor Nanostructures
by Camille Ndebeka-Bandou, Francesca Carosella & Gérald Bastard

An Introduction to Particle Dark Matter
by Stefano Profumo

More information on this series can also be found at https://www.worldscientific.com/series/atip

Advanced Textbooks in Physics

An Introduction to Plasmonics

Olivier Pluchery
Sorbonne University, France
Jean-François Bryche
CNRS, France & Sherbrooke University, Canada

World Scientific
NEW JERSEY • LONDON • SINGAPORE • BEIJING • SHANGHAI • HONG KONG • TAIPEI • CHENNAI • TOKYO

Published by

World Scientific Publishing Europe Ltd.
57 Shelton Street, Covent Garden, London WC2H 9HE
Head office: 5 Toh Tuck Link, Singapore 596224
USA office: 27 Warren Street, Suite 401-402, Hackensack, NJ 07601

Library of Congress Cataloging-in-Publication Data
Names: Pluchery, Olivier, author. | Bryche, Jean-François, author.
Title: An introduction to plasmonics / Olivier Pluchery, Sorbonne University, France,
Jean-François Bryche, CNRS, France & Sherbrooke University, Canada.
Description: New Jersey : World Scientific, 2024. | Series: Advanced textbooks in physics,
2059-7711 | Includes bibliographical references and index.
Identifiers: LCCN 2022040620 | ISBN 9781800613393 (hardcover) |
ISBN 9781800613409 (ebook) | ISBN 9781800613416 (ebook other)
Subjects: LCSH: Plasmonics.
Classification: LCC TK8585 .P59 2024 | DDC 621.36/5--dc23/eng/20221115
LC record available at https://lccn.loc.gov/2022040620

British Library Cataloguing-in-Publication Data
A catalogue record for this book is available from the British Library.

First published 2024 (hardcover)
Reprinted 2026 (in paperback edition)
ISBN 9781800619784 (pbk)

For any available supplementary material, please visit
https://www.worldscientific.com/worldscibooks/10.1142/Q0395#t=suppl

Desk Editors: Aanand Jayaraman/Adam Binnie/Shi Ying Koe

Typeset by Stallion Press
Email: enquiries@stallionpress.com

To my father,
To my good friend Joseph.
— Olivier Pluchery

To my family for their unwavering support.
— Jean-François Bryche

Preface

The first time I discovered plasmonics was during my PhD thesis from my adviser, A. Tadjeddine, in 1998. He told me about the beauty of coupling the beam of a laser with a strange surface wave that he called *plasmon*, and described how this wave was able to reveal the presence of a few molecules on a surface. He was among the pioneers of plasmonics, with his first published article in 1977, when he generated a plasmon wave to investigate the electrochemical interface. He also explained the dispersion curve with tears in his eyes, which definitively struck me. A "plasmonic" seed was planted in my mind and ever since then I have met other colleagues and fellow researchers, who seem to have developed the same kind of sentimental relationship with plasmonics.

Plasmonics emerged as a distinct field in the research community during the 1990s, which flourished at the frontier between optics and condensed matter physics. It spans chemistry, biophysics, and signal processing, reinforcing the idea that interdisciplinarity is a trademark of plasmonics. However, there is no well-established plasmonics course: It is sometimes mentioned in physics courses at the university level but usually as an illustrative example of electromagnetic surface waves in undergraduate courses (third year of *License* in the European system). Although the foundations of plasmonics can be found in reference textbooks, such as those by Born and Wolf[1] or Jackson,[2] there is not a single mention of plasmon wave propagation.

Since I was "bitten" by the passion for plasmonics, I decided to tackle the question from the experimental perspective, and I prepared lab work for my students. I was a freshly recruited Associate Professor at Sorbonne University (at that time known as University Pierre et Marie Curie), and I wanted to lead my students into modern and active research fields. In 2006, I started to assemble the first lab work illustrating the surface plasmon polariton. It evolved over the years, culminating in an article published in 2011 by two master's students. Eventually, a commercial setup was proposed in collaboration with a small company, *Dida Concept.* In parallel, I also became interested in gold nanoparticles and the localized surface plasmon resonance, which then emerged into lab work for my second-year master's students. It also progressed, and with my colleague and friend Catherine Louis, we launched a very active research network called Or-Nano (*Gold-Nano* in French), devoted to interdisciplinary research on gold nanoparticles. At that stage, I decided to propose a structured course on plasmonics, which proved difficult, since there was no core textbook.

In terms of books dealing with plasmonics, an impressive 80 books were published between 2004–2021, along with more than 700 book chapters. So, why write a new one? Despite the large number of publications, it is very difficult to find a unified presentation of this field that is accessible to university students. Most of the books are compilations of authoritative works by renowned researchers and illustrate how creative and attractive this field can be regarding research development. *An Introduction to Plasmonics* aims at tackling this deficiency, with our goal being to provide a self-standing course on plasmonics.

In conclusion, I would like to thank the many colleagues and friends who have accompanied me on my plasmonic journey. Thanks to Jean-Michel Courty and Agnès Maître who encouraged me in developing the lab work

[1] Born and Wolf, *Principle of Optics*, 1st edition 1959, 7th edition 1999.
[2] Jackson, *Classical Electrodynamics*, 1st edition 1965, 3rd edition 1998.

on plasmonics. Thanks to Souhir Boujday for the passionate discussions on LSPR. I am also deeply grateful to Catherine Louis for having urged me to embark on the gold nanoparticle adventure. Thanks to Anna Lévy and many other colleagues for critically evaluating some of the chapters, and thanks to all my students for their feedback on the early stages of this course.

Olivier Pluchery
Paris, France

Nature is full of mysterious phenomena that science studies and then, once understood, reproduces and uses to create new phenomena. Among them, plasmonics is an exciting field that I discovered during a lecture given by E. Boer-Duchemin in 2011. Plasmonics phenomena are observed at interfaces that, under appropriate conditions, enable the control of light at the nanoscale.

With this idea in mind, I started my PhD in 2013 in plasmonic biosensors under the supervision of two great directors, Michael Canva from the Laboratoire Charles Fabry and Bernard Bartenlian from the Centre de Nanosciences et de Nanotechnologies. Each of them, through their different personalities, brought me their vision of research: a dynamism associated with interdisciplinary and collaborative approach as well as a quiet strength related to the meticulousness of a specific field. Their teams were pioneers in the realization of plasmonic biosensors. Michael Canva has reached a limit in the detection of SPR biosensors with uniform gold films. This thesis project only strengthened my interest in plasmonics—an interdisciplinary field where one can combine fundamental approaches, exploratory studies, and fascinating experiments. After my PhD, wishing to continue in the field of plasmonics, I started a postdoctoral position in thermoplasmonics and ultrafast phenomena. This work, done in Sherbrooke, Canada, was a worthwhile challenge, undertaken with the teams of professors Paul Charette and Denis Morris. This challenge turned into a great success with the transition in 2021 to a CNRS researcher position at the *Laboratoire de Nanotechnologies et Nanosystèmes.*

I have been involved in teaching through mentoring and lectures throughout the years, through which I aim to impart my passion for this subject. It is always highly enriching to give a lecture and to stimulate and encourage students' curiosity while at the same time developing their critcal thinking. In this way, I hope that this book on plasmonics will allow students to discover or deepen their knowledge and see the wide potential of this field.

Thanks to Michael Canva and Bernard Bartenlian for their trust through all these years and for mentoring and introducing me to plasmonics research. Thanks to Éric Le Moal and Julien Moreau for the wonderful discussions around plasmonics excitation and with whom I enjoyed working on many collaborations. Thanks to Paul Charette and Denis Morris for welcoming me to the thermoplasmonics project. Thanks to Philippe Gogol for the passionate discussions on plasmonics and mode coupling, and for the critical proofreading of some of the chapters. Thanks to Olivier Pluchery for his trust in my participation in this work, which was quite an adventure and an impressive challenge in my young career.

I conclude these greetings by indicating that there are still many people who deserve a few words of acknowledgment because science is the fruit of common work, and I am sure that they are aware, through the exchanges that I had with them, of my gratitude.

Jean-François Bryche
Sherbrooke, Canada

About the Authors

Olivier Pluchery is currently Professor of Physics at Sorbonne University in Paris, where he has taught optics, electromagnetism, and plasmonics since 2002. He leads a research group at the Institute of Nanosciences in Paris, which focuses on the understanding of the singular properties of gold nanoparticles studied at the single-object level: charge transfer induced by molecular layers, conductivity, and plasmonic properties. He spent sabbaticals at UC Berkeley and at The University of Texas at Dallas. He is Director of the national research network Or-Nano, dealing with the many applications of gold nanoparticles. He is the editor, with Pr. C. Louis, of *Gold Nanoparticles for Physics, Chemistry and Biology* (two editions), and has authored more than 60 research articles. He received his PhD from University of Paris-Saclay in 2000 in laser physics and did a postdoctoral research stay at Bell Labs (New Jersey) https://www.linkedin.com/in/olivier-pluchery-0703718/.

Jean-François Bryche has been a CNRS researcher since 2022 at the *Laboratoire de Nanotechnologies et Nanosystèmes* (LN2), an international research laboratory funded by France and Canada, at the University of Sherbrooke. Previously, he was Research Assistant Professor at Sherbrooke University, where he taught optics, electronics, and plasmonics. He obtained his PhD in 2016 at the Université Paris-Saclay in France in the field of nanostructured plasmonic biosensors. During a first postdoctoral stay at the University of Sherbrooke, he contributed to the field of thermoplasmonics.

Then, he focused on electrically driven nanosources of light during a second postdoc at the Institute for Molecular Science of Orsay in France. His research interests include plasmonics, nanophotonics, surface-enhanced Raman scattering, biosensors, near-field optics, biophotonics, nanofabrication, and thermoplasmonics. He has authored more than 20 research articles.

Contents

Introduction

Plasmonics in a Few Simple Words

Plasmonics deals with the plasma formed by the free electrons in a conductive solid. To be precise, a plasma is a special state of matter in which electrons are detached from the nucleus of the atoms. This is the field of plasma physics, and it requires very high temperatures, similar to those found in the sun. At several millions degrees, the thermal energy can break the cohesion usually ensured by electrostatic forces in normal matter. As a result, electrical charges behave very differently from those in normal matter, and this rapid motion of charges induces an electromagnetic field. A plasma is therefore a medium with a strong mutual influence (called a coupling) between free charges and electromagnetic waves. This is indeed a violent world.

Plasmonics is not a part of plasma physics. However, there is a strong analogy between a plasma and a conductive solid (metal), and this is what gave its name to our topic. In a metal, some electrons are not tightly bound to the ionic cores. These electrons behave as if they were free electrons, solely forced to remain within the boundary of the piece of metal under consideration. The free electrons are very sensitive to the electromagnetic field of light. They oscillate and also re-emit a wave like an antenna. In the majority of cases, this interaction between light and free electrons prevents the light from penetrating the metal, and this explains the reflective power of pure metals in the visible range. It explains the shiny appearance of metals. However, there are two cases where the coupling of electrons with an electromagnetic wave generates a surface wave called the plasmon wave.

A Strange Surface Wave that Mixes Electrons and Photons

The first case is surface plasmon polariton (SPP), and it occurs at the interface between a flat metallic surface and a dielectric medium (i.e. insulating and transparent). Under special illumination conditions, a surface wave is generated that crawls along the surface. This wave is not radiative and is restricted to the near-field region (~100 nm away from the surface). The surface plasmon wave probes the proximity of an interface and is used to detect very localized surface events (see biosensor applications). The SPP waves result from a synchronized oscillation of the free electrons of the surface that couples with an electromagnetic surface wave. As a proof of this strong mutual coupling, the SPP waves can be either excited by the photons of a light beam or by the electrons of an electron beam. The first detection of an SPP wave was made using electron excitations by Ritchie in 1957, before the term "plasmon" had been invented [1].

Plasmonics and Nanotechnology

The second case of coupling between electrons and electromagnetic waves is the localized surface plasmon resonance (LSPR). It occurs when light impinges on a metallic object whose size is much smaller than the optical wavelength. Given that the average wavelength of the visible light is 500 nm, the object should have a dimension of approximately 50 nm. Therefore, this category of localized plasmons is generated in metallic nanostructures. Here, plasmonics joins two length scales: the typical length scales of electron phenomena, which is the ångström (0.1 nm), and the typical length scales of optical phenomena, which is 500 nm. This leads to the scale of plasmonics being between these two values, at about 10–100 nm. This means that plasmonics drives optics into the sub-wavelength scale. From that perspective, plasmonics is an entry point to nano-optics and a way to explore the sub-diffraction limit beyond the *Rayleigh criterion*. We show in the following chapters how the interaction of the optical wave with the free electrons of metals depends essentially on the boundary conditions of the nanostructure. The shape of the nanostructure controls the resonance wavelengths, the intensity of the local electric field that is sometimes amplified by several orders of magnitude.

Plasmonics: A Historical Link Between the Medieval Era and the Future

A vivid illustration of plasmonics is given by the intriguing colors of glasses produced by glassmakers and alchemists of antiquity through to the medieval age. In particular, they inserted gold into certain glasses, following delicate and sometimes secret protocols, to produce ruby-red glasses. The historical description of the adventure of gold nanoparticles is told in Chapter 1 of Ref. [2]. We have known since the experiments conducted by Faraday, and his related article in 1857, that this red color is due to the interaction of light with spherical gold nanoparticles [3]. But the mathematical derivation and the systematic scientific experiments had to wait until the Mie theory in 1908 [4], then for the chemical protocol of Turkevich in 1951 [5], and the demonstrative experiments of Van Duyne in 1977 [6] to prove this. The curiosity of the scientific community toward plasmonics thus began to grow. Since then, the interest had not waned, and plasmonics has grown into a solid scientific field, where the exploration of the laws of nature is very active among scientists. Applications have also reached the general public. In terms of applications, the LSPR has led to numerous proofs of concept in biosensors. One universally known biosensor based on gold nanoparticles is the antigen test kits used during the COVID-19 pandemic in 2021. The red color band used for checking the presence of the virus receives its color from gold nanoparticles that concentrate at the barrier placed on the cellulose paper [7]. The SPP has also found applications in high-speed optical signal processing. All these perspectives will be the topic of the last chapter of the book, which will continue to sustain the scientific curiosity of the research community.

Scope of This Book

The current textbook establishes the foundations of plasmonics and is aimed at master's or PhD students with a progressive explanation of the topic. It is divided into four parts. **The first part** consists of three chapters and is an overview of the main physical properties of electromagnetic waves interacting with metals. The focus is set on the properties useful for plasmonics and will be helpful for those who need to refresh their knowledge on the consequences of the Maxwell equations. **The second part** deals with the propagating plasmon, also called the SPP, and spans through Chapter 4, where the most important equations are derived; Chapter 5, with the consequences for plasmonic biosensing; and Chapter 6, with the applications to waveguides and signal processing. **The third part** treats the LSPR linked to metallic nanoparticles and starts with Chapters 7 and 8 by establishing the main properties of the LSPR. It is followed by a detailed discussion on biosensing applications in Chapter 9. Finally, **the last part** is a broad overview of the principal active research fields in plasmonics today, which will be the vibrant research fields of tomorrow. Twelve emerging topics are condensed into this Chapter 10. Let's stress that, to our knowledge, the books published on plasmonics during the last decade have set their focus on these emerging fields pushed by plasmonics. We have the conviction that the step-by-step introduction of our nine subsequent chapters is missing in the scientific literature. We hope that our textbook will contribute to filling this gap in the universal shelf of scientific knowledge.

References

[1] Ritchie R. H. 1957. Plasma losses by fast electrons in thin films. *Physical Review* **106**, 874–881.
[2] Louis C. and Pluchery O. (eds.) 2017. *Gold Nanoparticles for Physics, Chemistry and Biology (2nd Ed.)* (London: World Scientific).
[3] Faraday M. 1857. *Philosophical Transactions* **147**, 145.

[4] Mie G. 1908. Beiträge zur Optik trüber Medien, speziell kolloidaler Metallösungen. *Annalen der Physik* **25**, 377–445.
[5] Turkevich J., Stevenson P. C. and Hillier J. 1951. A study of the nucleation and growth processes in the synthesis of colloidal gold. *Discussion of the Faraday Society* **11**, 55–75.
[6] Jeanmaire D. L. and Van Duyne R. P. 1977. Surface raman spectroelectrochemistry: Part I. Heterocyclic, aromatic, and aliphatic amines adsorbed on the anodized silver electrode. *Journal of Electroanalytical Chemistry and Interfacial Electrochemistry* **84**, 1–20.
[7] Testard F. 2022. As measured by SAXS: The size of gold nanoparticles COVID tests is 28 nm (Official web page of CEA/IRAMIS). https://iramis.cea.fr/Phocea/Vie_des_labos/News/index.php?id_news=8386.

1

Key Concepts of Electromagnetism for Plasmonics

The foundations of the electromagnetic waves are summarized in this chapter. Starting from *the four Maxwell equations*, we show how the notion of propagation emerges from these equations, and we emphasize the case of *plane waves*. The *complex notation* for plane waves is detailed since it greatly simplifies the mathematical treatment. Wave propagation inside a material medium is also treated by introducing the *dielectric permittivity* that will be crucial for understanding plasmonics. Finally, the electromagnetic density of energy and the *Poynting vector* are defined.

1.1 What Is Light?

The question of the nature of light has puzzled humankind for a long time. Some intuitive approaches considered light to be a *simulacra* emitted by the surface of objects (Lucretius, Roman philosopher, first century BC). Then, the notion of light source was clarified: Light is emitted by some primary light sources (the sun, a light bulb, an LED) that have to be distinguished from secondary light sources (light diffused by objects that make them visible). But what is light really? Actually, it is easier to provide answers about the behavior of light and leave the question of its nature to philosophers. Three main approaches were elaborated over the centuries, as illustrated in Fig. 1.1: Light can be viewed as optical beams produced by light sources and deviated by objects: this is geometrical optics; or light

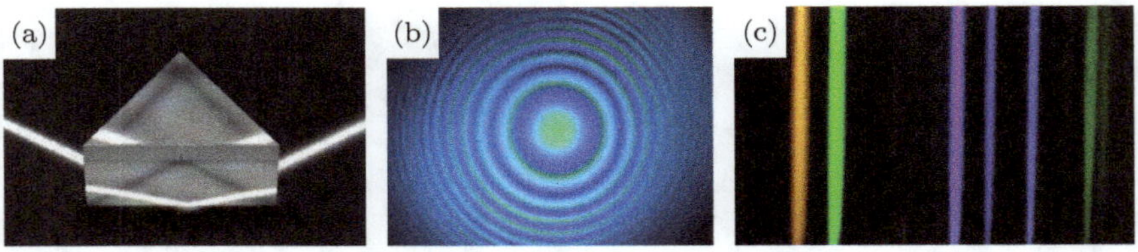

Fig. 1.1. The three descriptions of light: (a) *geometrical* optics based on light beams, where a glass prism is illuminated under conditions where total reflection occurs; (b) *wave* optics, where the interference is pattern generated by a Michelson interferometer in the parallel plates adjustment; (c) *photonic*, where the granular nature of the photons is illustrated by the spectral lines of a mercury lamp. Copyright © O. Pluchery.

can be modeled as electromagnetic waves propagating through vacuum or matter: this is the wave approach; or finally, it can be described as particles known as photons: this is photonics. These last two concepts of light might look incompatible: Can light be simultaneously a nonmaterial wave and a particle? This question gave rise to what is known as the wave–particle duality, which is again a philosophical issue. Anyway, in physics, we aim to describe how light behaves rather than discussing what it actually is. Therefore, when dealing with plasmonics, the most accurate description of light is based on the concept of waves, which will be used throughout this book. This chapter is a summary of the main concepts of electromagnetic waves, essential for understanding plasmonics. More details can be found in reference textbooks of electromagnetism, such as *Berkeley Physics Course* by Purcell [1], the book by Griffiths [2], or the one in French by Pérez [3]. The reference textbooks from Born and Wolf [4] and Jackson [5] are more demanding, and the work by Novotny and Hecht [6] focuses on nano-optics with some short discussions on plasmonics.

1.2 Maxwell Equations in Vacuum

Light, when considering its propagation and its interaction with matter, is accurately described by the oscillation of an *electric field* $\boldsymbol{E}$ and a *magnetic field* $\boldsymbol{B}$ (sometimes called *magnetic flux density*). These two fields are vectors, and their three components (E_x, E_y, and E_z for the electric field) depend on time t and space (x, y, z). Moreover, $\boldsymbol{E}$ and $\boldsymbol{B}$ are tightly

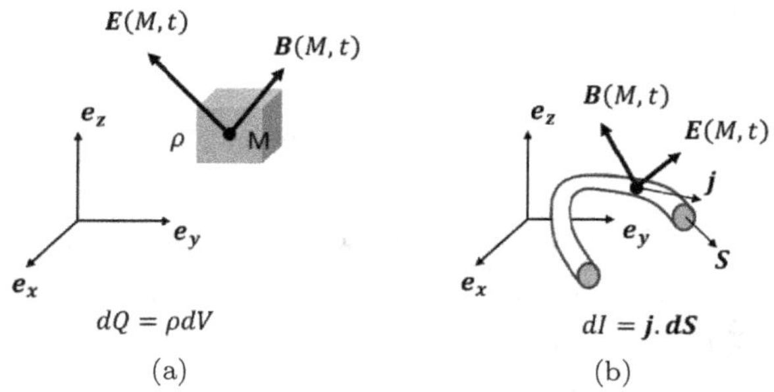

Fig. 1.2. (a) For a given charge density ρ at point M and time t, Maxwell equations allows deriving the electromagnetic fields $\boldsymbol{E}$ and $\boldsymbol{B}$. (b) The current density $\boldsymbol{j}$ also gives rise to similar electromagnetic fields. This current density is usually driven by a wire whose section is S so that the current passing through the wire is given by $I = \boldsymbol{j}.\boldsymbol{S}$.

bound and constitute the *electromagnetic field*. It took about two centuries to understand their mutual dependence, following the first experiments on magnetism (Gilbert, *De Magnete*, 1603) and on electricity (Galvani, *De virtutibus electricitatis*, 1791). The interplay between $\boldsymbol{E}$ and $\boldsymbol{B}$ is captured by the four Maxwell equations compiled between 1865 and 1884 by James Clerk Maxwell, Oliver Heaviside, and Josiah Willard Gibbs. We start from these founding equations.

We first consider the case of vacuum with free charges ρ and a possible current density denoted by $\boldsymbol{j}$. This could be vacuum or air outside of any solid medium, where there are a few fixed charges and an electric wire for driving the electric current $\boldsymbol{j}$. ρ and $\boldsymbol{j}$ are the *source terms* of the fields: Without charges and current, there is no generation of any fields (see Fig. 1.2), but these fields can propagate far away from these sources. This was the discovery by Heinrich Hertz in 1888 which then set the foundations of wave propagation.

Here are the four Maxwell equations, expressed in the SI system of units, in vacuum:

$$\boldsymbol{\nabla} \cdot \boldsymbol{E} = \frac{\rho}{\varepsilon_0} \quad \text{Maxwell–Gauss,} \tag{1.1}$$

$$\boldsymbol{\nabla} \cdot \boldsymbol{B} = 0 \quad \text{Maxwell–Thomson,} \tag{1.2}$$

$$\nabla \times \boldsymbol{E} = -\frac{\partial \boldsymbol{B}}{\partial t} \quad \text{Maxwell–Faraday,} \tag{1.3}$$

$$\nabla \times \boldsymbol{B} = \mu_0 \left(\boldsymbol{j} + \varepsilon_0 \frac{\partial \boldsymbol{E}}{\partial t} \right) \quad \text{Maxwell–Ampère.} \tag{1.4}$$

ε_0 is the dielectric permittivity of vacuum and μ_0 the magnetic permeability of vacuum. They are fundamental constants: $\varepsilon_0 = 8.854 \times 10^{-12}\,\text{F/m}$ and $\mu_0 = 1.257 \times 10^{-6}\,\text{H/m}$. Equation (1.1) is the basis of electrostatics, where the charge density ρ is the source of the electrostatic field $\boldsymbol{E}$. Equation (1.2) shows that there is no point source of magnetic field. Equation (1.3) states that the variation in the magnetic field over time generates an electric field, and Equation (1.4) states that a current $\boldsymbol{j}$ creates a magnetic field and so does a time-varying electric field. Note that Maxwell equations *locally* establish a relationship between the source terms ρ and $\boldsymbol{j}$ and the electromagnetic field. The integral form of this relationship allows calculating the values of $\boldsymbol{E}$ and $\boldsymbol{B}$ away from the sources. These integral forms are the topics of electrostatics and magnetostatics in the case of static charges, and they are covered by specific textbooks. We refrain from commenting further on these foundational equations.

We now aim to derive some important properties related to wave propagation. For this, we seek $\boldsymbol{E}$ and $\boldsymbol{B}$ that would obey the four Maxwell equations. Since $\boldsymbol{E}$ and $\boldsymbol{B}$ are coupled, it is sufficient to solve for one of them. Usually, in optics, we focus on the propagation of the electric field $\boldsymbol{E}$.

1.2.1 *The wave equation*

At this stage, we consider that there is no charge and no current (no sources in the free space): $\rho = 0$, $\boldsymbol{j} = 0$. Let's calculate the quantity $\nabla \times (\nabla \times \boldsymbol{E})$. The properties of vectorial operations state that $\nabla \times (\nabla \times \boldsymbol{E}) = \nabla(\nabla \cdot \boldsymbol{E}) - \boldsymbol{\Delta E}$. From Equation (1.1), we conclude that $\nabla \cdot (\nabla \cdot \boldsymbol{E}) = 0$. Using Equations (1.3) and (1.4), we write

$$\begin{aligned} \nabla \times (\nabla \times \boldsymbol{E}) = \nabla \times \left(-\frac{\partial \boldsymbol{B}}{\partial t} \right) = -\frac{\partial}{\partial t} \nabla \times \boldsymbol{B} = -\frac{\partial}{\partial t} \left(\mu_0 \varepsilon_0 \frac{\partial \boldsymbol{E}}{\partial t} \right) \\ = -\mu_0 \varepsilon_0 \frac{\partial^2 \boldsymbol{E}}{\partial t^2}. \end{aligned}$$

Finally, by equating the two forms obtained for $\boldsymbol{\nabla} \times (\boldsymbol{\nabla} \times \boldsymbol{E})$, we can write the **wave equation** in free space:

$$\boldsymbol{\Delta E} - \mu_0 \varepsilon_0 \frac{\partial^2 \boldsymbol{E}}{\partial t^2} = 0. \tag{1.5}$$

In the general case, $\boldsymbol{E}$ is a vector with three coordinates: $\boldsymbol{E} = E_x \boldsymbol{e}_x + E_y \boldsymbol{e}_y + E_z \boldsymbol{e}_z$, and each of these three coordinates depends on time t and space $\boldsymbol{r}$. This often leads to cumbersome calculations, so we are going to look at specific simplifying cases.

Maxwell equations and the wave equation are linear; therefore, any linear combination of solutions remains a solution. If we can find a set of easy solutions, we will be able to linearly combine these solutions to fabricate more advanced solutions. This is a strong motivation for focusing on plane wave harmonic solutions. These solutions can be expressed as a simple cosine function characterized by one frequency ω (circular frequency in rad/s).

1.2.2 *The plane wave*

We restrict ourselves to *propagative solutions* for Equation (1.5). Propagative functions are of the kind $f(x - ct)$, where c is the phase velocity of the wave. Figure 1.3(a) shows an example of the propagation of an arbitrary function f along the x direction. In a three-coordinate space, the propagative argument is written as $f(\alpha x + \beta y + \gamma z - ct)$, which can be condensed into $f(\boldsymbol{k} \cdot \boldsymbol{r} - \omega t)$. The vector $\boldsymbol{k}$ is the wave vector and plays a central role in wave propagation. Instead of any general function, we usually restrict ourselves to periodic functions, and we seek solutions of the form $\boldsymbol{E}_0 \cos(\boldsymbol{k} \cdot \boldsymbol{r} - \omega t)$. We also assume that the amplitude vector $\boldsymbol{E}_0$ is independent of time and space, and we write $\boldsymbol{E}_0 = E_0 \boldsymbol{e}_z$ (bold $\boldsymbol{E}_0$ is a vector, and normal E_0 is a scalar and is also constant). Therefore, $\boldsymbol{E}(\boldsymbol{r}, t) = E_0 \cos(\boldsymbol{k} \cdot \boldsymbol{r} - \omega t)\boldsymbol{e}_z$. Figure 1.3(b) shows such a wave when the electric field is along the vector $\boldsymbol{e}_z$. We now apply this possible solution to the wave equation (1.5), and we calculate $\boldsymbol{\Delta E} = (\partial^2 E_z/\partial x^2 + \partial^2 E_z/\partial y^2 + \partial^2 E_z/\partial z^2)\boldsymbol{e}_z$. Since $\cos(\boldsymbol{k} \cdot \boldsymbol{r} - \omega t) = \cos(k_x x + k_y y + k_z z - \omega t)$, we obtain $\boldsymbol{\Delta E} = -E_0 \boldsymbol{k}^2 \cos(\boldsymbol{k} \cdot \boldsymbol{r} - \omega t) = E_0(k_x^2 + k_y^2 + k_z^2)\cos(\boldsymbol{k} \cdot \boldsymbol{r} - \omega t)$. Finally,

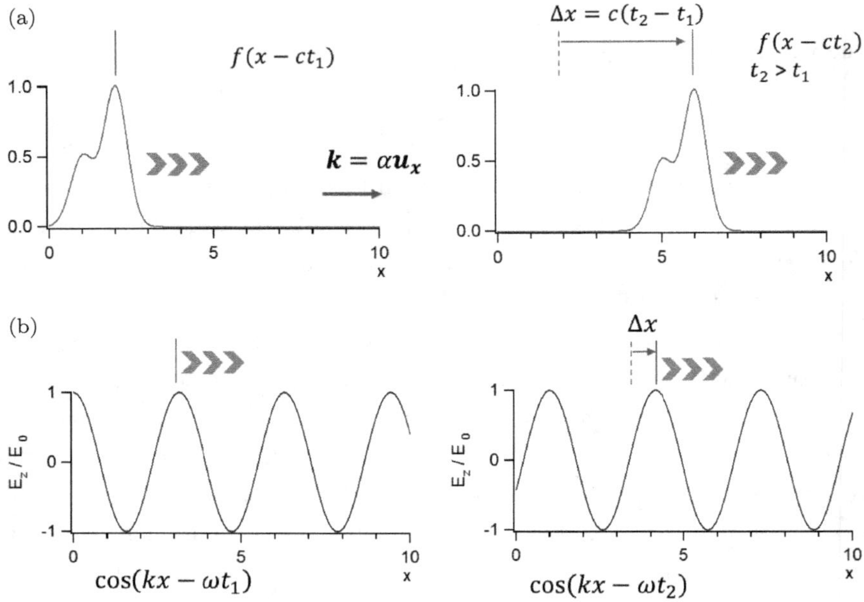

Fig. 1.3. The propagation of a wave shown at two different times: t_1 and t_2. (a) An arbitrary function f can describe a propagation provided the argument has the form $x - ct$. c is the phase velocity of the wave. During a time lapse of $t_2 - t_1$, the wave moves over a distance $\Delta x = c(t_2 - t_1)$. (b) The wave is often a periodic wave that is written with a cosine function. In this example, the electric field is $E_Z = E_0 \cos(kx - \omega t)$.

Equation (1.5) becomes

$$(-\boldsymbol{k}^2 + \mu_0 \varepsilon_0 \omega^2) E_0 \cos(\boldsymbol{k} \cdot \boldsymbol{r} - \omega t) = 0.$$

This relation should be verified for any t and any $\boldsymbol{r}$, which is the case only if

$$\boldsymbol{k}^2 = \frac{\omega^2}{c^2}, \quad \text{with } c^2 = \frac{1}{\mu_0 \varepsilon_0}, \tag{1.6}$$

where c is the velocity of light linked to the fundamental constants μ_0 and ε_0 (magnetic permeability and dielectric permittivity, respectively). Relation (1.6) is the *dispersion relation* in free space. Dispersion relation plays a pivotal role in optics, and it shows that the time period $T = 2\pi/\omega$ and the space period (wavelength $\lambda = 2\pi/|\boldsymbol{k}|$) are not independent but are linked by c, which is called the *phase velocity* of the wave.

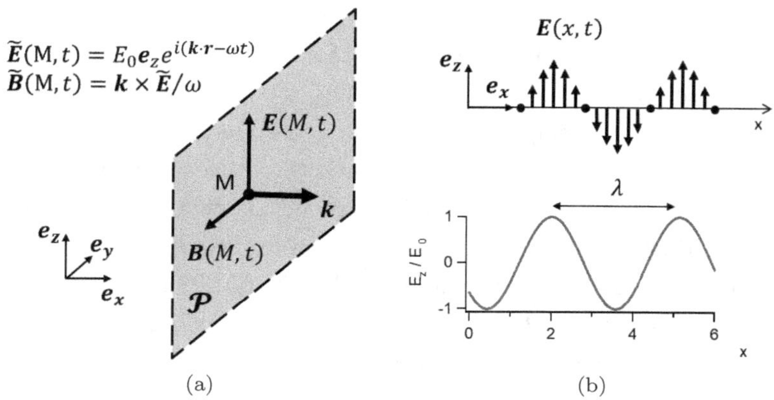

Fig. 1.4. (a) Representation of a monochromatic plane wave linearly polarized along the z direction and propagating along the x direction. The wave vector $\boldsymbol{k}$ sets the direction of propagation. The plane of the wave $\mathcal{P}$ is a plane perpendicular to $\boldsymbol{k}$. The values of $\boldsymbol{E}$ and $\boldsymbol{B}$ are uniform across this plane. (b) In the $(\boldsymbol{e_x}, \boldsymbol{e_z})$ plane, at an instant t, the electric field is an oscillation characterized by the wavelength λ.

1.2.3 *The vectorial structure of the electromagnetic wave*

Finally, a periodic plane wave in free space that obeys the Maxwell equations is written as

$$\boldsymbol{E}(\boldsymbol{r},t) = E_0 \cos(\boldsymbol{k}\cdot\boldsymbol{r} - \omega t)\boldsymbol{e}_z, \quad \text{with } \boldsymbol{k}^2 = \frac{\omega^2}{c^2}. \tag{1.7}$$

The wave vector $\boldsymbol{k}$ defines the *direction of propagation* of the wave, as shown in Fig. 1.4(a). Due to its propagative nature, this wave is called a *progressive wave* (as opposed to a *standing wave*). E_0 is the amplitude, and ω is the circular frequency or, simply, the frequency of the wave. The frequency f of oscillations is linked to ω as $f = \omega/2\pi$, and the wavelength λ is linked to the wave vector as $\lambda = 2\pi/|\boldsymbol{k}|$ (see Fig. 1.4(b)).

This wave is a *plane wave* because at a given time t, the value of $\boldsymbol{E}(\boldsymbol{r},t)$ is identical throughout any plane perpendicular to the wave vector $\boldsymbol{k}$. This plane is denoted as $\mathcal{P}$ in Fig. 1.4(a). This is, of course, an ideal case because a wave does not fill the whole space simultaneously (think of the beam from a laser pointer).

This wave is *monochromatic*. It is characterized by a single frequency or, in other words, by a single color. It could be the wave of a laser.

In the case of relation (1.7), the wave is *linearly polarized* along the direction $\boldsymbol{e}_z$. It means the electric field adopts a fixed direction that does not change either in space or with time. Other polarization states could be circular polarization, elliptic polarization, or *natural light* (unpolarized).

Another important feature of the solutions to Maxwell equations in free space is that waves are *transverse*, which means that the vector of the electric field is always perpendicular to the wave vector, say the direction of propagation. This property derives directly from the Maxwell–Gauss equation: $\boldsymbol{\nabla} \cdot \boldsymbol{E} = \partial E_x/\partial x + \partial E_y/\partial y + \partial E_z/\partial z = 0$, which in the case of a periodic monochromatic wave becomes $-k_x \cdot E_x - k_y \cdot E_y - k_z \cdot E_z = 0$, or also, $\boldsymbol{k} \cdot \boldsymbol{E} = 0$. In our example, $\boldsymbol{E}$ has only one z component, which means $\boldsymbol{k} = k_x\boldsymbol{e}_x + k_y\boldsymbol{e}_y$. We see later that in solid media, such as metals, the wave is not necessarily transverse and can be longitudinal.

1.2.4 *The complex notation for monochromatic waves*

The exact description of the monochromatic plane wave is given by relation (1.7) with cosine or sine functions. However, the math with these functions can easily become delicate (addition, factorization, etc.). This is one reason why a preferred representation is provided by the complex notation. The wave in relation (1.7) is then expressed as

$$\tilde{\boldsymbol{E}}(\boldsymbol{r}, t) = E_0 \exp[i(\boldsymbol{k} \cdot \boldsymbol{r} - \omega t)]\boldsymbol{e}_z, \tag{1.8}$$

where i is the *imaginary unit* and the tilde ($\sim$) over $\tilde{\boldsymbol{E}}$ indicates that the complex notation is used. Relation (1.8) can also be written as $\tilde{\boldsymbol{E}}(\boldsymbol{r}, t) = E_0 e^{i\boldsymbol{k}\cdot\boldsymbol{r}} e^{-i\omega t}\boldsymbol{e}_z$, and when the case of a monochromatic wave is discussed, the $e^{-i\omega t}$ shows up systematically everywhere and is eventually dropped. Therefore, with the complex notation, a monochromatic plane wave can be simply written as $\tilde{\boldsymbol{E}}(\boldsymbol{r}, t) = E_0 e^{i\boldsymbol{k}\cdot\boldsymbol{r}}\boldsymbol{e}_z$. This makes all the calculations much lighter and easier. The time dependence is therefore implicit.

When one needs to go back to the real notation, for example to plot the time evolution of the amplitude of an electric field, one needs to multiply the complex amplitude with $e^{-i\omega t}$ and use

$$\boldsymbol{E}(\boldsymbol{r}, t) = \frac{1}{2}\Re\left[E_0 e^{i\boldsymbol{k}\cdot\boldsymbol{r}} e^{-i\omega t}\right]\boldsymbol{e}_z, \tag{1.9}$$

where $\Re(\tilde{x})$ is the real part of the complex number $\tilde{x}$. In this example, the wave under consideration is polarized along $\boldsymbol{e}_z$, but this formula also applies when the electric field decomposes over all three components $(\boldsymbol{e}_x, \boldsymbol{e}_y, \boldsymbol{e}_z)$ and also when the amplitude is a complex number, $\tilde{E}_0$.

In the following, we mostly use the complex notation and the tilde is dropped for simplifying our notations.

1.2.5 *Magnetic field B*

In the case of monochromatic plane waves, the magnetic field $\boldsymbol{B}$ is easily obtained from $\boldsymbol{E}$ using the Maxwell–Faraday equation (1.3):

$\tilde{\boldsymbol{B}}(\boldsymbol{r}, t) = E_0/\omega(k_y \boldsymbol{e}_x - k_x \boldsymbol{e}_y) \exp[i(\boldsymbol{k} \cdot \boldsymbol{r} - \omega t)]$, which simplifies into

$$\tilde{\boldsymbol{B}} = \boldsymbol{k} \times \tilde{\boldsymbol{E}}/\omega. \tag{1.10}$$

In free space, the magnetic field is perpendicular to both the wave vector and the electric field, as depicted in Fig. 1.4(a).

1.2.6 *Energy of a plane wave*

The energy transported by a wave is linked to the Poynting vector, which is defined by

$$\boldsymbol{\Pi} = \boldsymbol{E} \times \boldsymbol{B}/\mu_0, \tag{1.11}$$

where $\boldsymbol{\Pi}$ is the electromagnetic power flow (in W/m^2). It is not exactly equivalent to the *intensity*, which is power per unit solid angle (in W/sr).

$\boldsymbol{E}$ and $\boldsymbol{B}$ oscillate at a frequency ω, and their product also exhibits a high-frequency oscillating behavior in the order of 10^{15} Hz. Detectors, such as a photodiode or the human eye, systematically measure the time-averaged Poynting vector. In complex notation, the time average of a harmonic electromagnetic wave is simply given by

$$\langle \boldsymbol{\Pi} \rangle = \frac{1}{2} \Re \left[\frac{\boldsymbol{E} \times \boldsymbol{B}^*}{\mu_0} \right]. \tag{1.12}$$

If the Poynting vector is homogeneous over the surface S crossed by the wave, the power flux going through the surface (in W) is given by

$$\Phi = \boldsymbol{\Pi} \cdot \mathbf{S}. \tag{1.13}$$

The local electromagnetic density of energy is denoted by u_{em} (in $\mathrm{W/m^3}$) and is given by the following expression with the two contributions (magnetic energy and electric energy):

$$u_{em} = \frac{1}{2}B^2/\mu_0 + \frac{1}{2}\varepsilon_0 E^2.$$

1.3 Wave Propagation in Matter

The propagation of an electromagnetic wave is modified when it passes through a medium. In some materials, the propagation is stopped, e.g. waves of sufficiently low frequencies interacting with a metal. This causes the wave to be reflected. In other materials, propagation is possible, but the wave is modified compared to what happens in vacuum: Its direction may be modified (as in the case of refraction of a beam of light), its polarization state may change (e.g. in an anisotropic crystal), or its intensity may be attenuated (in an absorbing media). When a medium allows the electric field to penetrate, it is called a *dielectric medium*. Usually, this corresponds to insulating media (e.g. polymers, a volume of water, or a piece of wood). The interaction between a wave and matter is the result of microscopic interactions at the atomic level between the electrons and the electromagnetic field. To be more precise, under the influence of the fields $\boldsymbol{E}$ and $\boldsymbol{B}$, the charge density ρ and the local current density $\boldsymbol{j}$ are modified, and therefore, the source terms of the fields in the Maxwell equations are also modified. This change in the source terms changes the fields, and this, in turn, affects the source terms again. This interaction acts like a feedback loop. The purpose of this section is to elucidate this interaction.

1.3.1 *Polarization of dielectric media*

The Maxwell equations, as presented in Equations (1.1)–(1.4), are not only valid in vacuum but also in solid or liquid media provided that the meanings of the density of charge ρ and of the local current density $\boldsymbol{j}$ are clarified. Let's consider a medium of volume $\mathcal{V}$, where ρ and $\boldsymbol{j}$ are continuous (one type of material, no domain boundary, and no interfaces). The charge density ρ has two contributions: the external charges ρ_{ext} injected into the medium and the local charges ρ_{local} released by the medium itself so that[1]

$$\rho = \rho_{ext}+\rho_{local}. \tag{1.14}$$

Local charges are precisely the charges induced by the fields when the electron cloud of the medium is deformed by the electric and magnetic fields. The global charge neutrality is maintained, but locally, negative charges are moved away from the positive core of the atoms, and this gives rise to the local density of charge ρ_{local}. Typically, the existence and the value of ρ_{local} strongly depends on the nature of the material, and this is given by the *constitutive relation*. Qualitatively, the electric field within the medium induces a deformation of the electron clouds in the atoms and moves their center of gravity (which is negatively charged) away from the center of gravity of the positively charged nucleus (which are supposed to remain immobilized by the atomic lattice). This is schematically shown in Fig. 1.5, where the electric field gives rise to a more pronounced variation in ρ_{local} (Fig. 1.5(b)). This induces a local electric dipole. When all these local dipoles are averaged over the whole volume $\mathcal{V}$, they are expressed as polarization per unit volume $\boldsymbol{P}$.

We now consider the case of *linear, homogeneous, isotropic dielectrics*, abbreviated LHID. This means that under the excitation of a given electric field $\boldsymbol{E}$, the polarization vector $\boldsymbol{P}$ is linear with $\boldsymbol{E}$, homogeneous over the whole volume $\mathcal{V}$, and isotropic (the linear coefficient does not depend on the direction of $\boldsymbol{E}$). Under these assumptions, $\boldsymbol{P}$ is written as

$$\boldsymbol{P} = \varepsilon_0 \chi \boldsymbol{E}, \tag{1.15}$$

where χ is a scalar called the *dielectric susceptibility*.

The *constitutive relation* provides a mathematical form for χ. The Drude model, for example, is an easy and efficient approach for deriving χ in the case of metals, and it will be developed in the next chapter.

---

[1] In some textbooks, the two contributions are distinguished as free and bound charges. However, in the case of metals, which is our focus, this distinction could be misleading since the local charge splits into free and bound electrons.

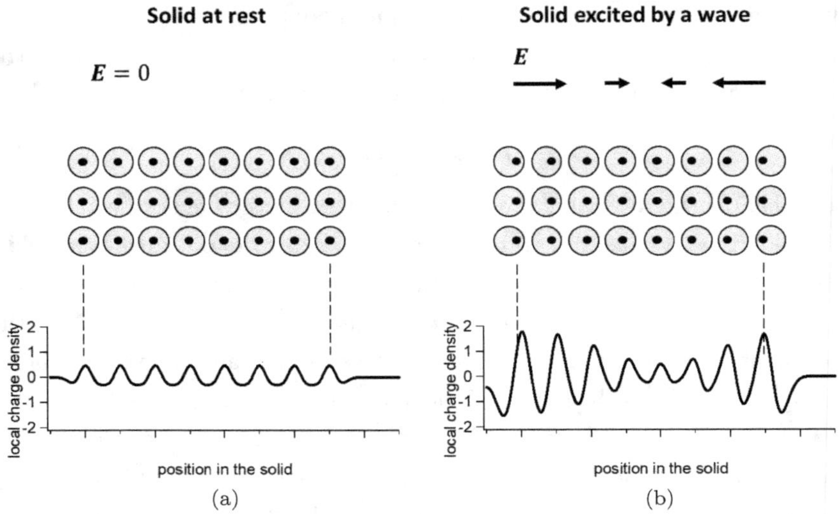

Fig. 1.5. (a) When the solid is at rest, the electron cloud and the ionic cores are centered at the same point. The variation in the local charge density in space is minimized (bottom graph). (b) However, when an electromagnetic wave disturbs this equilibrium, the centers of mass of the positive and negative charges are not superimposed anymore, and this generates a strong local charge density.

In textbooks on electromagnetism [1,2], it is shown that the local charge is linked to the polarization $\boldsymbol{P}$ by

$$\rho_{local} = -\boldsymbol{\nabla} \cdot \boldsymbol{P}. \tag{1.16}$$

If the local charge density depends on or fluctuates with time, there will be also the generation of a local current given by

$$\boldsymbol{j}_{local} = \partial \boldsymbol{P}/\partial t. \tag{1.17}$$

At the surface of the polarized medium, a surface charge density may appear and is linked to the polarization by

$$\sigma = -\boldsymbol{P} \cdot \boldsymbol{n}_s, \tag{1.18}$$

where $\boldsymbol{n}_s$ is the unit vector perpendicular to the surface and pointing outward.

1.3.2 *Maxwell equations in dielectric media*

As said above, the Maxwell equations (1.1)–(1.4) are also valid in a dielectric medium (LHID). However, using relations (1.13) and (1.14) to express the new source terms generated by the polarization of the medium, the Maxwell–Gauss equation becomes

$$\nabla \cdot \boldsymbol{E} = \frac{\rho_{ext}}{\varepsilon_0} + \frac{\rho_{local}}{\varepsilon_0} = \frac{\rho_{ext}}{\varepsilon_0} - \nabla \cdot \boldsymbol{P},$$

$$\nabla \cdot (\boldsymbol{E} + \boldsymbol{P}/\varepsilon_0) = \frac{\rho_{ext}}{\varepsilon_0}. \tag{1.19}$$

Maxwell–Ampère equation (1.4) transforms into

$$\nabla \times \boldsymbol{B} = \mu_0 \left(\boldsymbol{j}_{ext} + \boldsymbol{j}_{local} + \varepsilon_0 \frac{\partial \boldsymbol{E}}{\partial t} \right)$$

$$= \mu_0 \left(\boldsymbol{j}_{ext} + \varepsilon_0 \frac{\partial}{\partial t} [\boldsymbol{E} + \boldsymbol{P}/\varepsilon_0] \right). \tag{1.20}$$

In the preceding two equations, a new quantity shows up: $\boldsymbol{D} = \varepsilon_0 \boldsymbol{E} + \boldsymbol{P}$. $\boldsymbol{D}$ is called the *dielectric displacement*. The two Maxwell equations can be rewritten in the following way:

$$\nabla \cdot \boldsymbol{D} = \frac{\rho_{ext}}{\varepsilon_0}, \tag{1.21}$$

$$\nabla \times \boldsymbol{B} = \mu_0 \left(\boldsymbol{j}_{ext} + \varepsilon_0 \frac{\partial \boldsymbol{D}}{\partial t} \right). \tag{1.22}$$

They have the same form as before, but $\boldsymbol{E}$ is now replaced with $\boldsymbol{D}$, where the polarization of the medium is taken into account.

In an LHID medium, the expression of $\boldsymbol{D}$ can be clarified and greatly simplified by taking into account the constitutive relation (1.14):

$$\boldsymbol{D} = \varepsilon_0 \boldsymbol{E} + \varepsilon_0 \chi \boldsymbol{E} = \varepsilon_0 (1 + \chi) \boldsymbol{E} = \varepsilon_0 \varepsilon \boldsymbol{E}. \tag{1.23}$$

Here, we have introduced the *dielectric permittivity* $\varepsilon_0 \varepsilon$ and the *relative dielectric permittivity* ε. It is basically a constant of proportionality between the electric field and the *dielectric displacement*. ε plays a pivotal role in optics and will be used and discussed throughout the book. Note that ε is a real number in the simplest cases (isotropic medium) but can also be a (3×3) matrix for anisotropic media, such as birefringent crystals, liquid crystals, and oriented molecules at a metallic interface.

1.3.3 *About magnetic media*

So far, we have not mentioned magnetic media. They play only a minor role in optics (except for metamaterials). Nevertheless, a few words should be added. When a magnetic medium is immersed in an electromagnetic wave, it retrofits into the field in a similar way to a dielectric medium. In this case, a magnetization $\boldsymbol{M}$ arises from the material, and in case of a magnetic linear homogeneous isotropic (MLHI) medium, $\boldsymbol{M}$ is linked to $\boldsymbol{B}$ with a scalar, linear relationship. The *magnetic susceptibility* χ_m is defined by the magnetic constitutive relation:

$$\boldsymbol{M} = \chi_m \boldsymbol{B}/\mu_0. \tag{1.24}$$

The local current induced by the magnetization is written as

$$\boldsymbol{j}_{local} = \boldsymbol{\nabla} \times \boldsymbol{M} + \partial \boldsymbol{P}/\partial t. \tag{1.25}$$

This is a more complete form of the local current and can be added to the Maxwell equations. It is convenient to introduce the vector $\boldsymbol{H}$, called the *magnetic field strength*, so that $\boldsymbol{B} = \mu_0 \mu \boldsymbol{H}$, where μ is the *magnetic permeability* linked to the magnetic susceptibility by $\mu = 1/(1 - \chi_m)$. Regarding B and H, there can be a confusion in their names: $\boldsymbol{B}$ is the magnetic field but is sometimes called magnetic induction, and $\boldsymbol{H}$ is sometimes called the magnetic excitation. If $\chi_m > 0$ ($\mu < \mu_0$), the medium is paramagnetic, and if $\chi_m < 0$ ($\mu > \mu_0$), the medium is diamagnetic.

In the Maxwell–Ampère equation, if the term $\boldsymbol{\nabla} \times \boldsymbol{M}$ from relation (1.24) is introduced to take into account the local current induced by the magnetization, it is easy to verify that the equation becomes

$$\boldsymbol{\nabla} \times (\boldsymbol{B} - \mu_0 \boldsymbol{M}) = \mu_0 \left(\boldsymbol{j}_{ext} + \varepsilon_0 \frac{\partial \boldsymbol{D}}{\partial t} \right).$$

Taking into account the constitutive relation (1.23), it yields

$$\boldsymbol{\nabla} \times \left(\boldsymbol{B} - \frac{\boldsymbol{B}}{\chi_m} \right) = \mu_0 \boldsymbol{\nabla} \times \boldsymbol{H} = \mu_0 \left(\boldsymbol{j}_{ext} + \varepsilon_0 \frac{\partial \boldsymbol{D}}{\partial t} \right),$$

and

$$\boldsymbol{\nabla} \times \boldsymbol{H} = \boldsymbol{j}_{ext} + \varepsilon_0 \frac{\partial \boldsymbol{D}}{\partial t}. \tag{1.26}$$

Relation (1.26) is the form of the Maxwell–Ampère equation in a dielectric, magnetic medium.

Finally, in such dielectric, magnetic media, the Maxwell equations are written as a function of $\boldsymbol{H}$ and $\boldsymbol{D}$, *with their most general form* as follows:

$$\boldsymbol{\nabla} \cdot \boldsymbol{D} = \rho_{ext} \quad \text{Maxwell–Gauss,} \tag{1.27}$$

$$\boldsymbol{\nabla} \cdot \boldsymbol{B} = 0 \quad \text{Maxwell–Thomson,} \tag{1.28}$$

$$\boldsymbol{\nabla} \times \boldsymbol{E} = -\frac{\partial \boldsymbol{B}}{\partial t} \quad \text{Maxwell–Faraday,} \tag{1.29}$$

$$\boldsymbol{\nabla} \times \boldsymbol{H} = \boldsymbol{j}_{ext} + \frac{\partial \boldsymbol{D}}{\partial t} \quad \text{Maxwell–Ampère,} \tag{1.30}$$

with the two relations $\boldsymbol{D} = \varepsilon_0 \varepsilon \boldsymbol{E}$ and $\boldsymbol{B} = \mu_0 \mu \boldsymbol{H}$. ε and μ are physical quantities that contain the electrical and magnetic properties of the material under consideration, respectively. They are the *constitutive relations* for LHID and MLHI media, respectively. They describe how electrons react to an electromagnetic field, yield a specific dielectric polarization, and generate local currents. The most accurate description of these properties is provided by solid state physics, explained in detail in textbooks by authors such as Kittel [7] and Ashcroft [8]. In our case, we remain at the elementary level of this approach and use the most elementary, yet relevant, Drude model (see Chapter 3). The Drude model provides a simple analytical expression for ε and its dependency on the frequency ω.

The Maxwell equations are valid for any kind of electromagnetic field: propagative fields, non-propagative fields (near fields), monochromatic waves, soliton waves, etc. In the particular case of interest for this chapter, we restrict ourselves to propagative, harmonic plane waves, and we use the complex notation. Maxwell equations can be written for the complex amplitudes, knowing that the differentiations turn into simple multiplications. For example, the Maxwell–Faraday equation becomes (see Exercises for the other three Maxwell equations)

$$\boldsymbol{k} \times \boldsymbol{E} = i\omega \boldsymbol{B}. \tag{1.31}$$

1.3.4 *Wave propagation in an LHID medium*

We now return to a dielectric and nonmagnetic medium. This means that $\varepsilon \neq 1$ and $\mu = 1$. We are interested in harmonic propagative solutions of the

Maxwell equations, oscillating at a frequency ω and propagating throughout this medium. Therefore, we use the complex notation for the fields:

$$\tilde{\boldsymbol{E}}(\boldsymbol{r},t) = E_0 \exp[i(\boldsymbol{k}\cdot\boldsymbol{r} - \omega t)]\boldsymbol{e}_z. \tag{1.32}$$

The constitutive relation may also be complex:

$$\tilde{\boldsymbol{P}} = \varepsilon_0 \tilde{\chi} \tilde{\boldsymbol{E}}. \tag{1.33}$$

The meaning of a complex $\tilde{\chi}$ is that a dephasing is possible between $\tilde{\boldsymbol{P}}$ and $\tilde{\boldsymbol{E}}$: The amplitudes of the polarization and the electric field are not simultaneously maximum, or in other words, the polarization of matter reacts with a time delay to the electric excitation. Of course, this time delay depends on the frequency of the electric field. Subsequently, the dielectric permittivity is also complex: $\tilde{\varepsilon} = 1 + \tilde{\chi}$.

The dispersion relation is established in the same way as in Section 1.2.1. We derive it as

$$\boldsymbol{k}^2 = \tilde{\varepsilon}\frac{\omega^2}{c^2}. \tag{1.34}$$

Since $\tilde{\varepsilon}$ is a complex number, $\boldsymbol{k}$ is also complex, and this leads to some important consequences. We review in the following two consequences of propagation in dielectric media.

1.3.5 *Poynting vector and energy in LHID*

The Poynting vector that expresses the power flow (power per surface unit) is identical in a LHID as in vacuum. The flux averaged over time is given by

$$\langle \boldsymbol{\Pi} \rangle = \frac{1}{2}\Re\left[\frac{\boldsymbol{E}\times\boldsymbol{B}}{\mu_0}\right] = \frac{1}{2}\Re\left[\frac{\boldsymbol{E}\times(\boldsymbol{k}\times\boldsymbol{E})}{\mu_0\omega}\right]. \tag{1.35}$$

The difference from vacuum is the fact that $\boldsymbol{k}$ is a complex vector.

1.4 Two Important Consequences of the Dispersion Relation in LHID

1.4.1 *First consequence: The optical index*

Let's first consider the case of a medium characterized by a noncomplex permittivity. This is the case of glass or any transparent medium (such as Plexiglas and water at first approximation). ε is a real number, greater than 1, and can depend on ω. We introduce the optical refractive index (a positive quantity):

$$n = \sqrt{\varepsilon}.$$

The dispersion relation (1.34) becomes

$$\boldsymbol{k} = \pm n\frac{\omega}{c}\boldsymbol{e}_x, \tag{1.36}$$

which can also be written as

$$\boldsymbol{k} = \pm n\frac{2\pi}{\lambda_0}\boldsymbol{e}_x = \pm n\boldsymbol{k}_0,$$

where $\boldsymbol{k}_0$ and λ_0 are the wave vector and the wavelength of a plane wave of frequency ω in vacuum, respectively. The electric field can be written as

$$\tilde{E}(\boldsymbol{r}, t) = E_0 \exp\left[i(n\boldsymbol{k_0} \cdot \boldsymbol{r} - \omega t)\right] \boldsymbol{e}_z. \tag{1.37}$$

The positive $\boldsymbol{k} = +n\boldsymbol{k}_0$ is a wave propagating along x-axis in the positive direction, and the propagation is reversed for $\boldsymbol{k} = -n\boldsymbol{k}_0$.

The propagation, therefore, looks very similar to the propagation in vacuum, except for a few changes in its characteristics: The wavelength in a transparent medium is now $\lambda = \lambda_0/n$. The phase velocity of the wave is obtained from the propagation argument in (1.36): $i(n\boldsymbol{k_0} \cdot \boldsymbol{r} - \omega t)$. It is the ratio of the quantity in front of the time coordinate t over the quantity in front of the space coordinate $\boldsymbol{r}$. It yields $c' = \omega/nk_0 = c/n$. The speed of light c' decreases in transparent media and so does the wavelength (see Fig. 1.6(a)). It also corresponds to the definition of the optical index used in geometrical optics.

1.4.2 *Second consequence: Absorbing media*

Let's now assume that $\tilde{\varepsilon}$ is complex. A complex optical index is also defined as $\tilde{n}^2 = \tilde{\varepsilon}$ and is written with its real and imaginary parts: $\tilde{n} = n' + in''$. The dispersion relation can be expressed as[2] (with the direction of propagation

[2]When calculating the square root $n = \pm\sqrt{\varepsilon}$, we only keep the positive solution because it corresponds to a wave propagating along the x-axis in the positive direction with an energy decay. The other solution would correspond to a medium generating energy.

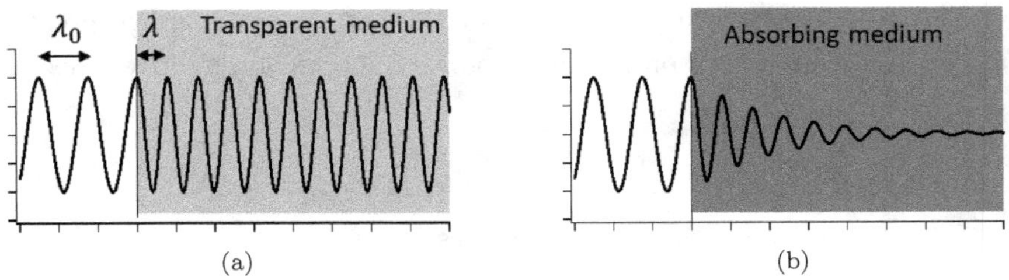

Fig. 1.6. (a) At the interface between vacuum and a transparent medium, the wavelength changes from λ_0 to λ. (b) When the second medium is absorbing, the transmitted wave will undergo a change in wavelength and an intensity decay.

along $\boldsymbol{e}_x$)

$$\tilde{\boldsymbol{k}} = \tilde{n}\frac{\omega}{c}\boldsymbol{e}_x. \tag{1.38}$$

The complex wave vector $\tilde{\boldsymbol{k}}$ decomposes into real and imaginary parts: $k' = n'\omega/c$ and $k'' = n''\omega/c$. The resulting plane wave in the medium is

$$\begin{aligned}\tilde{\boldsymbol{E}}(\boldsymbol{r},t) &= E_0 \exp[i(k'x + ik''x - \omega t)]\boldsymbol{e}_z,\\ \tilde{\boldsymbol{E}}(\boldsymbol{r},t) &= E_0 \exp(-k''x)\exp[i(k'x - \omega t)]\boldsymbol{e}_z.\end{aligned} \tag{1.39}$$

Relation (1.39) is a wave with the usual plane wave propagating term along the x-axis affected by a decay term $\exp(-k''x)$. The propagating term is identical to the case of transparent media explained above and occurs with a wavelength $\lambda = \lambda_0/n'$.

It is more meaningful to consider the energy of a wave rather than its amplitude. The former is given by the Poynting vector. According to relations (1.35) and (1.39), one obtains

$$\langle \boldsymbol{\Pi} \rangle = \frac{n' E_0^2}{2c\mu_0}\exp(-2k''x)\boldsymbol{e}_x = \frac{n' E_0^2}{2c\mu_0}\exp\left(-\frac{2\pi n''}{\lambda_0}x\right)\boldsymbol{e}_x. \tag{1.40}$$

This relation shows that the energy of the plane wave decays in a medium with a decay length of $l = 1/2k'' = \lambda_0/4\pi n''$. The energy is divided by a factor $e \approx 2.7$ after a propagation length of l. If the medium is not absorbing, $n'' = 0$ and this length is infinite. If the medium is slightly absorbing, like water, for which $n = 1.34 + 9.2 \times 10^{-10}$ (in case of pure water) at $\lambda_0 = 500$ nm, then the decay length is calculated to be 4.3 m.

An important application: the Beer–Lambert law: The Beer–Lambert law is an experimental law used for describing how a solution containing a given chemical compound absorbs light. The important parameters are the thickness x of the container (usually the cuvette of a spectrometer), the nature of the dissolved molecule, which is characterized by its molar extinction coefficient α, and the concentration C:

$$I = I_0 \exp(-\alpha C x).$$

This relation becomes equivalent to (1.39) by equating $\alpha C = 2\pi n''/\lambda_0$.

1.5 Conclusion

This chapter has provided a "survival kit" for understanding the propagation of electromagnetic waves based on the four Maxwell equations. Of course, in 20 pages or so, this overview does not have the pretension of offering the same in-depth developments as reference textbooks of 600 pages. However, we have stressed the primary role played by transverse plane waves and the dispersion equation. These concepts have been revealed to be essential for understanding surface plasmon waves, even if these waves are neither plane nor transverse. This will be developed in Chapter 4.

References

[1] Purcell E. M. and Morin D. J. 2013. *Electricity and Magnetism* (Cambridge: Cambridge University Press).
[2] Griffiths D. J. 2017. *Introduction to Electrodynamics* (New York: Cambridge University Press).
[3] Pérez J.-P., Carles R. and Fleckinger R. 2020. *Électromagnétisme: Fondements et applications* (Paris: Dunod).
[4] Born M. and Wolf E. 1999. *Principles of Optics: Electromagnetic Theory of Propagation, Interference and Diffraction of Light* (Cambridge: Cambridge University Press).
[5] Jackson J. D. 1998. *Classical Electrodynamics. 3rd Ed.* (New York: Wiley).
[6] Novotny L. and Hecht B. 2012. *Principles of Nano-Optics (2nd Ed.)* (Cambridge: Cambridge University Press).
[7] Kittel C. 2004. *Introduction to Solid State Physics* (New York: John Wiley & Sons, Inc.).
[8] Ashcroft N. and Mermin D. 1976. *Solid State Physics* (Philadelphia: Saunders College Publishing).

Exercises

(Exercises marked with an asterisk (*) are difficult.)

(1) **Gauss law:** Derive the Gauss law from the Maxwell–Gauss equation (1.1). We recall that the electric flux through a closed surface S is $\Phi = \oint_S \boldsymbol{E} \cdot d\boldsymbol{S}$. Use the Green–Ostrogradski theorem.

(2) **Electric field of a point charge:** Consider a point charge q at a position O. Use the Gauss law to calculate the electric field $\boldsymbol{E}(\boldsymbol{r})$ at a position M situated at a distance $\boldsymbol{r}$ from O ($\boldsymbol{r} = \vec{OM}$). You will need to use the fact that the field has the same symmetry as the charge distribution, which is a simple spherical symmetry.

(3) (*) **Electric field of a dipole:** Consider now an electrostatic dipole made of two charges of opposite values placed at a distance d from each other. Say the positive charge q is at point P with the coordinate $x = +d/2$ and the negative charge $-q$ is at point N at $x = -d/2$. The electrostatic dipole is defined as $\boldsymbol{p} = q\vec{NP}$.

(a) Consider a closed surface S containing the two charges. What is the value of the electric flux? Why is it not helpful to apply the Gauss law in this situation?

(b) Instead, calculate the electric potential $V(\vec{r})$ at a point M situated at a distance r from the center O of the dipole ($\vec{r} = \vec{OM}$). Use the fact that the electric potential is additive: The potential created by the dipole is the summation of the potentials created by each charge individually. Perform the calculation at the first order in d/r, and calculate the electrostatic potential of the dipole at a distance $r \gg d$.

(c) Derive the electric field $\vec{E}(\vec{r})$. Use the spherical coordinates (r, θ, φ).

(4) **Plane wave:** Consider a plane wave with the form: $\boldsymbol{E}(x, y, t) = \boldsymbol{E}_0 \exp i(ax - 2ay - \omega_0 t)$. This wave propagates in vacuum in the (x, y) plane, and a is a constant with the dimension of the inverse of a distance: $1/a = 500$ nm. Determine the wave vector $\boldsymbol{k}_0$ and the

direction of propagation. Make a drawing. Calculate the wavelength of this wave.

(5) **A plane electromagnetic wave** of frequency 20 GHz moves in the positive x-axis direction such that its electric field is pointed along the z-axis. The amplitude of the electric field is 10 V/m. The start of time is chosen so that at $t = 0$, the electric field has a value 10 V/m at the origin.

(a) Write the wave function that will describe the electric field wave, and calculate numerically the values of all the parameters.

(b) Give the expression of the wave function that describes the associated magnetic field wave and calculate the field amplitude B_0.

(6) **Damped plane wave:** Consider a damped wave propagating along the x-axis. In complex notation, it is written as $\boldsymbol{E}(x,t) = \boldsymbol{E}_0 \exp(-\alpha x) \exp i(kx - \omega_0 t)$. Represent this wave at a time t_0 if we assume that $\alpha = 1/2\lambda$. Using the wave equation, show that this wave cannot propagate in vacuum.

(7) **(*) Dispersion relation in a waveguide** (inspired by Sophocles Orfanidis): Consider a rectangular waveguide made of metal and filled with air, as depicted in the adjacent figure. Due to this geometry and applying the Maxwell equations, it is possible to show that one set of solutions is a transverse electric (TE) wave, which can be written as

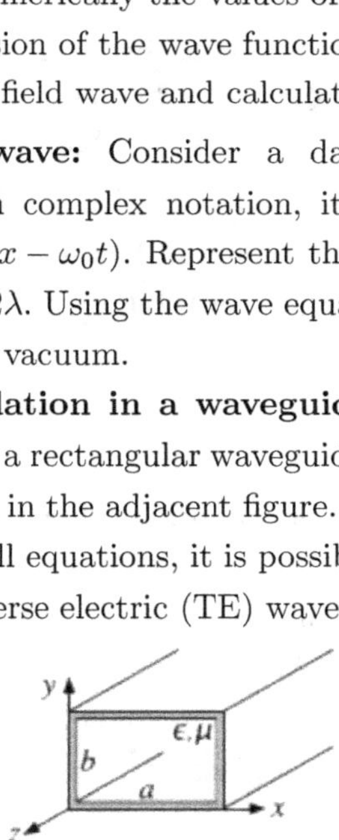

$$H_x = H_1 \sin(k_c x) \exp i(\omega t - \beta z),$$
$$H_z = H_0 \cos(k_c x) \exp i(\omega t - \beta z),$$
$$E_y = E_0 \sin(k_c x) \exp i(\omega t - \beta z),$$

while all the remaining components of $\boldsymbol{E}$ and $\boldsymbol{H}$ are zero.

(a) Derive the dispersion relationship. The parameter $\omega_c = ck_c$ is the cut-off frequency. Plot this relationship (ω as a function of β).

(b) Apply the boundary conditions for the electric field on the walls of the waveguide and determine the TE_{n0} modes allowed in this waveguide.

(8) What is the physical significance of the Poynting vector? A 2.0 mW helium–neon laser transmits a continuous beam of red light of cross-sectional diameter 2 mm at 60 cm from the laser aperture. The beam has a divergence of 3 mrad. Express the dependence of the amplitude of the Poynting vector as a function of distance x.

(9) **Laser safety:** In the visible range and for a continuous laser, the damage threshold of the human eye is $P_{eye} = 1$ mW. The diameter of the pupil is 7 mm. Consider the helium–neon laser of the previous exercise. At what distance does this laser cease to be harmful?

(10) **(*)** Consider a sphere of radius R with a uniform polarization $\boldsymbol{P}$ along the z-axis. We want to know the resulting charge distribution at a point M. Use the spherical coordinates (r, θ, φ) for M.

(a) Derive the expression of the charge density inside the sphere $\rho(r, \theta, \varphi)$ and of the surface charge $\sigma(\theta, \varphi)$ at the sphere surface.

(b) Show that this charge distribution can be produced by two interpenetrating spheres, uniformly charged with $+\rho_0$ and $-\rho_0$ and positioned at points P ($z = d/2$) and N ($z = -d/2$), respectively. Express the relationship between P, ρ_0, and d.

(c) Show that a metallic sphere of dielectric susceptibility χ placed in a uniform electric field $E_0\boldsymbol{e}_z$ behaves like two point charges of opposite charges and placed at P ($z = d/2$) and N ($z = -d/2$).

2

Optical Properties of Metals

Why is a metal not transparent to light? What makes the electron cloud of an insulator transmit light but that of a metal induce *metallic reflection*? The interaction of an optical wave with a metal is described by the **Drude model**. This chapter derives the **dielectric permittivity** of metals using this model and discusses how their response is dominated by free electrons, a phenomenon which is also called **intraband transition**. For metals in which **interband transitions** make an important contribution, the **Drude–Lorentz model** is used. This chapter also provides numerical values for dielectric permittivity of gold so that accurate calculations can be carried out and compared with experimental data.

2.1 How Can We Describe Metals?

The most advanced description of metals is provided by the concepts of condensed matter physics, which is based on the quantum description of atoms, applied to the case where these atoms are highly organized into a crystalline network [1–3]. Condensed matter physics mainly describes how electrons behave in these crystalline lattices, and this helps to understand various properties, such as the bonds between atoms, electrical conductivity, magnetic properties, and optical properties. The metal aluminum, for example, possesses 13 electrons separated into 10 core electrons and 3 valence electrons. Copper has 29 electrons and just 1 valence electron. Valence electrons in a metal are nearly free electrons, and they make the metal conductive.

They are also the principal contributors to a metal's optical properties. It is therefore particularly important to describe their behavior as precisely as possible. As far as optical properties are concerned, we can restrict our vision of a metal to a network of fixed ionic cores (with each ionic core consisting of a nucleus surrounded by its bound electrons) surrounded by freely moving electrons. The ionic core is positively charged, and together with the negatively charged electrons, the charge neutrality of the whole solid is ensured. In this book, we mainly confine ourselves to a classical description of the motion of free electrons. This approach is sufficient to capture the fundamental optical properties of metals. This is represented by the Drude model, which provides a qualitative agreement with experimental data. If a quantitative agreement is needed, we either implement refinements to the Drude model, thus extending it to the Drude–Lorentz model, or use parameters obtained from measurements (experimental values of the dielectric function). An excellent overview of the optical properties of solids can be found in the textbook by Fox [4].

2.2 The Drude Model for Metals

The Drude model was established in 1900 by Paul Drude in order to explain the conductivity of metals when submitted to a DC electric field. It is still the basis of all approaches to describing the dielectric permittivity of metals. The driving idea was that the valence electrons of metals can be considered as a *gas of electrons*, where electrons are free to move inside the solid. These electrons collide with the ionic core, and between two collisions, they do not suffer any other interaction with the solid (neither with other electrons nor with the core). When considering an optical beam, the electric field oscillates at a high frequency ($\sim 10^{15}$ Hz). Consequently, the motion of electrons is an oscillation, and it gives rises to a set of coordinated oscillating dipoles. One idea of the Drude model is to represent the effects of an optical wave on metal by these dipoles. Indeed, the oscillating dipole was described by Henrick A. Lorentz in 1878. Let us now convert these ideas into equations.

We consider an electromagnetic wave oscillating at a frequency ω at the point $\boldsymbol{r} = 0$. It is written as $\boldsymbol{E}(0,t) = E_0 \exp(-i\omega t)\boldsymbol{e}_x$. The electron oscillates along the x-axis under an electric force, and the second law of Newton is written as[1]:

$$m\frac{d^2x}{dt^2} + m\gamma\frac{dx}{dt} = -eE_0 \exp(-i\omega t), \tag{2.1}$$

where x describes the spacing between the fixed positive charges and the moving electrons. In relation (2.1), m is the mass of the electron γ and represents the frictional damping in the medium, which is directly linked to the time between two collisions.

We are interested in the forced solution once the steady-state regime is reached. Therefore, we use the complex notation introduced in Chapter 1. The electron also oscillates at the frequency ω: $x = \tilde{x}_0 \exp(-i\omega t)$, where $\tilde{x}_0$ is the complex amplitude of this motion. Relation (2.1) yields

$$\tilde{x}_0 = \frac{e}{m(\omega^2 + i\gamma\omega)}E_0. \tag{2.2}$$

The coordinate x corresponds to a local electric dipole that is given by $p = -ex$. By noting N the density of free electrons (number of electrons per unit volume), the polarization of the electron gas is given by

$$\boldsymbol{P} = -Nex\,\boldsymbol{e}_x = -\frac{Ne^2}{m(\omega^2 + i\gamma\omega)}E_0 \exp(-i\omega t)\,\boldsymbol{e}_x. \tag{2.3}$$

Relation (2.3) shows that the polarization is proportional to the electric field, and this proportionality coefficient is known as the dielectric susceptibility χ. χ was introduced in the previous chapter in relation (1.15), which stated $\boldsymbol{P} = \varepsilon_0\tilde{\chi}\boldsymbol{E}$. Therefore, we deduce the following:

$$\tilde{\chi} = -\frac{Ne^2}{\varepsilon_0 m(\omega^2 + i\gamma\omega)} = \frac{\omega_p^2}{\omega^2 + i\gamma\omega}. \tag{2.4}$$

[1] Here, since we deal with free electrons, we do not include the elastic restoring force on the left side of the equality, which would be $m\omega_0^2 x$. This term would play an important role in discussing the polarization in dielectric media and shows up in the complete derivation of the *dipole oscillator model.*

In relation (2.4), ω_p is the *plasma frequency*, which is an important characteristic of each metal:

$$\omega_p^2 = \frac{Ne^2}{\varepsilon_0} m. \tag{2.5}$$

The plasma frequency is computed from the electronic density N, with the other parameters being physical constants: $\varepsilon_0 = 8.85 \times 10^{-12}\,\mathrm{F/m}$, $e = 1.60 \times 10^{-19}\,\mathrm{C}$, and $m = 9.11 \times 10^{-31}\,\mathrm{kg}$.[2] The electronic density is the product of the density of the metal atoms and their respective valency. The value of N for some metals are given in Table 2.1.

The dielectric function is straightforwardly obtained from $\tilde{\chi}$ since $\tilde{\varepsilon} = 1 + \tilde{\chi}$, and this gives rise to the Drude model for free electrons in metals:

$$\tilde{\varepsilon} = 1 - \frac{\omega_p^2}{\omega^2 + i\gamma\omega} = 1 - \frac{\omega_p^2}{\omega^2 + \gamma^2} + i\frac{\omega_p^2\gamma}{(\omega^2 + \gamma^2)\omega}. \tag{2.6}$$

Relation (2.6) shows that $\tilde{\varepsilon}$ separates into its real and imaginary parts: $\varepsilon = \varepsilon' + i\varepsilon''$, with

$$\varepsilon' = 1 - \frac{\omega_p^2}{\omega^2 + \gamma^2} \quad \text{and} \quad \varepsilon'' = \frac{\omega_p^2\gamma}{(\omega^2 + \gamma^2)\omega}. \tag{2.7}$$

Table 2.1 shows a set of values of the plasma frequency ω_p for different metals and compares the calculated values with the experimental ones. The agreement for alkali metals, such as lithium, sodium, and potassium,

Table 2.1. Values for the Drude model for metals obtained directly from relation (2.5). Data taken from Ref. [2]. The last two columns contain the experimental values for the plasma frequency and the damping coefficients obtained by a fit to the actual dielectric function. Data taken from Ref. [5].

Metal	Valency	N $10^{28}\,\mathrm{m}^{-3}$	$\hbar\omega_p$ eV	λ_p nm	$\hbar\omega_p$ exp eV	γ exp eV
Li (77 K)	1	4.70	8.00	155		
Na (5 K)	1	2.65	6.00	207	**5.7**	
K (5 K)	1	1.40	4.40	282	**3.7**	
Cu	1	8.45	10.80	115	**9.8**	**0.096**
Ag	1	5.85	8.95	139	**9.6**	**0.023**
Au	1	5.90	9.10	136	**8.6**	**0.018**
Al	3	18.10	15.70	79	**15.3**	**0.600**

[2] m is the mass of an electron in vacuum, but for most of the metals, m should be replaced by m^*, the effective mass of an electron whose value depends on the material.

presented in the first three lines is rather good. Nevertheless, even if the behavior of free electrons in metals are well described, they are not the only ones playing a role in the optical properties. This is what happens for noble metals, such as gold and silver.

2.3 The Drude Model and the Optical Reflectivity of Metallic Surfaces

Optical waves have wavelengths situated between 400 and 800 nm, corresponding to frequency $\omega/2\pi$, ranging between 7.5×10^{14} and 1.5×10^{15} Hz. γ has typical values of 10^{14} Hz. $\gamma = \tau^{-1}$, where τ is the average time between two collisions on the electron trajectory. The shorter the collision time, the higher the damping coefficient. Therefore, considering that $\gamma \ll \omega$, the Drude formula simplifies into a real quantity:

$$\varepsilon = 1 - \frac{\omega_p^2}{\omega^2}. \tag{2.8}$$

Plasma frequency plays a critical role. When the optical frequency ω is smaller than ω_p, ε is negative. For example, in the case of aluminum, the metal has three valence electrons and therefore a very high electronic density N of $18.1 \times 10^{28}\,\mathrm{m}^{-3}$. The plasma frequency is computed using relation (1.5) and equals 15.7 eV. Figure 2.1(a) displays the dielectric function calculated with the Drude model using relation (1.6). The real and imaginary parts are plotted on the same graph as a function of the photon energy $\hbar\omega$. It shows that the real part $\Re(\tilde{\varepsilon})$ changes sign for $\omega = \omega_p$ and is negative for $\omega < \omega_p$. The visible range is 1.5–3.1 eV, and $\Re(\tilde{\varepsilon})$ varies from -95 to -22 in this range. The imaginary part remains much smaller.

The optical index is given by $\tilde{n} = \sqrt{\varepsilon}$, and using relation (1.48), it turns out to be a purely imaginary number. The reflectivity of a surface in air and at normal incidence is given by (see Chapter 3)

$$R = \left|\frac{\tilde{n}-1}{\tilde{n}+1}\right|^2. \tag{2.9}$$

Therefore, for $\omega < \omega_p$, $R = 1$ and the metal is perfectly reflective. Thus, the Drude model explains the reflective character of metals. On the other hand,

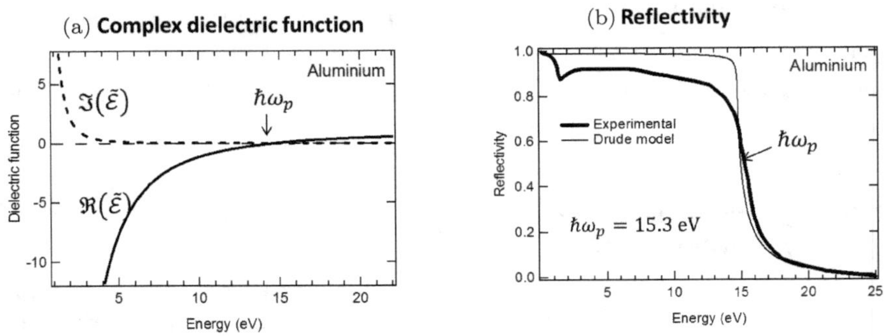

Fig. 2.1. The Drude model applied to aluminum: (a) Drude complex dielectric function; (b) reflectivity of aluminum as a function of photon energy. Experimental data are compared with the values calculated using the Drude model.

the reflectivity drops if $\omega > \omega_p$, and the metal tends to become transparent. For most metals, this cutoff frequency is in the UV range, and this property is known as the *ultraviolet transparency of metals.*

For aluminum and for a photon energy smaller than 15.3 eV, the metal is expected to be 100% reflective. This value corresponds to a theoretical cutoff wavelength of 79 nm, which is in the deep UV range. Figure 2.1(b) compares the reflectivity of a bulk piece of aluminum calculated according to the Drude model and the experimental values [6]. The global shape is well reproduced except that the value of reflectivity for $\omega < \omega_p$ is overestimated (experimental reflectivity is in the range 80–90% instead of 100%), and the feature at 1.5 eV is not reproduced by the model. This small dip in the reflectivity spectrum is due to the transition of an electron from the 2d band to the 3s band. This is called an interband transition. The existence of a marked feature at 1.5 eV is proof of a bound state that could be described as an oscillator with a defined oscillation frequency ω_0 (i.e. we would need to add a term $m\omega_0^2 x$ on the left side of Equation (2.1)). The electrons in bound state behave very differently from free electrons. Moreover, in the range 1.5–15.3 eV, a portion of the photon energy is damped into this bound state, and this explains the lower value of reflectivity. This example of aluminum shows that the optical behavior of metals is explained by a combination of the contributions from the free electrons (intraband transitions) and the bound electrons (interband transitions). The Drude model can satisfactorily account for the intraband transitions, but the interband transitions are more difficult to model.

2.4 Optical Reflectivity of Metals with Strong Interband Contribution

The Drude model is effective for metals whose optical properties are governed by the free electron gas. However, photons also excite the interband transitions, and the simple Drude model is no longer sufficient. The most iconic example is the case of gold. Figure 2.2 compares the reflectivity of gold calculated using the Drude model and the experimental values. It is clear that the Drude model is not adapted. The discrepancy is explained by the presence of two strong interband transitions from the 5d band to the 6s band situated at $\lambda \sim 470$ and 330 nm. Note also that the low reflectivity

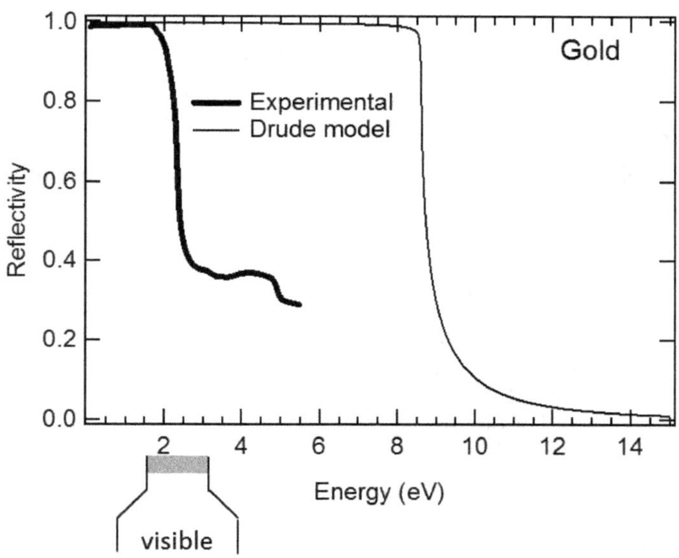

Fig. 2.2. Reflectivity of gold. The experimental reflectivity taken from Ref. [7] is compared with that obtained using the Drude model. For gold, the reflectivity is strongly affected by the interband transitions, and this generates a strong discrepancy with the model.

of gold above 2.5 eV is perceived as a lack of the blue contribution in the optical spectrum, which therefore gives gold its yellow color.

2.5 The Drude–Lorentz Model for Metals with Strong Interband Contribution

In order to predict the optical properties of metals with a strong contribution from the interband transitions, such as gold, copper, or silver, it is crucial to use more accurate values for the dielectric functions. The most popular model is a combination of the Drude model described above and the Lorentz oscillators, which include the interband transitions [8, 9]. In case of N oscillators, the relative dielectric permittivity is written as

$$\tilde{\varepsilon} = 1 - \frac{\omega_p^2}{\omega^2 + i\gamma\omega} + \omega_p^2 \sum_{j=1}^{N} \frac{f_j}{\omega_j^2 - \omega^2 - i\omega\gamma_j}, \tag{2.10}$$

where f_j is the oscillator strength of oscillator j, ω_j its resonance frequency, and γ_j its damping constant. Rakić, in 1998, applied the Drude–Lorentz model to 11 metals and discussed the number of oscillators needed to better reproduce the experimental value of optical reflectivity [8].

Nevertheless, the best values are those obtained by ellipsometric measurements, where the reflectivity of the metal is measured in certain conditions (under vacuum or under a controlled atmosphere). This makes it possible to evaluate the complex dielectric function as a function of photon energy. Such measures strongly depend on the actual chemical state of the surface of the metal, and one has to pay attention when using such sets of data.

2.5.1 *Analytical model for the dielectric function of gold*

In the case of gold, the most reliable data were published by Johnson and Christy in 1972 [10]. These data have been used by Etchegoin and Le Ru in 2006 to design an analytical model for $\varepsilon(\lambda)$ that efficiently reproduces the measured values [11]. Their Drude–Lorentz model includes two interband transitions, and the dielectric function of gold is given by

$$\varepsilon_{Au} = \varepsilon_\infty - \frac{1}{\lambda_p^2(1/\lambda^2 + i/\gamma_p\lambda)} + \sum_{i=1,2} \frac{A_i}{\lambda_i}\left[\frac{e^{i\phi_i}}{1/\lambda_i - 1/\lambda - i/\gamma_i} + \frac{e^{-i\phi_i}}{1/\lambda_i + 1/\lambda + i/\gamma_i}\right]. \quad (2.11)$$

The parameters are given in Table 2.2.

Figure 2.3 shows a plot of the Drude model for gold and the more elaborate Drude–Lorentz model using the parameters proposed by Etchegoin [11]. In the visible range (1.5–3.1 eV), the two curves are clearly different.

Table 2.2. Parameters of an analytical model for gold by Etchegoin and Le Ru [11] that reproduces the experimental data of Johnson and Christy [10].

Drude contribution		Interband 1		Interband 2	
ε_∞	1.53	A_1	0.94	A_2	1.36
		ϕ_1 (rad)	$-\pi/4$	ϕ_2 (rad)	$-\pi/4$
λ_p (nm)	145	λ_1 (nm)	468	λ_2 (nm)	331
γ_p (nm)	17,000	γ_1 (nm)	2300	γ_2 (nm)	940

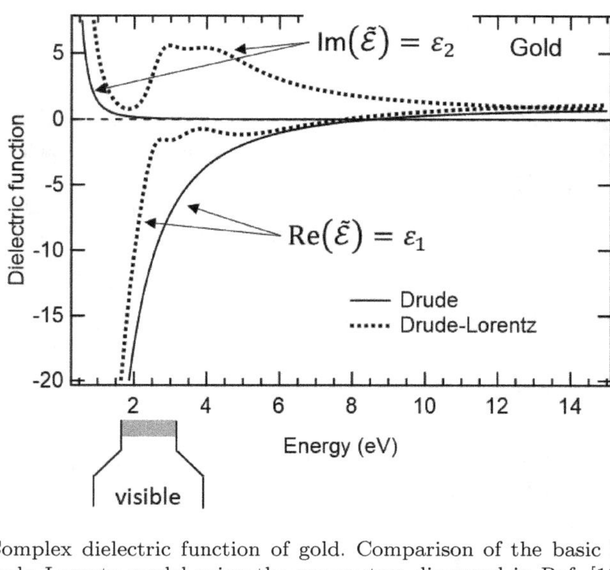

Fig. 2.3. Complex dielectric function of gold. Comparison of the basic Drude model with the Drude–Lorentz model using the parameters discussed in Ref. [11]. The latter accounts for the interband transitions.

2.5.2 *Tabulated values of the dielectric function of gold*

The most accurate values for the dielectric function of gold are obtained by ellipsometry on a freshly deposited gold layer. We give in Table 2.3 some values measured by a Swedish group, who became famous for their contribution to plasmonics [12].

Table 2.3. Values of the complex dielectric function of gold: $\varepsilon = \varepsilon_1 + i\varepsilon_2$, measured by H. Arwin. The gold layer (200 nm thickness) was sputtered and cleaned and the optical index was measured in air.

λ (nm)	ε_1	ε_2	λ	ε_1	ε_2	λ	ε_1	ε_2
400	−1.11	6.26	535	−5.14	2	670	−14.9	1.11
405	−1.13	6.25	540	−5.51	1.91	675	−15.3	1.11
410	−1.15	6.23	545	−5.88	1.84	680	−15.7	1.11
415	−1.16	6.19	550	−6.24	1.78	685	−16.1	1.1
420	−1.18	6.15	555	−6.6	1.72	690	−16.5	1.1
425	−1.19	6.09	560	−6.96	1.67	695	−16.9	1.11
430	−1.21	6.02	565	−7.32	1.62	700	−17.3	1.12
435	−1.23	5.95	570	−7.68	1.58	705	−17.7	1.12
440	−1.24	5.86	575	−8.03	1.54	710	−18.1	1.13
445	−1.26	5.75	580	−8.38	1.5	715	−18.5	1.15
450	−1.27	5.62	585	−8.74	1.47	720	−18.9	1.16
455	−1.29	5.48	590	−9.09	1.43	725	−19.3	1.17
460	−1.32	5.31	595	−9.45	1.4	730	−19.7	1.19
465	−1.36	5.11	600	−9.81	1.38	735	−20.1	1.2
470	−1.4	4.88	605	−10.2	1.35	740	−20.5	1.21
475	−1.48	4.63	610	−10.5	1.32	745	−20.9	1.23
480	−1.59	4.34	615	−10.9	1.3	750	−21.3	1.24
485	−1.73	4.04	620	−11.2	1.28	755	−21.7	1.26
490	−1.94	3.72	625	−11.6	1.26	760	−22.1	1.28
495	−2.2	3.4	630	−12	1.24	765	−22.5	1.3
500	−2.52	3.11	635	−12.3	1.22	770	−22.9	1.31
505	−2.87	2.86	640	−12.7	1.2	775	−23.3	1.32
510	−3.24	2.65	645	−13.1	1.18	780	−23.7	1.35
515	−3.62	2.47	650	−13.4	1.17	785	−24.2	1.37
520	−4	2.32	655	−13.8	1.15	790	−24.6	1.39
525	−4.39	2.19	660	−14.2	1.14	795	−25	1.4
530	−4.77	2.09	665	−14.6	1.13	800	−25.4	1.42

2.6 Conclusion

Despite its relatively simplistic approach, the Drude–Lorentz model presented in this chapter is highly effective and applies to an immense number of situations in physics: It explains the Hall effect for a conductor under a magnetic field; it serves to calculate the optical indices of metals and the reflection coefficient of metallic coatings for optical lenses; it allows us to calculate the skin depth of a metal; and it is, of course, the foundation for deriving the key features of plasmonics. This model also has well-identified shortcomings, such as in the prediction of conductivity of metals, which the model underestimates. Moreover, it does not properly explain the changes in this conductivity when temperature decreases. However, it remains unbeatable for putting into equations the dielectric function of plasmonic metals, such as gold, silver, copper, and aluminum. In the derivation of this model, the only input parameters are electronic density and the mass of an electron. It does not include any length parameter. The length scale is introduced by the optical wavelength λ. Yet, this model predicts a penetration depth of 25 nm for an optical wave in gold. When considering metallic nanoparticles and applying the oscillating dipole model, the Drude model also predicts the extension of the electromagnetic near field to distances of 10–50 nm. This simple and old model has served as the foundation for most of the modern nano-optics. The following chapters will discuss the plasmonic side of nano-optics.

References

[1] Ashcroft N. and Mermin D. 1976. *Solid State Physics.* (Pacific Grove, CA: Brooks Cole).
[2] Kittel C. 2004. *Introduction to Solid State Physics* (New York: John Wiley & Sons, Inc.).
[3] Singleton J. 2001. *Band Theory and Electronic Properties of Solids* (New York: Oxford University Press).
[4] Fox M. 2001. *Optical Properties of Solids* (Oxford: Oxford University Press).
[5] Blaber M. G., Arnold M. D. and Ford M. J. 2009. Search for the ideal plasmonic nanoshell: The effects of surface scattering and alternatives to gold and silver. *J. Phys. Chem. C* **113**, 3041–3045.

[6] Ehrenreich H., Philipp H. R. and Segall B. 1963. Optical properties of aluminum. *Phys. Rev.* **132**, 1918–1928.
[7] Wolfe W. L. and Zissis G. J. 1985. *The Infrared Handbook* (Ann Arbor, Michigan: Environmental Research Institute of Michigan).
[8] Rakić A. D., Djurišić A. B., Elazar J. M. and Majewski M. L. 1998. Optical properties of metallic films for vertical-cavity optoelectronic devices. *Appl. Opt.* **37**, 5271–5283.
[9] Alabastri A., Tuccio S., Giugni A., Toma A., Liberale C., Das G., Angelis F. D., Fabrizio E. D. and Zaccaria R. P. 2013. Molding of plasmonic resonances in metallic nanostructures: Dependence of the non-linear electric permittivity on system size and temperature. *Materials* **6**, 4879–4910.
[10] Johnson P. B. and Christy R. W. 1972. Optical constants of the noble metals. *Phys. Rev. B* **6**, 4370.
[11] Etchegoin P. G., Le Ru E. C. and Meyer M. 2006. An analytic model for the optical properties of gold. *J. Chem. Phys.* **125**, 164705–164707.
[12] Johansen K., Arwin H., Lundstrom I. and Liedberg B. 2000. Imaging surface plasmon resonance sensor based on multiple wavelengths: Sensitivity considerations *Rev. Sci. Instrum.* **71**, 3530–3538.
[13] Chakrabarti S. 2019. Determination of the damping co-efficient of electrons in optically transparent glasses at the true resonance frequency in the ultraviolet from an analysis of the Lorentz-Maxwell model of dispersion. *arXiv:1907.04499v1 [physics.optics].*

Exercises

(Exercises marked with an asterisk (*) are difficult.)

(1) **The Drude model and interband transitions:** (a) Use the Drude formula (2.9) and the data from Table 2.1 to plot the dielectric permittivity ε_r for gold in the visible region (1.5–3.2 eV). Compare with the experimental data provided in Table 2.3, which are also reproduced in the following figure. ε' and ε'' are the real and imaginary parts of ε_r, respectively. (b) A discrepancy arises from not taking into account the interband transitions of gold, although they occur in the UV region. A first correction consists in adding a constant contribution ε_∞ so that $\varepsilon = \varepsilon_\infty - \omega_p^2/\omega^2$. Fit the experimental data and calculate the best value for ε_∞. Optimize the region around 2.0 eV.

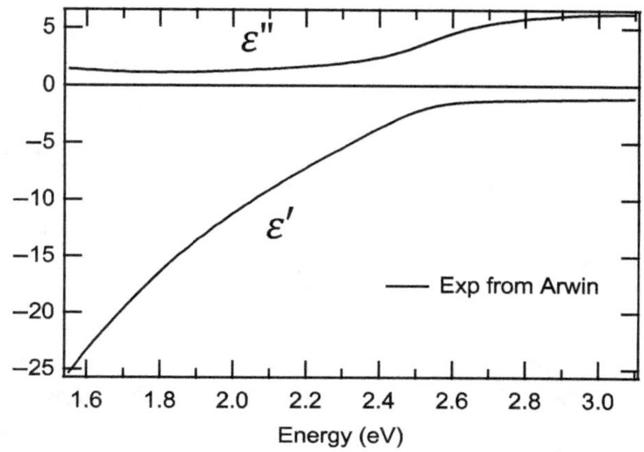

(2) **The Drude–Lorentz model for interband transition:** Following the previous exercise, improve the Drude model into the Drude–Lorentz model of Equation (2.10). Use only one Lorentzian oscillator ($N = 1$) and $\varepsilon_\infty = 7.7$. Use a drawing software and make a plot of ε' and ε''. Find the optimal values of f_1, ω_1, and γ_1. Comment on the use of one single oscillator.

(3) **Conductivity:** $\boldsymbol{j} = -en\boldsymbol{v}$ is the current density along a conductor of section S containing a density of n free electrons that move with a velocity $\boldsymbol{v}$. The conductor is biased so that a uniform (but not constant) electric field $\boldsymbol{E}(t)$ is established along this conductor. The conductivity σ is defined as $\boldsymbol{j} = \sigma\boldsymbol{E}(t)$. Using a similar approach to that of Equation (2.1), write the expression of σ using the complex notation. Give the relationship between the relative dielectric permittivity ε_r and σ.

(4) **The Lorentz model:** In transparent media, the electrons are not free and undergo an elastic restoring force that appears as the supplemental term $m\omega_0^2 x$ in the left member of Equation (2.1). Include this force when writing the second law of Newton and establish the expression of the complex dielectric permittivity ε_r. Express its real and imaginary parts: $\tilde{\varepsilon}_r = \varepsilon' + \varepsilon''$. Plot the results.

(5) A transparent medium like glass is characterized by its optical index n. Write the relationship between n and ε_r. We assume that n and ε_r are both real. The visible range corresponds to the limit $\omega \to 0$. Using the result in Exercise (4), show that $\lim \varepsilon_r = 1 + \omega_p^2/\omega_0^2$. Here, ω_p is the plasma frequency and ω_0 is the resonance frequency of the dominant oscillator. For *Flint F2* glass, $\omega_p = 18.0 \times 10^{15}$ rad/s. Use the following dispersion plot and find the value of ω_0 [13].

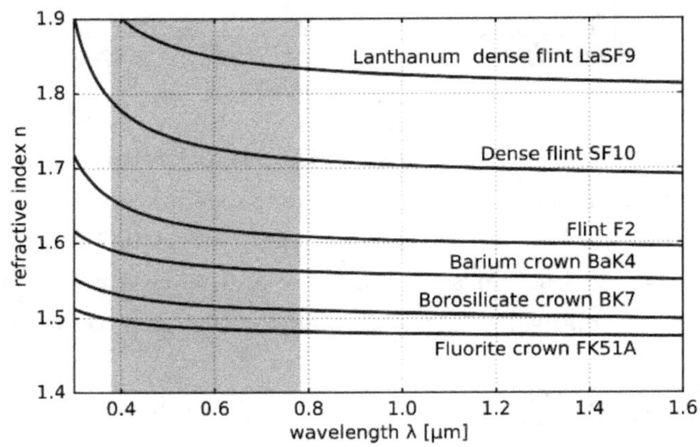

Variation of refractive index vs. vacuum wavelength for various glasses. The wavelengths of visible light are shaded in grey.
Source: Dispersion (optics), Wikipedia, https://commons.wikimedia.org/wiki/File:Mplwp_dispersion_curves.svg.

(6) The **Sellmeier dispersion relation** is a commonly used formula for expressing the dispersion of transparent materials. It expresses the value of n^2 as a function of λ^2. Use the Lorentz model established in Exercise (4) and consider a non-absorbing medium ($\gamma = 0$). Derive the expression of the Sellmeier equation considering three resonances. Include the six parameters: λ_p^i, λ_0^i for $i = 1, 2$.

In case of fused silica, the coefficients are given in the following table. Calculate the value of optical index at the three wavelengths: $\lambda_d = 587.6$ nm, $\lambda_f = 486.1$ nm, and $\lambda_d = 656.3$ nm. C_i is in μm^2.

	$i = 1$	$i = 2$	$i = 3$
$B_i = (\lambda_0^i/\lambda_p^i)^2$	0.6961663	0.4079426	0.8774794
$C_i = (\lambda_0^i)^2$	0.0046791481	0.0135120631	97.934

(7) The **Abbe number** describes the chromatic dispersion of transparent materials: A high Abbe number (value around 60) corresponds to high chromatic dispersion. It is defined by $\nu_d = (n_d - 1)/(n_f - n_e)$, where n_d, n_f, and n_e are the refractive indices at the wavelengths 587.6, 486.1, and 656.3 nm, respectively. Calculate the Abbe number for fused silica using the result of Exercise (6).

(8) **Dielectrics as perfect reflector:** Write down the reflection coefficient in terms of energy for a medium of optical index $\tilde{n}$ (see Equation (2.9)). The optical index $\tilde{n}$ of this dielectric can be described by the Lorentz model, as discussed in Exercise (4). For the sake of simplicity, consider $\omega_p = \omega_0$. In the case $\gamma = 0.2\omega_0$, plot the reflection coefficient R around ω_0. Draw the plots for $\gamma = 0.2\omega_0$ and $\gamma = 0$. Discuss the case of perfect reflectors.

3

Wave Propagation at Interfaces

This chapter initially derives a pivotal relationship called the **Fresnel equations**, which plays an important role in plasmonics, especially in the surface plasmon polariton. From these Fresnel equations, several familiar phenomena emerge, such as the Snell–Descartes law and the Brewster angle. The Fresnel equations are initially meant for non-absorbing surfaces, but here, they will be generalized for dealing also with absorbing surfaces. Next, the behavior of an optical wave across **thin films** is also treated with the **three-layer model**. Finally, we propose a simple **numerical approach based on recursivity** to treat complex interfaces with more than three layers.

3.1 What Is an Interface?

An interface is a surface separating two media. From a mathematical point of view, an interface has a thickness equal to zero. In physics, however, parameters, such as the density of charge or the electric field, have to be continuous when crossing an interface, and they undergo a rapid but continuous transition. At this frontier between two media, two length scales have to be considered. The first one is the length scale of the variation of charge at the edge of the crystal. This is a local charge, and it is linked to the spatial extension of the electron cloud. It decays within a distance of the lattice parameters, which is around 0.5 nm. The second length scale is set by the electric field associated with the optical wave: It changes over distances of the order of magnitude of $\lambda/10$, which is 50 nm or so. It corresponds to the decay length of an electromagnetic wave inside a metal,

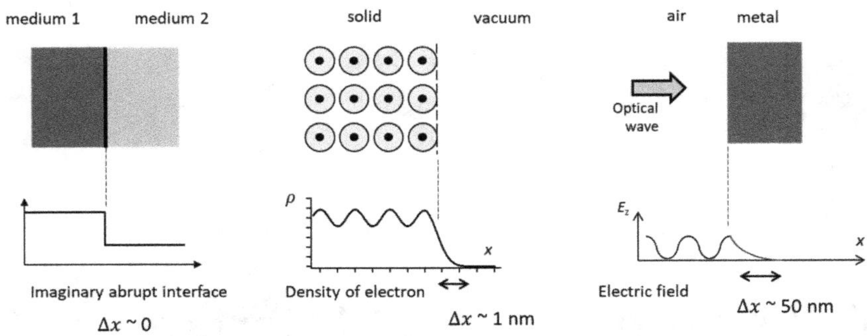

(a) (b) (c)

Fig. 3.1. Schematic view of the length scales to be considered at an interface. (a) A mathematical interface is characterized by an abrupt transition between the two media with a transition distance $\Delta x = 0$. (b) At the scale of the constitutive atoms of a solid, the electronic density decays to zero at the edge of the solid over a distance of the interatomic distance $\Delta x \approx 1$ nm. (c) For an optical wave, the transition distance when impinging on a solid interface is of the order of a fraction of the wavelength $\Delta x \approx 50$ nm.

which was given by Equation (1.39) in Chapter 1 and equals λ_0/n''. This length scale is 100 times greater than the first one and can be viewed as the *optical thickness* of an interface. Figure 3.1 shows a schematic description of these length scales. The optical phenomena that we are going to discuss in this chapter occur within this *optical layer* with a thickness that spans between 10 and 200 nm depending on the values of λ_0 and n''.

Since we are dealing with the effect of an interface on an optical wave, we expect changes in electromagnetic fields over length scales of 10 nm or so. Therefore, it is reasonable to consider the interface between two media as an abrupt change in the electronic density and also as an abrupt change in the optical index n. Our consideration does not depend directly on the lattice parameters of the material or on the crystalline orientation of the surface.

3.2 Light at Interfaces

We are now interested in predicting how light is reflected and transmitted at an interface. Even transparent interfaces produce a reflected beam: How does this happen? How much light is transmitted and how much does it depend on the incidence angle? The answers are given by the Fresnel coefficients. We first consider the case where the two media, **1** and **2**, are dielectrics. This problem is used in numerous textbooks by authors such as Born and Wolf [1], Griffiths [2], or more recently by Ware and Peatross [3]. However, most of them confine their discussion to the interface between two dielectrics, which corresponds to the case of a light beam impinging on glass or any insulating medium. In this chapter, we begin with this simple case of non-absorbing media, and we generalize to any media so that metallic surfaces (metals are highly absorbing media) can be treated. We also treat the case of three-medium interfaces that correspond to the case of thin films and will be crucial for treating the case of plasmon polariton.

Let's start by clarifying our notation. We consider a beam of light impinging on an interface. Light is described by its electric field. We define the *plane of incidence* as the plane containing the direction of the incident light and the normal of the interface (see Fig. 3.2). At this interface, the incident beam $\boldsymbol{E}_i$ gives rise to two other beams: the transmitted and the reflected beams: $\boldsymbol{E}_t$ and $\boldsymbol{E}_r$, respectively. The field vectors can all be decomposed into the two polarization states. p and s, where the two letters stand for *parallel* and perpendicular (*senkrecht* in German) to the plane of incidence, respectively. The p-polarized waves are also called *transverse*

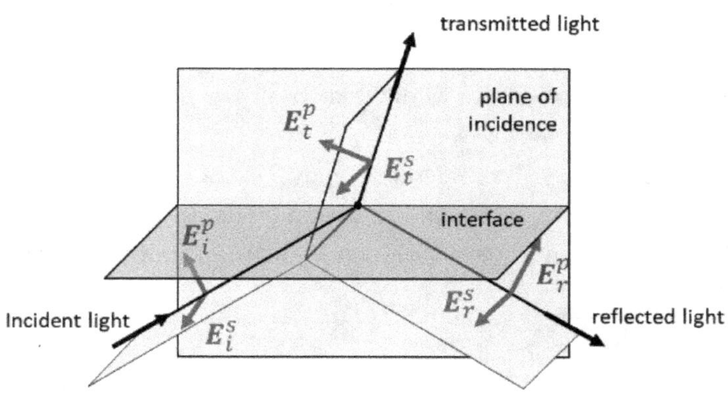

Fig. 3.2. An incident wave impinging on a surface defines the plane of incidence as the plane containing the incident direction and the normal of the interface plane. The electric fields can be decomposed into the two polarization states s and p.

magnetic (TM) and the s-polarized waves, *transverse electric* (TE). The corresponding fields are depicted in Fig. 3.2.

3.3 Boundary Conditions for Electromagnetic Waves at the Interface Between Two Media

We now consider the case of an interface between two media, denoted by **1** and **2**. Each medium is characterized by its dielectric permittivity, ε_1 and ε_2, and its magnetic permeability, μ_1 and μ_2. An interface is a geometric boundary, and we expect the dielectric permittivity to be discontinuous when crossing this boundary. From a physical standpoint, ε undergoes a rapid change from ε_1 to ε_2 on a distance scale much smaller than other typical distances and often smaller than the wavelength, as discussed in Section 3.1. When crossing the interface, the electromagnetic field obeys the so-called *boundary conditions.*

The boundary conditions derive directly from the Maxwell equations for dielectric media (see Chapter 1). The fields in medium **1** are denoted by subscript 1 (and similarly by subscript 2 for the fields in medium **2**). They are decomposed into two components: one component perpendicular to the interface and the second one parallel to this surface. For example, as shown in Fig. 3.3, the electric field in medium 2 is written as

$$\boldsymbol{E}_2 = (\boldsymbol{E}_2)_{\parallel} + (\boldsymbol{E}_2)_{\perp}.$$

Note that this decomposition is different from the p and s decomposition explained in the previous section.

The interface under consideration may host a surface charge, whose density is denoted by σ and a surface current denoted by $\boldsymbol{j}_s$. The boundary conditions are written as (see Born and Wolf [1] for a detailed derivation)

$$(\boldsymbol{D}_2)_{\perp} - (\boldsymbol{D}_1)_{\perp} = \sigma. \tag{3.1}$$

The normal component of the electric displacement changes abruptly through the interface by an amount of σ:

$$(\boldsymbol{B}_2)_{\perp} - (\boldsymbol{B}1)_{\perp} = 0. \tag{3.2}$$

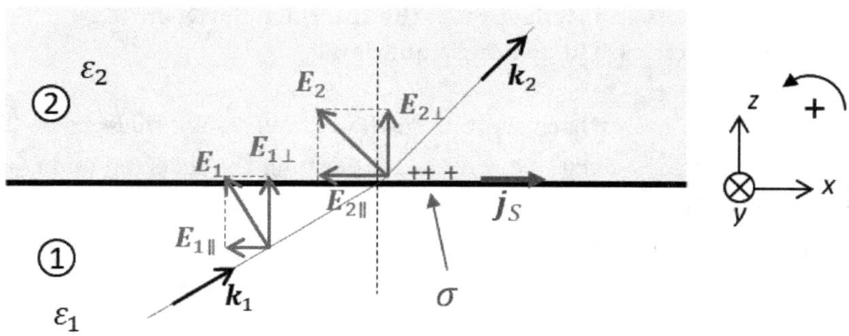

Fig. 3.3. Notation for the electric field at the interface between medium 1 and medium 2. Here, the electric field is supposed to be in the plane of incidence. The surface charge density σ and the surface current $\boldsymbol{j}_S$ are also represented.

The normal component of the magnetic induction is continuous across the interface:

$$(\boldsymbol{E}_2)_{\parallel} - (\boldsymbol{E}_1)_{\parallel} = 0. \tag{3.3}$$

The parallel component of the electric field is continuous across the interface:

$$(\boldsymbol{H}_2)_{\parallel} - (\boldsymbol{H}_1)_{\parallel} = j_s. \tag{3.4}$$

The parallel component of the magnetic field undergoes a discontinuity across the interface by an amount j_s.

In these cases, we are going to investigate the following: media **1** and **2** are nonmagnetic ($\mu_1 = \mu_2 = 1$) and the interface is deprived or free of charges ($\sigma = 0$) and of surface external current ($j_s = 0$). Consequently, the boundary conditions simplify. In particular, condition (3.1) becomes $(\boldsymbol{D}_2)_{\perp} = (\boldsymbol{D}_1)_{\perp}$, and since $\boldsymbol{D}_1 = \varepsilon_0\varepsilon_1\mathbf{E}_1$ and $\boldsymbol{D}_2 = \varepsilon_0\varepsilon_2\boldsymbol{E}_2$, it can be rewritten as

$$\varepsilon_1(\boldsymbol{E}_2)_{\perp} = \varepsilon_2(\boldsymbol{E}_1)_{\perp}. \tag{3.5}$$

Using the fact that $\boldsymbol{B}_1 = \mu_0\mu_1\boldsymbol{H}_1 = \mu_0\boldsymbol{H}_1$ and $\boldsymbol{B}_2 = \mu_0\boldsymbol{H}_2$, relation (3.4) also simplifies to

$$(\boldsymbol{B}_2)_{\parallel} - (\boldsymbol{B}_1)_{\parallel} = 0. \tag{3.6}$$

3.4 Transmission and Reflection at the Interface Between Two Dielectrics: The Fresnel Equations

We consider the case where the two media, **1** and **2**, are dielectrics. The boundary conditions given above allow for deriving the reflected and transmitted fields. At this stage, we do not consider the reflection by a metallic surface or by any absorbing layers. We also suppose that the wave arrives from medium **1**, which is purely transparent. Therefore, the dielectric permittivities ε_1 and ε_2 are real and positive numbers, so that the optical indices are defined without any ambiguity as $n_1 = \sqrt{\varepsilon_1}$ and $n_2 = \sqrt{\varepsilon_2}$.

The incident wave has a wave vector, denoted by $\boldsymbol{k}_{1i}$, impinging on the surface with an angle of incidence θ_1 (angles of incidence are measured relative to the normal of the interface). At this stage, we choose the orientation of the basis so that $\boldsymbol{e}_z$ is the normal of the surface and $(\boldsymbol{e}_x, \boldsymbol{e}_z)$ corresponds to the *plane of incidence*. Therefore, $\boldsymbol{k}_{1i}$ is written as $\boldsymbol{k}_{1i} = k_{1x}\boldsymbol{e}_x + k_{1z}\boldsymbol{e}_z$.

We consider linearly polarized electric fields, and their polarization state is expressed as a combination of the two polarization states p and s. We consider only isotropic media, which means that the polarization is not modified when the beam is traveling through the medium.

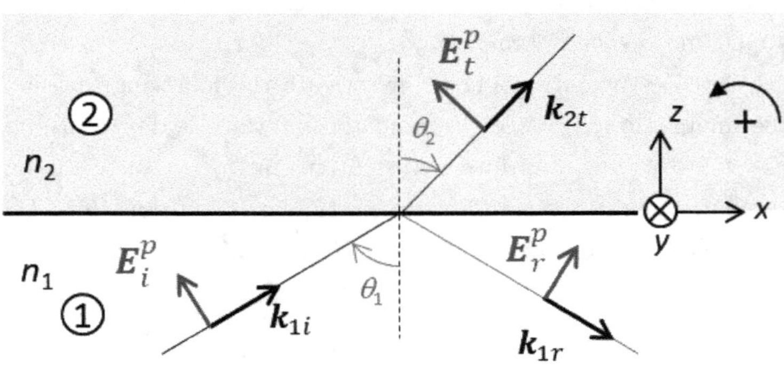

Fig. 3.4. Interface between medium **1** and medium **2** characterized by their indices n_1 and n_2, respectively. The incident, reflected, and transmitted electric fields are drawn in the case of a p-polarized incident field. This sketch defines all the sign conventions used in the formulas.

3.4.1 *Fresnel equations for p-polarized electric fields*

Here, we give a detailed derivation in the case of p polarization, and the corresponding quantities are indicated by a superscript p as shown in Fig. 3.4. The case of p-polarized electric field is also called the *TM mode.* The incident electric field is written as, with $\boldsymbol{E}_0^p$ being the incident amplitude,

$$\tilde{\boldsymbol{E}}_i^p(\boldsymbol{r}, t) = \boldsymbol{E}_0^p \exp[i(\boldsymbol{k}_1 \boldsymbol{i} \cdot \boldsymbol{r} - \omega t)].$$

At the interface, a reflected beam and a transmitted beam are generated, and they are characterized by their wave vectors $\boldsymbol{k}_{1r}$ and $\boldsymbol{k}_{2t}$ and their amplitudes $\tilde{\boldsymbol{E}}_{r0}^p$ and $\tilde{\boldsymbol{E}}_{t0}^p$, respectively. The wave vector in medium **2** is expressed as follows:

$$\boldsymbol{k}_{2t} = n_2 \frac{\omega}{c} \boldsymbol{u}. \tag{3.7}$$

The transmitted and reflected electric fields are written, respectively, as

$$\tilde{\boldsymbol{E}}_t^p(\boldsymbol{r}, t) = \boldsymbol{E}_t^p \exp[i(\boldsymbol{k}_{\mathbf{2t}} \cdot \boldsymbol{r} - \omega t)]$$

$$\text{and} \quad \tilde{\boldsymbol{E}}_r^p(\boldsymbol{r}, t) = \boldsymbol{E}_r^p \exp[i(\boldsymbol{k}_{\mathbf{1}r} \cdot \boldsymbol{r} - \omega t)]. \tag{3.8}$$

Our goal is to determine the wave vectors $\boldsymbol{k}_{1r}$ and $\boldsymbol{k}_{2t}$ and express the complex amplitudes $\boldsymbol{E}_{r0}^p$ and $\boldsymbol{E}_t^p$ as a function of the amplitude of the incident wave.

Before using the boundary conditions detailed in Section 3.3, we have to explicitly define the electric fields, $\boldsymbol{E}_1$ and $\boldsymbol{E}_2$. $\boldsymbol{E}_2$ is simply the transmitted field $\tilde{\boldsymbol{E}}_t^p$. However, $\boldsymbol{E}_2$ is the addition of the two fields present in medium **1**, $\tilde{\boldsymbol{E}}_r^p$ and $\tilde{\boldsymbol{E}}_t^p$.

Therefore, the continuity equation (3.3) can be projected on $\boldsymbol{e}_x$ and written as

$$E_{i0}^p \cos\theta_1 \exp[i(\boldsymbol{k}_{1i} \cdot \boldsymbol{r} - \omega t)] + E_{r0}^p \cos\theta_1 \exp[i(\boldsymbol{k}_{1\mathrm{r}} \cdot \boldsymbol{r} - \omega t)]$$

$$= -E_{t0}^p \cos\theta_2 \exp[i(\boldsymbol{k}_{2\mathrm{t}} \cdot \boldsymbol{r} - \omega t)]. \tag{3.9}$$

In this relation, the quantity $\exp(-i\omega t)$ can be factored in front of all terms. Therefore, we can simplify by dividing all terms by $\exp(-i\omega t)$. Moreover, the relation should be fulfilled for any position $\boldsymbol{r}$ at the interface. This is only possible if the x component of the three wave vectors are equal, which is written as

$$k_{ix} = k_{rx} = k_{tx}. \tag{3.10}$$

Note that relation (3.10) is the Snell–Descartes equation because it can also be written as

$$k_1 \sin\theta_1 = k_1 \sin\theta_1 = k_2 \sin\theta_2.$$

Since $k_1 = n_1\omega/c$ and $k_2 = n_2\,\omega/c$, it simplifies into

$$n_1 \sin\theta_1 = n_2 \sin\theta_2, \tag{3.11}$$

which is the well-known Snell–Descartes relation.

After the simplifications explained above, relation (3.10) becomes

$$E^p_{i0}\cos\theta_1 + E^p_{r0}\cos\theta_1 = -E^p_{t0}\cos\theta_2. \tag{3.12}$$

We now use the boundary condition (3.6), which states that the magnetic induction is continuous across the interface. Using the relation $\boldsymbol{B} = (\boldsymbol{k} \times \boldsymbol{E})/\mu_0$ and the fact that $\boldsymbol{E}_1 = \boldsymbol{E}_{1i} + \boldsymbol{E}_{1r}$, we obtain

$$k_i E^p_{i0} + k_r E^p_{r0} = k_t E^p_{t0},$$

which is written as $$n_1\frac{\omega}{c}E^p_{i0} + n_1\frac{\omega}{c}E^p_{r0} = n_2\frac{\omega}{c}E^p_{t0}. \tag{3.13}$$

We define the *reflection coefficient in amplitude* for polarization p as

$$r^p = E^p_{r0}/E^p_{i0}. \tag{3.14}$$

From Equations (3.12) and (3.13), we can eliminate E^p_{t0} and obtain the first Fresnel coefficient for reflection:

$$r^p = \frac{n_2\cos\theta_1 - n_1\cos\theta_2}{n_2\cos\theta_1 + n_1\cos\theta_2}. \tag{3.15}$$

Similarly, we define the *transmission coefficient in amplitude* for polarization p as

$$t^p = E^p_{t0}/E^p_{i0}. \tag{3.16}$$

By eliminating E^p_{r0} from Equations (3.12) and (3.13), we obtain the second Fresnel coefficient for the transmitted amplitude:

$$t^p = \frac{2n_1\cos\theta_1}{n_2\cos\theta_1 + n_1\cos\theta_2}. \tag{3.17}$$

Formula (3.15) is the foundation of the Brewster angle and the polarization of light at a dielectric interface. For example, it shows that a light beam reflected at an air–glass interface with an incidence angle of 57° is s-polarized because the coefficient r^p is zero at this angle.

3.4.2 *Fresnel equations in s-polarization*

The case of s-polarized electric field is very similar to the case of p-polarization. The electric and magnetic fields considered in this case correspond to the TE mode. The incident electric field is written as

$$\tilde{\boldsymbol{E}}_i^s(\boldsymbol{r},t) = \boldsymbol{E}_0^s \exp[i(\boldsymbol{k_{1i}}\cdot\boldsymbol{r} - \omega t)].$$

The reflected and transmitted beams are characterized by the same wave vectors as in the p-polarization case, $\boldsymbol{k}_{1r}$ and $\boldsymbol{k}_{2t}$, and the electric fields are denoted by $\tilde{\boldsymbol{E}}_{r0}^s$ and $\tilde{\boldsymbol{E}}_{t0}^s$.

The *reflection* and *transmission coefficients in amplitude* for polarization s are defined in a similar way as

$$r^s = E_{r0}^s/E_{i0}^s \quad \text{and} \quad t^s = E_{t0}^s E_{i0}^s.$$

By using boundary conditions (3.1) and (3.3), one can demonstrate the following:

$$r^s = \frac{n_1\cos\theta_1 - n_2\cos\theta_2}{n_1\cos\theta_1 + n_2\cos\theta_2}, \tag{3.18}$$

$$t^s = \frac{2n_1\cos\theta_1}{n_1\cos\theta_1 + n_2\cos\theta_2}. \tag{3.19}$$

3.5 Generalized Fresnel Equations at Metallic and Absorbing Surfaces

The Fresnel equations are very efficient for an interface between transparent media. However, if one wants to use this equation for calculating the reflected light by an absorbing media, a difficulty appears: In this case, $\tilde{n}_2$ is complex and the Snell–Descartes law (3.11) states that the angle θ_2 would also be complex, which is confusing. We resolve this issue and derive the case where an optical wave arrives from a transparent medium (n_1 is real) onto an absorbing layer (complex $\tilde{n}_2$). Although the incidence angle of the transmitted wave into medium **2** is not defined, the wave is likely to propagate through this absorbing medium. Remember from Chapter 1 that a complex optical index generates a wave having the usual propagating term accompanied by a decaying amplitude. It so happens that angles are not suited for treating absorbing media.

The angles were introduced when the wave vectors were projected onto the surface, which yielded k_{ix}, k_{rx}, k_{tx} in Equation (3.10). This relation shows that the projection of the wave vector k_x is conserved across the interface. Since $k_x = n_1 \sin\theta_1 \omega/c$, we define an invariant for the wave traveling across the interface as the complex quantity $\tilde{n}_2 \cdot s_2$ so that

$$\tilde{n}_2 \cdot s_2 = n_1 \sin\theta_1 = n_1 \cdot s_1. \tag{3.20}$$

s_2 is a complex quantity. The projection of the wave vector normal to the interface (analog to $\cos\theta_1$ and $\cos\theta_2$) is also generalized by the expressions $\sqrt{1-s_1^2}$ and $\sqrt{1-s_2^2}$. Note that in any medium, denoted by the index j, the wave vector components can be written as

$$k_{x,j} = n_j \frac{\omega}{c} s_j,$$

$$k_{z,j} = n_j \frac{\omega}{c} \sqrt{1-s_j^2}.$$

Therefore, the Fresnel equations for the reflected beams are generalized by replacing in (3.15–3.19) $\cos\theta_1$ and $\cos\theta_2$ with $\sqrt{1-s_1^2}$ and $\sqrt{1-s_2^2}$, respectively, and then s_2 by $(n_1/\tilde{n}_2)s_1$, with $s_1 = \sin\theta_1$:

$$\tilde{r}_{12}^P = \frac{\tilde{n}_2\sqrt{1-s_1^2} - n_1\sqrt{1-\left(\frac{n_1}{\tilde{n}_2}s_1\right)^2}}{\tilde{n}_2\sqrt{1-s_1^2} + n_1\sqrt{1-\left(\frac{n_1}{\tilde{n}_2}s_1\right)^2}}.$$

By dividing the two levels of the fraction by $\tilde{n}_2$, we see that $\tilde{r}_{12}^P$ depends only on $n_1/\tilde{n}_2$. We define the **reduced $\boldsymbol{n_{12}}$ index** as $\boldsymbol{n_{12} = n_1/\tilde{n}_2}$, and the formula becomes:

$$\tilde{r}_{12}^P = \frac{\sqrt{1-s_1^2} - n_{12}\sqrt{1-(n_{12}s_1)^2}}{\sqrt{1-s_1^2} + n_{12}\sqrt{1-(n_{12}s_1)^2}}. \tag{3.21}$$

In this generalized formula $s_1 = \sin\theta_1$, but we keep the notation s_1 because it could be complex. For example, in a three-layer system, where we seek the reflection coefficients between the two absorbing layers 2 and 3, the formula above can be used.

The other three Fresnel equations are written similarly:

$$\tilde{t}_{12}^{P} = \frac{2n_{12}\sqrt{1-s_1^2}}{\sqrt{1-s_1^2} + n_{12}\sqrt{1-(n_{12}s_1)^2}}. \tag{3.22}$$

For the s-polarized electric field, it yields

$$\tilde{r}_{12}^{s} = \frac{n_{12}\sqrt{1-s_1^2} - \sqrt{1-(n_{12}s_1)^2}}{n_{12}\sqrt{1-s_1^2} + \sqrt{1-(n_{12}s_1)^2}}, \tag{3.23}$$

$$\tilde{t}_{12}^{s} = \frac{2n_{12}\sqrt{1-s_1^2}}{n_{12}\sqrt{1-s_1^2} + \sqrt{1-(n_{12}s_1)^2}}. \tag{3.24}$$

These expressions of the Fresnel coefficients handle complex numbers since n_{12} is the complex, reduced optical index. It might also happen that $\sqrt{1-(n_{12}s_1)^2}$ is complex even if n_1 and n_2 are real optical indices. This is the case when $n_{12}s_1 = n_1/n_2 \sin\theta_1 > 1$, which corresponds to an incident wave impinging on the surface with an angle greater than the critical angle. This is the total internal reflection, which will be quickly reviewed in the next section.

These expressions are especially suited for programming the Fresnel coefficient. They apply to an immense variety of situations and not just visible light: X-ray radiations, infrared waves, etc.

3.6 Evanescent Waves

We now introduce evanescent waves, which are waves generated at some specific interfaces and which will be very important for understanding the propagating plasmon wave.

3.6.1 *Evanescent waves in the case of total internal reflection*

We consider an interface between two dielectric media characterized by their optical indices n_1 and n_2, which are both real numbers. The incident

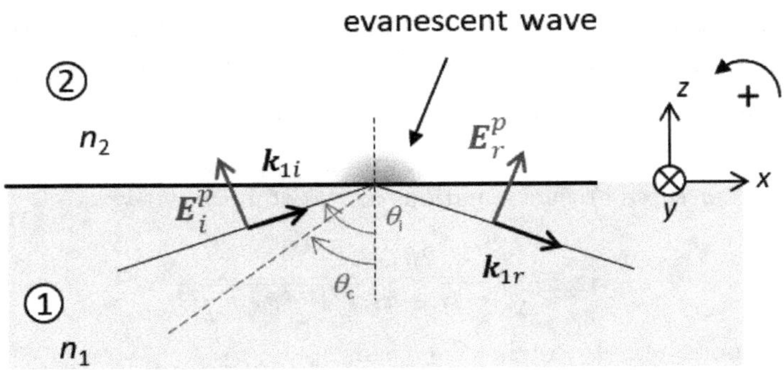

Fig. 3.5. Interface between medium 1 and medium 2, with $n_1 > n_2$. Here, the incidence angle θ_i is greater than the critical angle θ_c. This induces total reflection. In medium **2**, a particular wave hardly detectable is also generated: This is the evanescent wave.

wave reaches the interface from medium **1** with an incidence angle θ_i, and we suppose here that $n_1 > n_2$. The critical angle θ_c is defined by $\sin\theta_\mathrm{c} = n_2/n_1$. If $\theta_\mathrm{i} > \theta_\mathrm{c}$, this is the case of *total internal reflection*, and no light beam is transmitted, as represented in Fig. 3.5. However, we would like to know more about the electric field near this interface. We see that in medium **2**, there is an electric field $\boldsymbol{E}_2$ of a special nature.

In medium **2**, the amplitude of the electric field is given by

$$E_t^p = \tilde{t}_{12}^p E_i^p \text{ and } E_t^s = \tilde{t}_{12}^s E_i^s \text{ for } p \text{ and } s \text{ polarizations,}$$

where the transmission coefficients are given by relations (3.22) and (3.24). The wave in medium **2** is written as

$$E_t^p(\boldsymbol{r}, t) = E_t^p \exp[i(\tilde{\boldsymbol{k}}_2 \cdot \boldsymbol{r} - \omega t)]. \tag{3.25}$$

The key point here is that $\tilde{\boldsymbol{k}}_2$ is a vector with complex components $\tilde{\boldsymbol{k}}_{2x}$ and $\tilde{\boldsymbol{k}}_{2z}$ that determines the nature of the transmitted wave. Here, we focus our attention on the exponential term rather than on the amplitude E_t^p.

From relation (3.10), we know that

$$\tilde{k}_{2x} = k_{1x} = (\omega/c) n_1 \sin\theta_i = (2\pi/\lambda_0) n_1 \sin\theta_i. \tag{3.26}$$

Therefore, $\tilde{k}_{2x}$ is real and can be written as k_{2x} (without the *tilde* sign indicating that a number is complex). Moreover, the dispersion relation in medium **2** is

$$(\tilde{\boldsymbol{k}}_2)^2 = \left(n_2 \frac{\omega}{c}\right)^2. \tag{3.27}$$

Since $(\tilde{\boldsymbol{k}}_2)^2 = (k_{2x})^2 + (\tilde{k}_{2z})^2$, we can compute the following expression for $\tilde{k}_{2z}$:

$$(k_{2z})^2 = \left(n_2 \frac{\omega}{c}\right)^2 - \left(n_1 \frac{\omega}{c} \sin\theta_i\right)^2 = \left(\frac{\omega}{c}\right)^2 [n_2^2 - n_1^2 \sin^2\theta_i]. \tag{3.28}$$

Since $\theta_\mathrm{i} > \theta_\mathrm{c}$, the expression $n_2^2 - n_1^2 \sin^2\theta_i$ in (3.28) is negative, and the component $\tilde{k}_{2z}$ of the transmitted wave vector is a pure imaginary number.

Let's write $\tilde{k}_{2z} = i/\beta_2$. β_2 is real and has the dimension of length:

$$\beta_2 = \frac{c}{\omega} \cdot \frac{1}{\sqrt{n_1^2 \sin^2\theta_i - n_2^2}} = \lambda_0/2\pi \cdot \frac{1}{\sqrt{n_1^2 \sin^2\theta_i - n_2^2}}. \tag{3.29}$$

We can now substitute the expressions of k_{2x} and $\tilde{k}_{2z}$ into (3.25), which yields

$$E_t^p(\boldsymbol{r}, t) = E_t^p \exp\left(-z/\beta_2\right) \exp[i(k_{2x} \cdot x - \omega t)]. \tag{3.30}$$

This wave exhibits a propagative behavior along the x-axis, which is along the interface and has a vanishing character along the z direction away from the interface, with the decaying factor $\exp(-z/\beta_2)$. The typical decay length is given by β_2. This wave is an inhomogeneous wave and is not a plane wave anymore. It is called an *evanescent wave.*

An important feature of the evanescent wave shown in relation (3.30) is that it is not propagative along the z-axis. It is not an attenuated or damped wave, which would have a propagative factor of the form $\exp[i(k_{2z} \cdot z - \omega t)]$.

In the case of glass–air interface, the critical angle is $\theta_\mathrm{c} = 41.1°$. If we consider an incidence angle of $\theta_\mathrm{i} = 45°$ and $\lambda_0 = 500\,\mathrm{nm}$, the decay length is $\beta_2 = 201\,\mathrm{nm}$. It decreases to 78 nm with $\theta_\mathrm{i} = 70°$. These values show that the evanescent wave barely escapes from the interface. Born and Wolf show that no energy is conveyed by this wave and that the energy is entirely reflected as expected [1]. Nice illustrations of this evanescent wave have been produced using finite-difference time-domain (FDTD) calculations by the group of Hogan from the University of Reading (UK) [4] and represented in Fig. 3.6. The structure of the evanescent waves will be discussed in greater detail in Chapter 4.

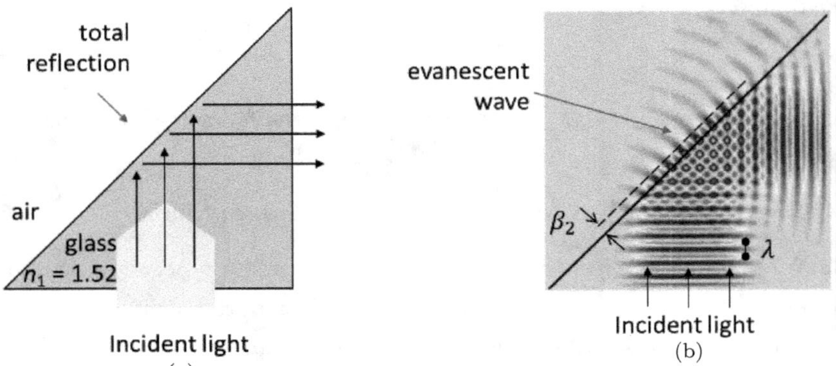

Fig. 3.6. Evanescent wave at a glass–air interface. (a) An incident light propagated into glass (optical index $n_1 = 1.52$) and reaches the glass–air interface with an angle of 45°, which is greater than the critical angle ($\theta_c = 41.1°$). The light undergoes total reflection. (b) The calculation performed by Hogan shows that the amplitude of the s-polarized component of the light. At the interface, the evanescent wave overlaps the frontier within a distance $\beta_2 = 200$ nm for visible light at $\lambda_0 = 500$ nm.

Source: Reproduced from Ref. [4].

3.6.2 *Waves in the case of metallic reflection*

We consider now the case of a wave impinging on a metallic surface with an incidence angle that can take any value between −90° and +90°, as shown in Fig. 3.7. Based on common experience, we know that the light beam is totally reflected, and this is the working principle of metallic mirrors. Nevertheless, we can investigate the wave on the metallic side of the interface. This wave has the same form as in the case of the total reflection above, given by Equation (3.25). The nature of the wave in medium **2** is controlled by the wave vector $\tilde{\boldsymbol{k}}_2$, whose two components are written as $\tilde{k}_{2x}\boldsymbol{e}_x$ and $\tilde{k}_{2z}\boldsymbol{e}_z$. $\tilde{k}_{2x}$ is actually a real number, as in (3.26): $k_{2x} = (\omega/c)n_1 \sin\theta_i$.

$\tilde{k}_{2z}$ is obtained similarly, and according to (3.28), it is written as

$$(\tilde{k}_{2z})^2 = \left(\tilde{n}_2\frac{\omega}{c}\right)^2 - \left(n_1\frac{\omega}{c}\sin\theta_i\right)^2 = (\tilde{\varepsilon}_2 - \varepsilon_1\sin^2\theta_i)\left(\frac{\omega}{c}\right)^2, \tag{3.31}$$

since $\tilde{n}_2^2 = \tilde{\varepsilon}_2$ and $n_1^2 = \varepsilon_1$. This allows computing the real and imaginary parts of $\tilde{k}_{2z}$:

$$\tilde{k}_{2z} = \frac{\omega}{c}\sqrt{\tilde{\varepsilon}_2 - \varepsilon_1\sin^2\theta_i}. \tag{3.32}$$

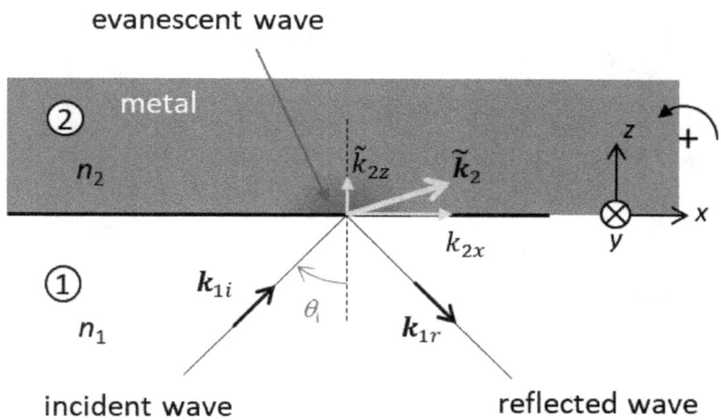

Fig. 3.7. Evanescent wave generated at the surface of a metal. The incoming wave arrives from below.

$\tilde{k}_{2z}$ is a complex number with real and imaginary parts: $\tilde{k}_{2z} = k'_{2z} + ik''_{2z}$. Therefore, introducing the expressions of $k_{2\text{x}}$ and $\tilde{k}_{2z}$ into (3.25) yields

$$E_t^p(\boldsymbol{r}, t) = E_t^p \exp(-k''_{2z} \cdot z) \exp[i(k_{2x} \cdot x + k'_{2z} \cdot z - \omega t)]. \tag{3.33}$$

This wave is inhomogeneous and has a propagation vector $\boldsymbol{k}_2 = k_{2x} = k_{2x} \cdot \boldsymbol{e}_x + k'_{2z}\boldsymbol{e}_z$ and a decay factor $\exp(-k''_{2z} \cdot z)$. This decay factor quickly prevents any propagation when moving away from the interface. For example, with an incidence angle of 45° and an air–gold interface, the decay length given by $1/k''_{2}z$ is 29 nm at $\lambda_0 = 633\,\text{nm}$. This very quick attenuation of the wave on the metal side is known as the skin depth and is discussed in the next section.

3.6.3 *Penetration of a wave into a metallic surface: Skin depth*

The dielectric permittivity of a metal is essentially a real and negative number (see Chapter 2). For example, $\tilde{\varepsilon}_2 = -11.7 + 1.26i$ for gold at $\lambda_0 = 633\,\text{nm}$, and we can approximate this value to $\tilde{\varepsilon}_2 \approx -11.7$. Moreover, if we consider the case of normal incidence, $\theta_i = 0$, and the square root in (3.32) simplifies into $\sqrt{\tilde{\varepsilon}_2} = in''_2$, where n''_2 is the imaginary part of the optical index of the metal. Finally, $\tilde{k}_{2z}$ at normal incidence is written as

$$\tilde{k}_{2z} = i\, n''_2 \frac{\omega}{c}. \tag{3.34}$$

Table 3.1. Calculations of the skin depth for various metals and various wavelengths.

	λ_0	520 nm	633 nm	1550 nm
Gold	n_2	$0.64 + i2.07$	$0.18 + i3.43$	$0.52 + i10.7$
	$\boldsymbol{\beta_2}$	**40 nm**	**29 nm**	**23 nm**
Silver	n_2	$0.05 + i3.32$	$0.06 + i4.28$	$0.14 + i11.4$
	$\boldsymbol{\beta_2}$	**25 nm**	**24 nm**	**22 nm**
Aluminum	n_2	$0.49 + i4.84$	$1.45 + i7.54$	$1.58 + i15.7$
	$\boldsymbol{\beta_2}$	**13 nm**	**13 nm**	**16 nm**

Substituting the values k_{2x} and $\tilde{k}_{2z}$ into (3.25) yields the same expression as (3.30) but with a different value for β_2:

$$E_t^p(\boldsymbol{r}, t) = E_t^p \exp(-z/\beta_2) \exp[i(k_{2x} \cdot x - \omega t)]. \tag{3.35}$$

This is again the structure of an evanescent wave.

Since β_2 is defined by $\tilde{k}_{2z} = i/\beta_2$, we obtain

$$\beta_2 = \frac{1}{n''_2} \frac{c}{\omega} = \frac{1}{n''_2} \frac{\lambda_0}{2\pi}. \tag{3.36}$$

β_2 is the penetration depth of the wave at normal incidence into the metal, also known as the *skin depth* (skin depth in amplitude here). The skin depth depends on the nature of the metal and on the wavelength. Table 3.1 shows, for example, that for gold at 633 nm, the skin depth is 29 nm. For aluminum, it is just 13 nm. This concept will be especially relevant when we deal with metallic nanoparticles of 20 nm diameter or so, since they indicate that the optical wave completely penetrates the nanoparticle. A nanoparticle will therefore handle the le electromagnetic field completely differently from a metal surface. This will be the topic of Chapter 7.

3.7 Transmission and Reflection Across Thin Films

Thin films are constituted by a series of parallel interfaces. The simplest case of thin films is a single layer of thickness d_2 placed between two semi-infinite media, e.g. a glass window in air. This is a three-medium system, and the problem lies in calculating how an electromagnetic wave crosses this film and how much light is reflected and transmitted. We would like to calculate the reflection ($\mathcal{R}_2$) and transmission ($\mathcal{T}_2$) coefficients. This calculation includes considering the interference effects due to multiple reflections within the film. For example, this is the situation for a thin piece of glass in air. If this glass is protected by an antireflection coating (usually a thin aluminum layer), it turns into a four-layer system. Analytical solutions exist for two-layer and three-layer systems only. Surprisingly, there is no analytical solution when considering four or more layers. In these cases, numerical solutions must be used. Solutions for two-layer systems are given by the Fresnel coefficients, and the solutions for three-layer systems will be the subject of this section.

3.7.1 *The three-layer system*

Let us now consider a system made of three layers **1**, **2**, and **3** of indices $\tilde{n}_1$, $\tilde{n}_2$, and $\tilde{n}_3$, respectively. Medium **2** has a thickness d_2, and the other two are considered semi-infinite in the z direction. Since there are two interfaces, two series of Fresnel coefficients can be calculated: one at interface **1–2** and the other at interface **2–3**. The reflection ($\mathcal{R}_2$) and transmission ($\mathcal{T}_2$) coefficients of this thin film are calculated as a function of these Fresnel coefficients (Fig. 3.8). $\mathcal{R}_2$ and $\mathcal{T}_2$ are the coefficients in amplitude for the electric field. The calculation considers the multiple reflections within layer **2** (interferences patterns) and yields the following results (see Exercise 6 at the end of this chapter for the derivation):

$$\mathcal{R}_2 = \tilde{r}_{12} + \frac{\tilde{t}_{12} \cdot \tilde{t}_{21} \cdot \tilde{r}_{23} \cdot e^{2ik_0\delta_2}}{1 + \tilde{r}_{12} \cdot \tilde{r}_{23} e^{2ik_0\delta_2}}, \tag{3.37}$$

$$\mathcal{T}_2 = \frac{\tilde{t}_{12} \cdot \tilde{t}_{23} \cdot e^{ik_0\delta_2}}{1 + \tilde{r}_{12} \cdot \tilde{r}_{23} e^{2ik_0\delta_2}}. \tag{3.38}$$

k_0 is the wavelength in vacuum, and δ_2 has the dimension of length with the following expression:

$$\delta_2 = \tilde{n}_2 d_2 / \sqrt{1 - s_2^2}. \tag{3.39}$$

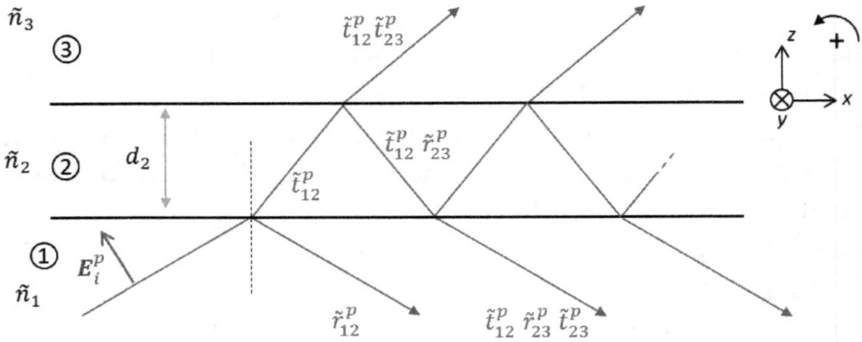

Fig. 3.8. A three-layer system of two parallel interfaces between three media **1**, **2**, and **3**, with the respective complex indices $\tilde{n}_1$, $\tilde{n}_2$, and $\tilde{n}_3$. Here, an incoming p-polarized wave impinges on interface **1–2**. The reflected and transmitted amplitudes are calculated with the transmission and reflection Fresnel coefficients r_{12} and t_{12}, respectively. The transmitted beam undergoes a series of multiple reflections and transmissions through the various interfaces.

More precisely, δ_2 is the optical path of light in medium 2 at oblique incidence. In (3.37) and (3.38), the superscripts p and s of the Fresnel coefficients are not indicated since the formulas for $\mathcal{R}_2$ and $\mathcal{T}_2$ are identical for both polarizations.

After some mathematical treatment (see Exercise 7 at the end of this chapter), formulas (3.27) and (3.28) can be simplified into the following [1], where δ_2 is still given by (3.39):

$$\mathcal{R} = \frac{\tilde{r}_{12} + \tilde{r}_{23}e^{2ik_0\delta_2}}{1 + \tilde{r}_{12}\cdot\tilde{r}_{23}e^{2ik_0\delta_2}}, \tag{3.40}$$

$$\mathcal{T} = \frac{\tilde{t}_{12}\cdot\tilde{t}_{23}e^{2ik_0\delta_2}}{1 + \tilde{r}_{12}\cdot\tilde{r}_{23}e^{2ik_0\delta_2}}. \tag{3.41}$$

These formulas handle the amplitude of the electric field. For example, the p-polarized electric field reflected by the three-layer system, shown in Fig. 3.8, is written as

$$\boldsymbol{E}^p_r = \mathcal{R}\boldsymbol{E}^p_i, \tag{3.42}$$

where $\mathcal{R}$ is calculated using the $\tilde{r}^p_{12}$ and $\tilde{r}^p_{23}$ coefficients. If one or more of the three media is absorbing, the generalized Fresnel coefficients introduced in Section 3.5 must be used.

3.7.2 *Thin films of more than three layers*

The formulas given above are robust enough to be applied to any type of layers characterized by their complex indices. However, for more than three layers, there is no analytical formula. Calculating the reflection and transmission coefficients through a many-layer system can be carried out by two distinct numerical approaches. The most common is based on a matrix method pioneered by F. Abeles in 1948 [5]. Each layer is characterized by a transfer matrix, and the resulting coefficients can be computed by multiplying the matrices. An elegant method was proposed earlier in 1934 by Rouard [6], which was explained by Heavens in 1955 [7], and is based on the recursion property. This property consists of a function that needs to call itself to proceed with the calculation until a break condition is reached. In programming, this was first introduced in computer languages, such as Algol and LISP, in the early 1970s and is now available in many computer codes. Actually, the recursion approach promoted by Rouard [6] seems to have been mostly overlooked except a few cases (see, for example, Refs. [8–10]). Let's briefly explain the Rouard method.

We consider a series of $m+1$ media of indices $\tilde{n}_l (l = 1, 2, \ldots, m+1)$ defining $m-1$ thin layers [7], as shown in Fig. 3.9(a). We are interested in the calculation of the transmitted amplitudes $\mathcal{T}_m$ and reflected amplitudes $\mathcal{R}_m$ through this set of thin layers. We first consider the last film (medium $m-1$): If we knew the amplitude of the field incident on this film, we would just apply the three-layer formula to this last layer. Therefore, according to Equation (3.41), the transmitted amplitude $\mathcal{T}_m$ through this system is written as

$$\mathcal{T}_m = \frac{t_{m-1,m} \cdot t_{m,m+1} \cdot e^{i\delta_m}}{1 + r_{m-1,m} \cdot r_{m,m+1} \cdot e^{2i\delta_m}}. \tag{3.43}$$

In this expression, $t_{m,m+1}$ and $r_{m,m+1}$ are the Fresnel coefficient of the last interface between medium m and medium $m+1$, which can be straightforwardly calculated using Equations (3.21) or (3.22). $t_{m-1,m}$ is more complex since this is the transmitted amplitude through the m previous media, which is given by $\mathcal{T}_{m-1}$ (see Fig. 3.9(b)). It shows that the $m+1$-layer problem can be expressed as a function of the same problem but with

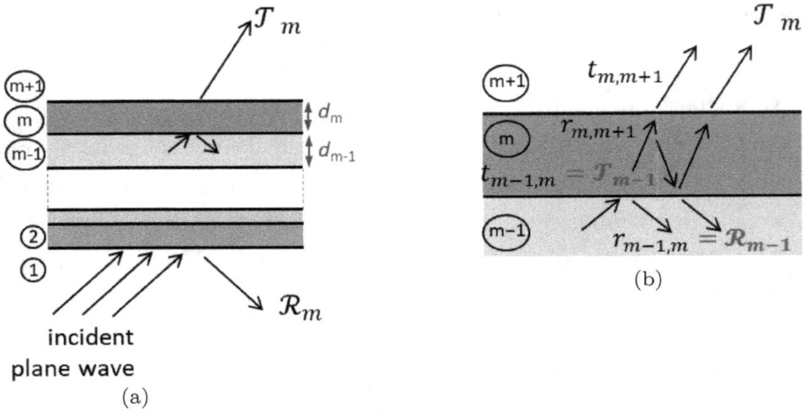

Fig. 3.9. (a) Recursion approach for calculating the transmission and reflection coefficients across a system made of $m - 1$ layers ($m + 1$ media). (b) The transmitted electric field into medium $m + 1$ is given by the coefficient $\mathcal{T}_m$ calculated using a three-layer model, where the incident field results from the transmission through the previous $m - 2$ layers (coefficient $\mathcal{T}_{m-1}$). This step-by-step calculation is part of the recursion capability of most modern programming languages.

only m layers. Similarly, $r_{m-1,m} = \mathcal{R}_{\mathrm{m}-1}$. This means that the calculation calls itself but with a decreasing index m and a decreasing complexity This recursion calculation proceeds until interface **1**–**2** is reached, where the reflection/transmission coefficients are given by the Fresnel equations. This calculation is carried out for each wavelength of interest. It also assumes that none of the media are anisotropic since the p and s polarizations are treated independently. If the complex indices of each layer are precisely known, the reflectivity and transmission spectra can be computed accurately. In principle, there is no limit in terms of the number of layers in recursion method other than that of computing memory.

3.7.3 *Reflected and transmitted energy*

A light detector is sensitive to the energy and not to the amplitude of the waves. In order to compare these calculations with experimental results, we have to calculate the energy per time and per unit area, which is the power flux, using the Poynting vector. The power flux of the reflected and transmitted beams is written as follows:

$$\text{Reflected energy:} \quad R = |\mathcal{R}_m|^2, \tag{3.44}$$

$$\text{Transmitted energy:} \quad T = \frac{\cos\theta_m}{\cos\theta_1}|\mathcal{T}_m|^2. \tag{3.45}$$

3.8 Examples of Application

3.8.1 *Case of a metallic layer on glass*

Here, we apply the formula based on the Fresnel coefficients developed previously to the case of a thin metallic gold film deposited on glass. This is a three-layer system of glass–gold–air. The gold film is 50 nm thick. We suppose in the present case that the beam is incident on the glass at an incidence angle θ, as shown in Fig. 3.10(a). If the light is s-polarized, the reflected energy is roughly independent of the incidence angle and close to 92% (see Fig. 3.10(b)). However, for the p-polarized beam, the reflectivity exhibits a pronounced minimum at 40.6°. This dip is the signature of the surface plasmon polariton (SPP) wave that is generated at the metal–air

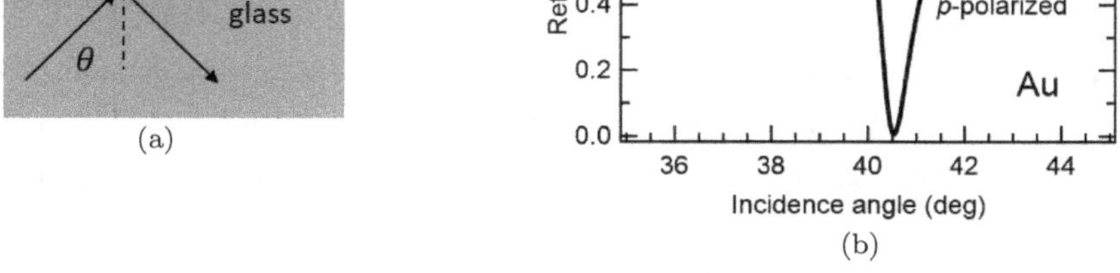

Fig. 3.10. (a) Geometry for the internal reflection of light on a thin gold film (50 nm) evaporated on glass ($n_1 = 1.61$). The optical index of gold is taken as $n_2 = 0.18+3.49i$ at 633 nm. (b) The p-polarized light undergoes a clear minimum when the incidence angle reaches 40.6°, which is the signature of the surface plasmon polariton. In the case of a bulk gold layer, there is no such minimum, nor is there one in the case of an s-polarized light.

interface and absorbs the energy of the incident wave. The SPP wave is the topic of the next three chapters and will not be discussed further here. Note that the same *p*-polarized wave impinging on a glass–gold interface that comprises only two layers does not generate any SPP wave, as shown in Fig. 3.10(b).

The calculations above can be carried out using analytical formulas (3.41) and (3.42) applicable to three-layer systems and using the generalized Fresnel equations (3.21)–(3.24) for the metallic layer.

3.8.2 *Case of an antireflection coating for lenses or glasses*

We consider now the simple case of a glass surface in air, such as the surface of a lens of a camera or of a glass. The air–glass interface corresponds to an abrupt change in index and a non-negligible amount of the incoming energy is reflected even if glass is transparent (see the *two-layer case* in Fig. 3.11(a)). For example, if the glass index is $n_1 = 1.65$ (it corresponds to *organic glasses* currently used for fabricating light-weight glasses) and at an incidence angle of 45°, the reflectivity is 1.6% (see Fig. 3.11(b)). The reflection or glare is clearly visible, as shown in Fig. 3.11(c). This glare is unpleasant because the eyes of the bespectacled person cannot be seen. In optical instruments, these reflections reduce the contrast.

An antireflection coating is a thin layer of dielectric with an adjusted thickness that will minimize this glare: See the *three-layer case* in Fig. 3.11(a). It works by generating a destructive interference between the two reflected waves: the wave reflected from the top surface and the wave reflected from the buried interface. Of course, the interference is optimized for one wavelength and is approached for the other wavelengths. The optimization of the antireflection efficiency goes through the layer thickness and its optical index. For example, if the glass ($n_1 = 1.65$) is coated with 120 nm of a dielectric of optical index $n_1 = 1.28$, the reflectivity is drastically decreased from 1.6 to 0.4%, as shown in Fig. 3.11(b). The curve shows that this coating is the most efficient at 480 nm and not uniform over the visible range. This calculation was carried out for a three-layer system using

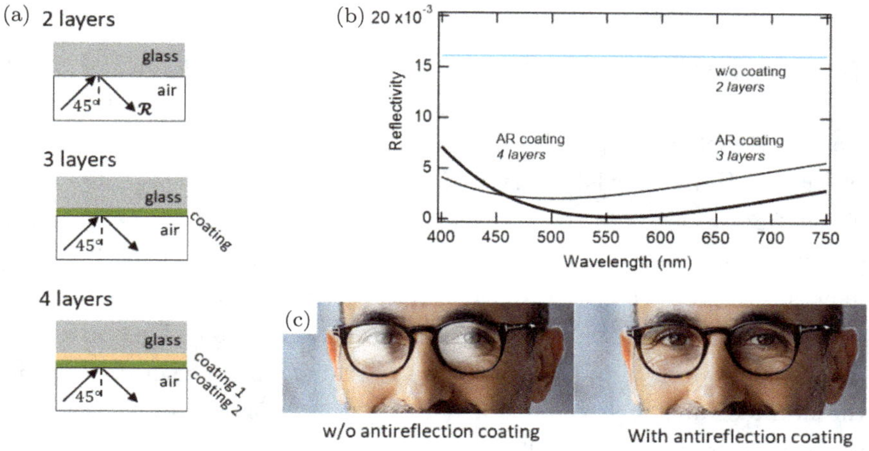

Fig. 3.11. (a) Description of the coating for minimizing the reflection on glass: *2 layers* correspond to the non-coated glass; *3 layers* correspond to simple coating, and *4 layers* to a coating with two thin layers. (b) Calculation of the reflectivity at an air–glass ($n = 1.65$) interface at 45° incidence. Without coating, 1.6% of the light is reflected. By adding a first coating ($n_1 = 1.28$ and thickness = 120 nm), the reflectivity drops by about 0.4%. With two antireflecting coatings properly adjusted ($n_1 = 1.25$, $d_1 = 30$ nm and $n_2 = 1.45$, $d_2 = 70$ nm), the reflectivity drops below 0.1%. (c) Illustration of the effect of the antireflective coating.

the analytical formula (3.41). See Exercises for further details on how to optimize the parameters of a dielectric layer.

In some cases, the antireflection efficiency can be improved by adding more coating layers. For example, if we consider two layers deposited on glass, we have a *four-layer* system depicted in Fig. 3.11(a). We consider the following optical indices and thicknesses: $n_1 = 1.25$, $d_1 = 30$ nm and $n_2 = 1.45$, $d_2 = 70$ nm. The reflection efficiency is shown in Fig. 3.11(b) and exhibits an even more reduced value, with only 0.1% reflectivity at 45°. This graph also shows that the reflectivity increases in the blue wavelength range (400–450 nm), and this explains why the antireflection coated lenses sometimes exhibit a faint blue tint.

These calculations for a *four-layer* system were carried out using the recursive method based on the Rouard approach, as explained in Section 3.7.2.

3.8.3 *Case of infrared transmitted beam through a silicon wafer*

The Fresnel coefficients and the related calculation methods can be applied not only to the visible light but to any radiation that is described with electromagnetic plane waves: This is the case for X-rays, UV light, and infrared radiations. However, what is crucial is to know the exact values of the complex refractive index in the wavelength range of interest. Sometimes, this complex index also undergoes significative variations due to specific resonances, such as electronic transitions in the UV or X-ray ranges, molecular absorption in the visible range, or phonon modes in the infrared range. As an example, we show the transmission spectrum through a silicon wafer in the infrared range between 400 and 2000 cm^{-1} which corresponds to the wavelengths between 5 and 25 μm [11].

Let's consider a piece of a silicon wafer of thickness 500 μm, with its two sides mirror polished, placed in air. This makes up a three-layer system of air–silicon–air. Here, we use low-doped silicon so that the infrared photons will not trigger electronic transitions. The refractive index of silicon is obtained from Palik's compilation of experimental refractive index data [12]. This real part of the index does not vary much over the range 400–2000 cm^{-1} and stays at 3.42. The imaginary part, however, undergoes a sudden change from 3×10^{-7} to 1×10^{-3}, indicating infrared resonances. Let's see how this impacts the transmission spectrum. The calculation is based on the three-layer model explained in Section 1.7 using the generalized Fresnel equations that can handle absorbing media (with a non-zero imaginary part of the refractive index). We consider *p*-polarized incoming plane waves, and the incidence angle was set to 72°, which corresponds to the Brewster angle of the air–silicon interface. It minimizes the multiple reflections and almost suppresses the interference patterns (parallel plates interference patterns). The simulated spectrum is shown in Fig. 3.12(a) and is compared with the experimental spectrum obtained with a Fourier-transform infrared (FTIR) spectrophotometer [11]. The agreement between the experimental and simulated spectra is remarkable.

The simulation spectrum displays slight interference patterns that are barely visible in the experimental one because of the smearing out due to

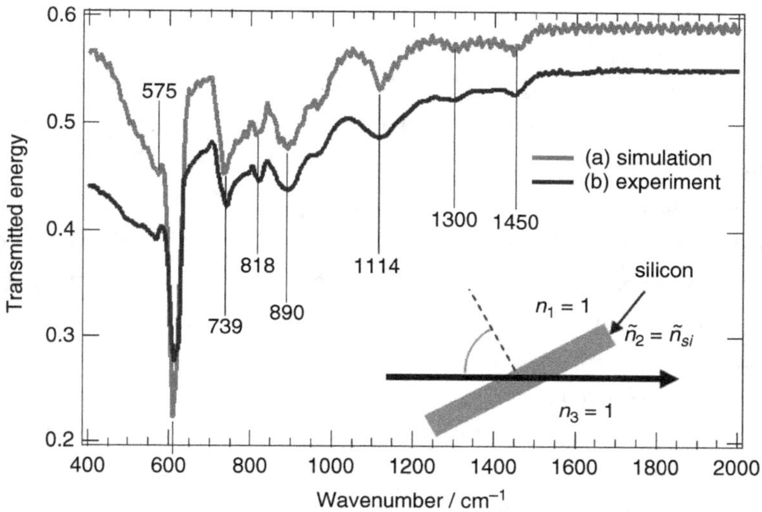

Fig. 3.12. (a) Simulated and (b) measured transmission spectra through a 500 μm thick low-doped silicon sample. In both cases, the incidence angle is 72°.

Table 3.2. Infrared absorption peaks of a crystalline silicon wafer with their assignment.

$\nu(\mathrm{cm}^{-1})$	575	609	739	818	890	1114	1300	1450
Intensity	Weak	Strong	Medium	Weak	Medium	Strong	Weak	Weak
mode	Si	Si	Si	Si	Si	Si-O	Si	Si
	LO+TA	TO+TA	LO+LA	TO+LA	TO+LO	stretching	2TO+LA	3TO

the experimental spectral resolution being limited to 4 cm^{-1}. Of the eight absorption peaks visible in Fig. 3.12, seven are due to phonon modes from the silicon crystal lattice [13] and one is due to the Si–O stretching mode. This latter peak is caused by a small amount of oxygen inserted into the silicon lattice. The seven other peaks are various harmonics resulting from combinations of the TO and LO modes of silicon at 521 cm^{-1} (threefold degenerate at zone center: LO, 2TO) with the TA and LA modes. These peaks will not be discussed here, but their assignment is given in Table 3.2.

This example shows how accurate the simulations based on the generalized Fresnel coefficients can be, provided the optical complex index is known.

3.9 Conclusion

This chapter has developed further the concepts explained in the first two chapters originating from the Maxwell equations. We have shown how electromagnetic plane waves are modified at interfaces because of the boundary conditions. We have extended the derivation of the Fresnel coefficients toward a generalized form so that even absorbing media can be handled. These Fresnel coefficients are meant for computing the transmitted and reflected electric fields through a single interface. But when it comes to thin layers, there are at least two interfaces. We presented a suitable model that can be used even in the case where more than three media are considered and where no analytical formula exists. These multilayer coefficients for transmission and reflection allow the explanation of many classical optical phenomena, such as the Brewster angle, skin depth of metals, and antireflection coating. We have also noted that in the case of a thin metallic layer placed on a dielectric, an anomaly in the reflected beam could be observed. This is the signature of an SPP wave, which will be treated in detail in the next three chapters.

References

[1] Born M. and Wolf E. 1999. *Principles of Optics: Electromagnetic Theory of Propagation, Interference and Diffraction of Light* (Cambridge: Cambridge University Press).

[2] Griffiths D. J. 2017. *Introduction to Electrodynamics* (New York: Cambridge University Press).

[3] Ware M. and Peatross J. 2020. *Physics of Light and Optics (Black & White)* (Brigham Young University, Department of Physics).

[4] Hogan R. 2022. Maxwell2D: Animations of electromagnetic waves. http://www.met.reading.ac.uk/clouds/maxwell/.

[5] Abeles F. 1948. Sur la propagation des ondes électromagnétiques dans les milieux stratifies. *Annales de Physique* **12**, 504–520.

[6] Rouard P. 1937. Etudes des propriétés optiques des lames métalliques très minces. *Annales de Physique* **11**, 291–384.

[7] Heavens O. S. 1991. *Optical Properties of Thin Solid Films* (North Chelmsford, Massachusetts: Courier Corporation).

[8] Crawford O. H. 1988. Radiation from oscillating dipoles embedded in a layered system. *The Journal of Chemical Physics* **89**, 6017–6027.

[9] Lecaruyer P., Maillart E., Canva M. and Rolland J. 2006. Generalization of the Rouard method to an absorbing thin-film stack and application to surface plasmon resonance. *Applied Optics* **45**, 8419–8423.

[10] Tomaš M.-S. 2010. Recursion relations for generalized Fresnel coefficients: Casimir force in a planar cavity. *Physical Review A* **81**, 044104.

[11] Pluchery O. and Costantini J.-M. 2012. Infrared spectroscopy characterization of 3C-SiC epitaxial layers on silicon. *Journal of Physics D* **45**, 495101.

[12] Lynch D. W. and Hunter W. R. 1985. (ed.) E. D. Palik (New York: Academic Press).

[13] Kittel C. 2004. *Introduction to Solid State Physics* (New York: John Wiley & Sons, Inc.).

Exercises

(Exercises with an asterisk (*) are difficult.)

(1) (*) **Brewster angle:** Consider an air–glass interface and a light beam incident at an angle θ_1.

  (a) Derive the Fresnel equations for the amplitude of the reflected electric field for the p and s polarizations as a function of θ_1 and identify a particular value of θ at which the p component of the electric field is zero. Does the s polarized electric field also go through zero? This particular angle is known as the Brewster angle, denoted by θ_B.

  (b) Derive the Brewster relationship: $\tan\theta_B = n_2/n_1$.

  (c) Demonstrate that at the Brewster angle, $\theta_1 + \theta_2 = \pi/2$.

  (d) Calculate θ_B for glass ($n_2 = 1.52$), water ($n_2 = 1.33$), and silicon ($n_2 = 3.42$).

(2) **Brewster angle and polarized light by reflection:** Consider an air–glass interface and a light beam incident at an angle θ_1. Assume that the glass has a refractive index $n_2 = 1.52$. Using the Fresnel equations, plot the amplitudes of the reflected electric fields for the p and s polarizations as a function of θ_1. Also plot the reflected energies. What is the Brewster angle? Explain the conditions under which the natural light reflected by this glass is polarized and what the type of polarization will be.

(3) Use the Brewster angle to propose a methodology to determine the polarization axis of an unknown polarizer when you have no specific optical bench at your disposal.

(4) Express the Fresnel reflection coefficient r at 0° incidence on a surface separating two dielectric media of indices n_1 and n_2.

(a) Calculate the reflection coefficient in energy $R = |r|^2$ for a glass surface ($n_2 = 1.52$) and for diamond ($n_2 = 2.46$). Explain why diamonds shine.

(b) The refractive index for a metal is a complex number ($\tilde{n}_2 = n'_2 + n''_2$). Obtain in this case the formula for R. Moreover, its value may also depend on the wavelength in the visible range. Compute the reflection coefficient R for the five main colors of the visible spectrum for gold, copper, aluminum, and platinum using the values of the refractive index given in the following table.

Complex refractive indices of four metals for the five main colors.

Color → Lambda →	Blue 450 nm	Green 520 nm	Yellow 580 nm	Orange 610 nm	Red 650 nm
Au	$1.38 + 1.92i$	$0.63 + 2.07i$	$0.29 + 2.84i$	$0.23 + 3.19i$	$0.16 + 3.60i$
Cu	$1.24 + 2.39i$	$1.18 + 2.60i$	$0.72 + 2.70i$	$0.38 + 3.11i$	$0.24 + 3.62i$
Al	$0.63 + 5.45i$	$0.89 + 6.28i$	$1.16 + 6.70i$	$1.31 + 7.30i$	$1.56 + 7.71i$
Pt	$0.63 + 3.75i$	$0.49 + 4.71i$	$0.46 + 5.48i$	$0.46 + 5.85i$	$0.47 + 6.33i$

Explain from these values why gold is yellow, copper is red-brown, and aluminum and platinum are grey.

(5) **Evanescent waves:** Consider a glass–water interface. A plane wave is incident on this interface from the glass side (medium **1**) at point A with an angle $\theta_1 = 65°$ greater than the critical angle ($\theta_c = 61°$). Point A is the origin of the (x, y, z) coordinates. The wave is p-polarized and has an amplitude E_0. We are interested in the electromagnetic wave at a point B placed at a distance $e = 1\,\mu$m from A (see the following Fig. (a) on the next page).

(a) What is the nature of the wave at point B? Give the expression of the wave vector $\tilde{\boldsymbol{k}}_2$ of this wave in medium **2** and its vectorial electric field using the complex notation. Calculate numerically the amplitude of the wave at point B as a function of E_0.

(b) We now consider another glass surface (medium **3**) at 1 μm from the first one, as shown in the following Fig. (b). Is there a wave in medium **3**? If yes, determine its nature: propagative, evanescent, transverse, or none of these. Give the expression of the electric field $\tilde{\boldsymbol{E}}_3$. Determine the direction of propagation $\tilde{\boldsymbol{k}}_3$.

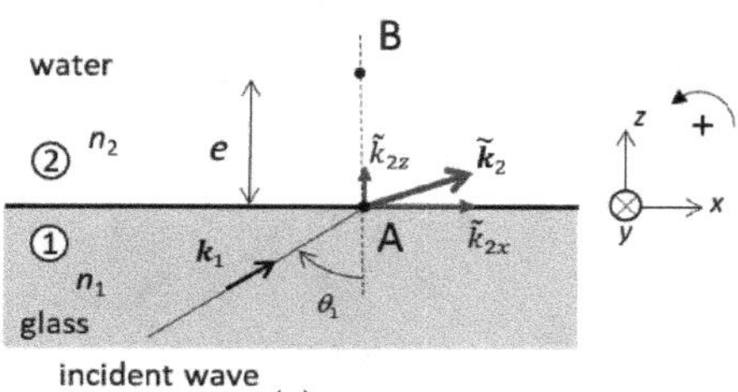

(a)

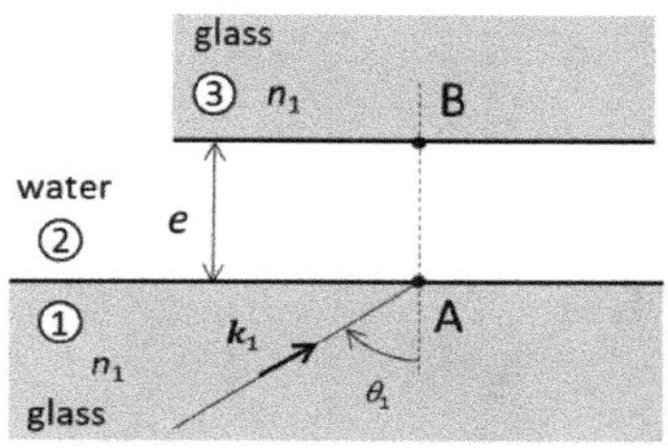

(b)

(6) Derive the three-layer formula (3.37) and (3.38) using the notation shown in Fig. 3.8. Use the mathematical results dealing with the convergence of infinite series: $\sum_{n=0}^{\infty} a^n = 1/(1-a)$ if $a < 1$.

(7) The formula (3.37) for calculating the electric field reflected by a thin layer can be simplified into formula (3.40). The goal is to prove (3.40). Start by establishing that $\tilde{r}_{12}^2 + \tilde{t}_{12}\tilde{t}_{21} = 1$. Then, prove that

$$\mathcal{R} = \frac{\tilde{r}_{12} + \tilde{r}_{23}e^{2ik_0\delta_2}}{1+\tilde{r}_{12} \cdot \tilde{r}_{23}e^{2ik_0\delta_2}}.$$

(8) A camera lens made of glass with $n = 1.6$ is covered by a thin layer with $n = 1.38$ to suppress reflections and maximize the transmittance at 540 nm wavelength. For the sake of simplification, assume that the lens surface is a plane surface. Also assume the thin layer to be of constant thickness.

(a) Derive the thickness d of the layer that suppresses the reflected beam.

(b) How is the reflectivity affected for $\lambda = 400$ nm and $\lambda = 700$ nm?

(c) What happens if the thickness of the thin layer is doubled and becomes $2d$?

(9) Demonstrate that the best antireflection efficiency is obtained for $n_1 = \sqrt{n_0 n_2}$.

4

Surface Plasmon Polaritons at Planar Interfaces

Surface plasmon polariton (SPP) is the result of the strong coupling between an electromagnetic wave and the oscillating charges of a metal. This is a **surface wave** propagating along the **metal–dielectric interface** and is **evanescent** perpendicular to this interface. The SPP wave is confined to 30–50 nm in the vicinity of the interface. It is controlled by a special **dispersion relation** that explains why launching an SPP wave requires special configurations, such as the **Kretschmann configuration**, or the use of a special grating. The analysis of the **SPP extinction curve** shows why a plasmonic biosensor can reach an extremely high sensitivity.

4.1 Introduction

Surface plasmon polariton (SPP) is a wave propagating at a planar interface between a metal and a dielectric medium (insulating medium). The slight penetration of the electromagnetic field into the metal induces the oscillation of conduction electrons (plasma oscillation) due to the Lorentz force. In return, the movement of electrons is the source of an electromagnetic field that will counter-react on the initial field. This strong coupling between the electromagnetic field and the oscillating charges is responsible for the peculiar structure of the SPP wave (see Fig. 4.1). It is a surface wave that propagates along the interface and has a vanishing behavior away from

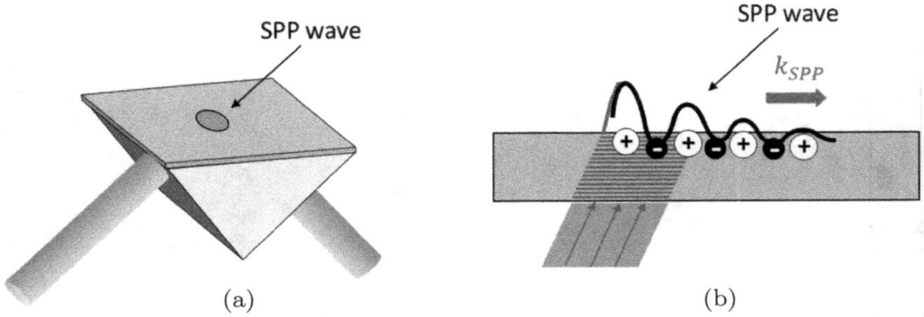

Fig. 4.1. Schematic representation of an SPP wave at the air–gold interface. (a) The SPP wave is excited within the Kretschmann configuration using a prism by total internal reflection. (b) The SPP wave develops at the metal–air interface and is an intricate coupling between a surface electromagnetic wave and a charge density wave. It decays away from the metal surface (not represented here) and is progressively damped in its propagation direction.

the surface. This confines the wave within 30 nm or so (for visible optical waves) close to the surface. This one-dimensional confinement within a distance much smaller than the wavelength is certainly a specificity of the SPP wave and the key to various applications in nano-optics. The wave is a combination between near-field confined optics and far-field propagation.

SPP waves are also called *propagating plasmon* or *surface plasmon resonance* (SPR).

For an SPP wave to be relevant, three conditions are required. First, a surface plasmon wave may exist only at specific interfaces that will be specified (condition of existence). The second condition is to assess the right conditions for its excitation, and the last condition is to be able to detect the SPP wave. In this chapter, we review these three aspects.

4.2 Conditions of Existence for an SPP Wave

4.2.1 *Surface wave and evanescent wave*

Since the SPP wave is a surface wave, we shortly discuss the structure of this kind of waves. The coupling of this wave with the dielectric medium at the other side of the interface is considered in the next section.

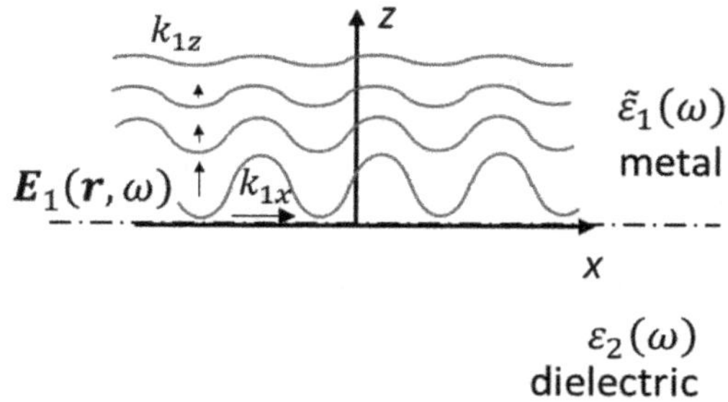

Fig. 4.2. Structure of a surface wave (evanescent wave). The wave propagates along the x direction with a wave vector k_{1x} and has an exponentially vanishing amplitude in the direction normal to the interface (direction z). In the z direction, the decay length is given by $1/k_z$, and there is no propagation but an oscillation of the electric field as a whole.

Let's consider the propagation of an electric field in a metallic medium of dielectric function $\tilde{\varepsilon}_1$ close to an interface defined by $z = 0$ (see Fig. 4.2). The wave vector is $\tilde{\boldsymbol{k}}_1$ and can be complex. This notation has to be examined closely because it will completely control the propagation. In the most general case, $\tilde{\boldsymbol{k}}_1$ is a vector with three coordinates $\tilde{k}_{1x}$, $\tilde{k}_{1y}$, and $\tilde{k}_{1z}$, and each of the coordinates is a complex number: $\tilde{k}_{1x} = k'_{1x} + ik''_{1x}$, etc. In our case, we seek solutions in the form of harmonic plane waves written as $\tilde{\boldsymbol{E}}(\boldsymbol{r},t) = \boldsymbol{E}_0 \exp[i(\tilde{\boldsymbol{k}}_1 \cdot \boldsymbol{r} - \omega t)]$. They obey the wave equation, as explained in Chapter 1, and the dispersion equation, which governs its propagation, is written as

$$\boldsymbol{k}_1^2 = \tilde{\varepsilon}_1 \frac{\omega^2}{c^2}. \tag{4.1}$$

By choosing the orientation of the axis such that $\tilde{\boldsymbol{k}}_1 = \tilde{k}_{1x}\boldsymbol{e}_x + \tilde{k}_{1z}\boldsymbol{e}_z$, relation (4.1) becomes

$$(\tilde{k}_{1x})^2 + (\tilde{k}_{1z})^2 = \tilde{\varepsilon}_1 \frac{\omega^2}{c^2}. \tag{4.2}$$

Now, the dielectric function $\tilde{\varepsilon}_1$ of a metal is predominantly a real number with a negative value (see the Drude model in Chapter 2). According to our notation, we now denote it as ε_1 (without the *tilde* sign). Hence, in order to fulfill relation (4.2), either k_{1x} or k_{1z} should be an imaginary number. The aim of the derivation that will be carried out in the next section is to find expressions for the wave vectors k_{1x} and k_{1z} at a given angular frequency ω.

Before that, we discuss the structure of the resulting wave. Let's assume that the z component of the wave vector is purely imaginary and the x component is real:

$$k_{1x} = k'_{1x} \text{ and } \tilde{k}_{1z} = ik''_{1z}. \tag{4.3}$$

Then, $\tilde{\boldsymbol{k}}_1 \cdot \boldsymbol{r} = k'_{1x}x + ik''_{1z}z$, and the resulting wave is an *evanescent wave*, or *surface wave*, given by

$$\tilde{\boldsymbol{E}}(\boldsymbol{r}, t) = \boldsymbol{E}_0 \exp(-k''_{1z}z) \exp[i(k'_{1x}x - \omega t)]. \tag{4.4}$$

This wave propagates along the x direction and has an exponentially decreasing amplitude along the z direction, as illustrated in Fig. 4.2. This means that the amplitude of the wave, $\boldsymbol{E}_0\exp(-k''_{1z}z)$, decays with a factor of $1/e = 0.37$ over a length of $1/k''_{1z}$ so that the wave is confined close to the surface at a distance of approximately $1/k''_{1z}$. In the next section, we demonstrate that the SPP wave is a wave of this kind. This section has shown that it fully depends on the structure of the wave vector.

4.2.2 *The dispersion relation of an SPP wave (TM polarization)*

Let's be more specific. In medium 1 (metal), the dielectric function is ε_1 and is ε_2 in medium 2 (dielectric). We first consider a p-polarized wave (also called transverse magnetic, TM). As shown in Fig. 4.3, the electric fields in medium 1 and medium 2 are written as:

$$\boldsymbol{E}_1 = \begin{bmatrix} E_{1x} \\ 0 \\ E_{1z} \end{bmatrix} \exp(i(\tilde{k}_{1x} \cdot x - \omega t)) \cdot \exp(i\tilde{k}_{1z} \cdot z), \tag{4.5}$$

$$\text{and } \boldsymbol{E}_2 = \begin{bmatrix} E_{2x} \\ 0 \\ E_{2z} \end{bmatrix} \exp\left(i(\tilde{k}_{2x} \cdot x - \omega t)\right) \cdot \exp(i\tilde{k}_{2z} \cdot z). \tag{4.6}$$

4.2.2.1 *Conservation of k_x across the interface*

The component of the electric field parallel to the interface must fulfill the *boundary conditions* described in Chapter 3 so that E_x is continuous when crossing the interface (see Fig. 4.4). Be aware that this condition of

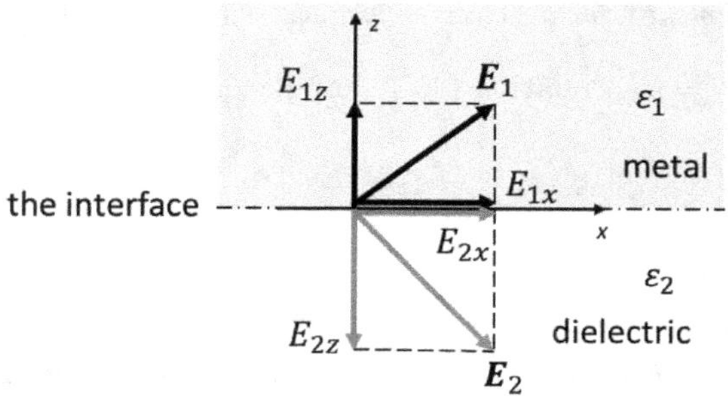

Fig. 4.3. Electric fields at the interface in the case of a p-polarized electric field (TM). The electric fields $\boldsymbol{E}_1$ and $\boldsymbol{E}_2$ are vectors that can be decomposed into components parallel to the interface, E_{1x} and E_{2x}, and into components normal to the interface, E_{1z} and E_{2z}.

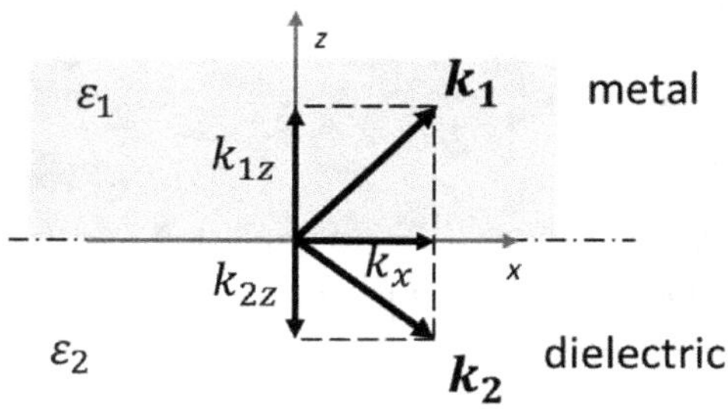

Fig. 4.4. Wave vectors $\boldsymbol{k_1}$ and $\boldsymbol{k_2}$ at the interface between a metal and a dielectric. This drawing is schematic and not intended to represent the exact direction of these vectors.

continuity has to be written for the complete complex wave and not just for the amplitudes E_{1x} and E_{2x}. At the interface, $z = 0$, and this continuity is written as

$$E_{1x}\exp(i(\tilde{k}_{1x}\cdot x - \omega t)) = E_{2x}\exp(i(\tilde{k}_{2x}\cdot x - \omega t)). \tag{4.7}$$

Relation (4.7) should be verified for any position x on the interface and at any time t. This is a very demanding condition, which requires that $\tilde{k}_{1x}$ and $\tilde{k}_{2x}$ be equal. This condition states that the wave vector parallel to the interface is preserved. In the following, this component will be simply denoted by $\tilde{k}_x$ irrespective of the medium, 1 or 2, under consideration (see Fig. 4.4).

4.2.2.2 *Relationships for the electric field across the interface*

Given that $\tilde{k}_{1x} = \tilde{k}_{2x}$, relation (4.7) simplifies into

$$E_{1x} = E_{2x}. \tag{4.8}$$

Regarding the components of the electric field perpendicular to the interface, the situation is different since they are discontinuous according to the *boundary conditions*. However, the normal component of the dielectric displacement $\boldsymbol{D}$ is continuous:

$$D_{1z} = D_{2z}. \tag{4.9}$$

From Chapter 1, we know that $\boldsymbol{D} = \varepsilon_0 \varepsilon \boldsymbol{E}$ so that condition (4.9) can be written as

$$\varepsilon_1 E_{1z} = \varepsilon_2 E_{2z}. \tag{4.10}$$

Moreover, using the Maxwell–Gauss equation (see Chapter 1) and the fact that the interface is free of sources ($\rho = 0$), we can write $\boldsymbol{\nabla} \cdot \boldsymbol{E} = 0$. Applying these conditions in medium 1 and medium 2, we obtain

$$k_{1x} E_{1x} + k_{1z} E_{1z} = 0, \tag{4.11}$$

$$k_{2x} E_{2x} + k_{2z} E_{2z} = 0. \tag{4.12}$$

Equations (4.8), (4.10), (4.11), and (4.12) form a set of four equations with the four field components as unknowns. We seek to express the components of the wave vectors k_x, k_{1z}, and k_{2z} as functions of the dielectric functions.

By combining Equations (4.8), (4.10), and (4.12), we easily obtain

$$\varepsilon_2 k_x E_{1x} + \varepsilon_1 k_{2z} E_{1z} = 0. \tag{4.13}$$

4.2.2.3 *Derivation of the dispersion equation of the SPP*

Equations (4.11) and (4.13) have an evident solution: $E_{1x} = E_{1z} = 0$, which is meaningless. We discard this trivial solution, which states that it is possible that there is no propagation at all at this interface. The equation can be expressed in a matrix form as

$$\begin{bmatrix} k_x & k_{1z} \\ \varepsilon_2 k_x & \varepsilon_1 k_{2z} \end{bmatrix} \begin{bmatrix} E_{1x} \\ E_{1z} \end{bmatrix} = 0. \tag{4.14}$$

In order for a solution different from zero to exist, the determinant of this system must be equated to zero: $\begin{vmatrix} k_x & k_{1z} \\ \varepsilon_2 k_x & \varepsilon_1 k_{2z} \end{vmatrix} = 0$, which can be written as $\varepsilon_1 k_x k_{2z} - \varepsilon_2 k_{1z} k_x = 0$.

This latter equation has two solutions: $k_x = 0$, which does not correspond to any wave at the interface, and the solution

$$\varepsilon_1 k_{2z} - \varepsilon_2 k_{1z} = 0. \tag{4.15}$$

The square of the components k_{1z} and k_{2z} can be both expressed as a function of k_x using the dispersion relations in medium 1 and medium 2 (relation (4.2)). Therefore, we first transform Equation (4.15) into

$$(\varepsilon_1 k_{2z})^2 = (\varepsilon_2 k_{1z})^2. \tag{4.16}$$

Then, $\varepsilon_1^2 \left[\varepsilon_2 \frac{\omega^2}{c^2} - k_x^2\right] = \varepsilon_2^2 \left[\varepsilon_1 \frac{\omega^2}{c^2} - k_x^2\right]$, which simplifies into

$$k_x^2 = \frac{\omega^2}{c^2} \cdot \frac{\varepsilon_1 \varepsilon_2}{\varepsilon_1 + \varepsilon_2}. \tag{4.17}$$

Equation (4.16) is the **dispersion equation** for the SPP, and it controls the propagation of the plasmon wave. We discuss all the consequences in the following sections.

However, relation (4.17) gives the expression of the x component of $\boldsymbol{k}$. Using again relation (4.2), we can also express the z components, which are different in media 1 and 2.

From relation (4.2), we have $k_{1z}^2 = \varepsilon_1 \frac{\omega^2}{c^2} - k_x^2$, then

$$k_{1z}^2 = \frac{\omega^2}{c^2} \cdot \frac{\varepsilon_1^2}{\varepsilon_1 + \varepsilon_2}, \tag{4.18}$$

and similarly for k_{2z}, $$k_{2z}^2 = \frac{\omega^2}{c^2} \cdot \frac{\varepsilon_2^2}{\varepsilon_1 + \varepsilon_2}. \tag{4.19}$$

Relations (4.17)–(4.19) completely define vectors $\boldsymbol{k}_1$ and $\boldsymbol{k}_2$.

4.2.3 What about the dispersion relation in TE modes?

The method described above can be applied to analyze the case of s-polarized waves (TE mode). As a general consideration, we do not expect an SPP wave in the TE mode since the essence of the plasmon wave is a coupling between an optical wave and a charge density wave, and these two kinds of wave have very different structure: An optical wave has a transverse structure ($\boldsymbol{k}$ and $\boldsymbol{E}$ are perpendicular), whereas a charge density wave has a longitudinal structure ($\boldsymbol{k}$ and $\boldsymbol{v}_e$, the velocity of charges, are parallel). An optical wave in the TE mode has no chance to excite the charge oscillations in the k_x direction. Let's now prove this from the electromagnetic equations.

In the TE mode, the electric fields oscillate along $\boldsymbol{e_y}$ and can be written as $\boldsymbol{E}_1 = E_1\boldsymbol{e_y}$ and $\boldsymbol{E}_2 = E_2\boldsymbol{e_y}$. Since we seek surface waves, the electric fields are written in the following form:

$$\boldsymbol{E}_1 = E_{10}\boldsymbol{e_y}\exp\left(i(\tilde{k}_x \cdot x - \omega t)\right) \cdot \exp(-k_{z1} \cdot z),$$

$$\boldsymbol{E}_2 = E_{20}\boldsymbol{e_y}\exp\left(i(\tilde{k}_x \cdot x - \omega t)\right) \cdot \exp(+k_{z2} \cdot z),$$

where $k_{z1} > 0$ and $k_{z2} > 0$ are positive wave vectors so that the field intensity decays away from the interface (since a "self-amplification" of these fields has no physical meaning).[1]

The parallel components of $\boldsymbol{E}$ are continuous across the interface ($z = 0$), which leads to $E_{10} = E_{20}$.

The magnetic field can be calculated from the Maxwell–Faraday equation: $\boldsymbol{\nabla}\times\boldsymbol{E} = -\partial\boldsymbol{B}/\partial t$. For the x components, it yields $B_{x1} = -k_{z1}/i\omega E_{10}$ and $B_{x2} = +k_{z2}/i\omega E_{20}$. Since $\boldsymbol{B}$ is continuous on crossing the interface ($B_{x1} = B_{x2}$) and since $E_{10} = E_{20}$, the relation transforms into

$$(k_{z1} + k_{z2})E_{10} = 0.$$

Since $k_{z1} > 0$ and $k_{z2} > 0$, $k_{z1} + k_{z2}$ cannot be null and the only solution is $E_{10} = 0$. In other words, no electromagnetic wave is allowed to travel as

[1] Here, we make the situation simple by considering k_{z1} and k_{z2} as real numbers. But we could also gain in generality by considering complex numbers. In this case, the conditions would be $\mathrm{Re}(k_{z1}) > 0$ and $\mathrm{Re}(k_{z2}) > 0$.

a surface wave at this interface in TE polarization. There is no SPP wave coupled with a TE optical wave.

4.2.4 *The meaning of the dispersion relation of the SPP*

The *dispersion relation* plays a central role in physics, and it is very important to derive this relation in the case of a plasmonic wave. The *dispersion relation* establishes the link between the energy of a wave $\hbar\omega$ and its momentum $\hbar\boldsymbol{k}$. For a photon of a given energy, if we want to generate an SPP wave, the allowed values for momentum (and therefore the allowed wave vectors) are controlled by the dispersion relation (4.17).

4.2.4.1 *Plasmon wavelength*

The dielectric functions ε_1 and ε_2 are functions of the frequency ω, and their dependence on ω is crucial to understanding the properties of an SPP wave: $\varepsilon_1 = \varepsilon_1(\omega)$ and $\varepsilon_2 = \varepsilon_2(\omega)$. In order to be more specific, we now consider a laser beam with a wavelength λ_0 in vacuum and, therefore, a circular frequency $\omega_0 = hc/\lambda_0$. We suppose $\lambda_0 = 633\,\text{nm}$ (helium–neon laser), and we consider a gold–air interface.

In this section, we derive the most important properties of an SPP wave, for which we use the values of the dielectric permittivity provided by the Drude model explained in Chapter 2. Therefore, the metal has a negative permittivity with no imaginary part (losses are discarded in a first approach):

$$\varepsilon_1 = 1 - \frac{\omega_p^2}{\omega^2}. \tag{4.20}$$

In the case of gold, $\hbar\omega_p = 8.99\,\text{eV}$, and in the visible range, $\omega < 8.99\,\text{eV}$ so that ε_1 is a negative number between -6 and -25 (see graph in Fig. 2.3). At $\lambda_0 = 633\,\text{nm}$, $\varepsilon_1 = -17.0$. Medium 2 is air and $\varepsilon_2 = 1$. Therefore, the term $\varepsilon_1\varepsilon_2/(\varepsilon_1 + \varepsilon_2)$ in relation (4.17) can be computed. Both the numerator and the denominator are negative numbers so that k_x^2 is positive. As a result, we expect that k_x takes real values. For example, at 633 nm, $\varepsilon_1\varepsilon_2/(\varepsilon_1 + \varepsilon_2) = 1.063$ and $k_x = 1.03k_0$, where $k_0 = \omega_0/c$, which is the wave vector in free space. By placing this value of k_x in Equation (4.5), we can demonstrate that the SPP wave has purely propagative behavior along the x direction. Moreover, the wavelength of the SPP is given by $\lambda_{SPP} = 2\pi/k_x = 1/1.03\lambda_0 = 0.97\lambda_0$. The SPP wavelength is always shorter than the corresponding wave in free space. In the next section, we look at what changes occur when losses are also included, such as when real metals are considered.

Let's be a bit more general. In order to generate an SPP wave, k_x should be a real number and the dielectric functions of the two media should fulfill the following conditions:

$$\varepsilon_1(\omega) \cdot \varepsilon_2(\omega) < 0, \tag{4.21}$$

$$\varepsilon_1(\omega) + \varepsilon_2(\omega) < 0. \tag{4.22}$$

If these relations are fulfilled, the dispersion relation (4.17) is written as

$$k_x = \pm\frac{\omega}{\mathrm{c}} \cdot \sqrt{\frac{\varepsilon_1\varepsilon_2}{\varepsilon_1 + \varepsilon_2}}. \tag{4.23}$$

Using the Drude relation (4.20), it is easy to plot the dispersion relation, as shown in Fig. 4.5(a).

Suppose the SPP wave has a frequency ω_0. According to the dispersion relation shown in Fig. 4.5, this wave would have a wave vector k_{SPP}, which is always greater than the wave vector k_0 of the corresponding wave traveling in free space. The free-space propagation is represented by the

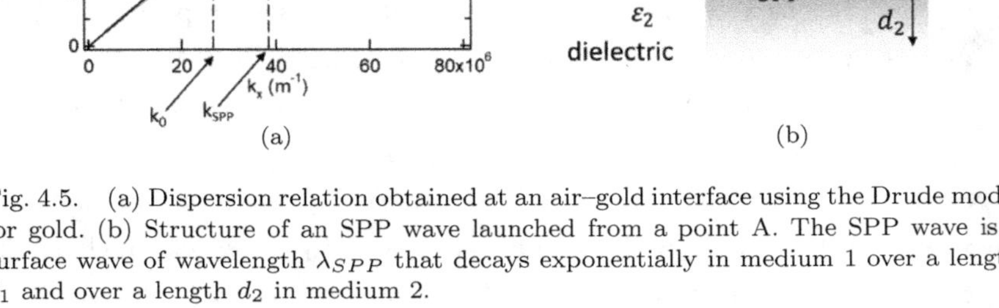

(a) (b)

Fig. 4.5. (a) Dispersion relation obtained at an air–gold interface using the Drude model for gold. (b) Structure of an SPP wave launched from a point A. The SPP wave is a surface wave of wavelength λ_{SPP} that decays exponentially in medium 1 over a length d_1 and over a length d_2 in medium 2.

light line, which is the plot of $\omega = k_x c$. Figure 4.5 also shows that the SPP wave has two branches. The high-energy branch is called the Brewster mode and is not a surface wave since the z component of the wave vector is not purely imaginary and contains a propagative term [1]. We will not discuss it further.

4.2.4.2 *Plasmon evanescent field decay length*

We now inspect the z components of the wave vectors in media 1 and 2. From relations (4.18) and (4.19), it is clear that k_{1z}^2 and k_{2z}^2 are negative; therefore, k_{1z} and k_{2z} are pure imaginary numbers. Placing these imaginary numbers in relations (4.5) and (4.6) demonstrates that the electric field is purely decaying along the z direction. In medium 2, (4.19) transforms into

$$k''_{2z} = \pm i\frac{\omega}{\mathrm{c}} \cdot \frac{\varepsilon_2}{\sqrt{-(\varepsilon_1 + \varepsilon_2)}}. \tag{4.24}$$

Using the values for the gold–air interface, as in the previous section, $\varepsilon_1 = -17.0$ for $\lambda_0 = 633\,\mathrm{nm}$, (1.24) yields

$$k''_{2z} = \pm i\frac{\omega}{\mathrm{c}} \cdot \frac{1}{\sqrt{16}} = \pm i\frac{2\pi}{4\lambda_0}. \tag{4.25}$$

The SPP wave in medium 2 is written as (using relation (4.6))

$$\boldsymbol{E}_2 = \begin{bmatrix} E_{2x} \\ 0 \\ E_{2z} \end{bmatrix} \exp\left(i(k_{SPP} \cdot x - \omega t)\right) \cdot \exp(k''_{2z} z). \tag{4.26}$$

Relation (4.26) describes the wave structure of the surface wave. The term $\exp(k''_{2z} z)$ corresponds to the decay of the SPP wave away from the interface ($z < 0$ in medium 2). It is a purely decaying term with no propagation in the z direction. The decay length is given by $d_2 = 1/k''_{2z}$, which is $d_2 = 4\lambda_0/2\pi = 400\,\mathrm{nm}$ in our example. This is the distance over which the amplitude of the wave is reduced by the factor e. The same calculation on the metallic side (medium 1) yields a decay length $d_1 = 1/k''_{1z} = 24\,\mathrm{nm}$. This demonstrates that the SPP wave is confined to a small volume of a typical thickness of 400 nm on the air side and 24 nm on the metal side (at $\lambda_0 = 633\,\mathrm{nm}$). These "twin" evanescent waves result in the SPP wave.

Be aware that the decay lengths apply to the amplitude of the electric field. The measurable quantity is the intensity which scales with the square of the electric field. Therefore, the decay lengths for the intensity are $D_1 = d_1/2$ and $D_2 = d_2/2$.

4.2.4.3 *Plasmon propagation length*

Since $\boldsymbol{k} = k_{1x}\boldsymbol{u_x} + k_{1z}\boldsymbol{u_z}$, and given that in our example we have modeled the metallic side by using a simple Drude formula, k_{1x} is purely real and k_{1z} is purely imaginary. The propagation occurs along the x-axis, whereas the wave is purely decaying along the z-axis. There is no decay in the x direction. In this case, the SPP wave would propagate indefinitely along the interface. However, in real metals, losses need to be included and the k_x is not a real number. We discuss this aspect in the next section.

4.2.5 *SPP wave in real metals*

4.2.5.1 *Effect of losses in a metal*

In real metals, the dielectric function ε_1 is not just a negative number but also a complex number: $\varepsilon_1 = \varepsilon'_1 + i\varepsilon''_1$. See, for example, relation (2.6), where the losses are included in the Drude model so that $\varepsilon_1 = 1 - (\omega_p^2/\omega^2 + i\gamma\omega)$. A more accurate value for ε_1 is obtained from ellipsometric measures (see Chapter 2), which yields $\varepsilon_1 = -12.3 + 1.22i$ at 633 nm. The losses in a metal are due to the collisions of free electrons of the metal with the ionic cores. Using the complex value for ε_1 in the dispersion relation (4.17) yields a complex value for k_x (also denoted by k_{SPP}) so that $k_x = k_{SPP} = k'_{SPP} + ik''_{SPP}$. Replacing ε_1 by $\varepsilon'_1 + i\varepsilon''_1$ in (4.17) and performing the calculation at the first order in $|\varepsilon''_1/\varepsilon'_1|$ yields the following expressions:

$$k'_{SPP} = \pm\frac{\omega}{c}\cdot\sqrt{\frac{\varepsilon'_1\varepsilon_2}{\varepsilon'_1+\varepsilon_2}}, \tag{4.27}$$

$$k''_{SPP} = \pm\frac{\omega}{c}\cdot\sqrt{\frac{\varepsilon'_1\varepsilon_2}{\varepsilon'_1+\varepsilon_2}}\frac{\varepsilon''_1\varepsilon_2}{2\varepsilon'_1(\varepsilon'_1+\varepsilon_2)}. \tag{4.28}$$

Expression (1.27) is identical to (1.17) and shows that the plasmon wavelength is identical to the value calculated in the absence of losses in a metal (Section 4.2.4.1).

Expression (1.28) of k''_{SPP} is responsible for the damping of the SPP wave along the x-axis and yields a decay term, $\exp(-k''_{SPP}x)$. The propagation length of the SPP wave is no longer infinite and is evaluated as $L_{SPP} = 1/k''_{SPP}$. Moreover, L_{SPP} corresponds to a decay in amplitude, but we are interested in the decay in intensity that scales as $\exp(-2k''_{SPP}x)$. Therefore, the decay in intensity L_{SPP} is simply $L_{SPP}/2$. Given that $\omega/c = 2\pi/\lambda_0$, L_{SPP} can be written as

$$L_{SPP} = \frac{1}{2}\frac{\lambda_0}{2\pi}\sqrt{\frac{\varepsilon_1' + \varepsilon_2}{\varepsilon_1'\varepsilon_2}}\frac{2\varepsilon_1'(\varepsilon_1' + \varepsilon_2)}{\varepsilon_1''\varepsilon_2}. \tag{4.29}$$

Note that L_{SPP} grows to infinity when $\varepsilon_1'' = 0$ as expected when losses are set to 0. Some of the calculated values are listed in Tables 4.1 and 4.2.

Table 4.1. Values of various parameters for the SPP at gold–air and gold–water interfaces at $\lambda_0 = 633\,\text{nm}$. k_{SPP} is the SPP wave vector. D_1 and D_2 are the decay lengths in the metal and dielectric, respectively. L_{SPP} is the propagation length along the interface. The decay lengths are given for intensity.

$\lambda_0 = 633\,\text{nm}$	Air–gold with the Drude model for ε_1	Air–gold with real values for ε_1	Water–gold with real ε_1
k_{SPP}	$1.03k_0$	$1.04k_0$	$1.44k_0$
$\lambda_{SPP} = 2\pi/k'_{SPP}$	$0.97\lambda_0$	$0.96\lambda_0$	$0.69\lambda_0$
$D_1 = 1/2k''_{1z}$	12 nm	14 nm	13 nm
$D_2 = 1/2k''_{2z}$	200 nm	170 nm	92 nm
$L_{SPP} = 1/2k''_{SPP}$	∞	11 μm	4.2 μm

Table 4.2. Values of SPP parameters at gold–air and gold–water interfaces at $\lambda_0 = 532\,\text{nm}$.

$\lambda_0 = 532\,\text{nm}$	Air–gold with the Drude model for ε_1	Air–gold with real values for ε_1	Water–gold with real ε_1
k_{SPP}	$1.03k_0$	$1.12k_0$	$1.66k_0$
$\lambda_{SPP} = 2\pi/k''_{SPP}$	$0.97\lambda_0$	$0.83\lambda_0$	$0.60\lambda_0$
$D_1 = 1/2k'_{1z}$	12 nm	17 nm	15 nm
$D_2 = 1/2k''_{2z}$	140 nm	84 nm	42 nm
$L_{SPP} = 1/2k''_{SPP}$	∞	700 nm	210 nm

At 633 nm, the SPP wave at the gold–air interface propagates over a length of 11 μm and its energy is confined to a depth of 170 nm on the air side. When considering the gold–water interface, the propagation length reduces to 4.2 μm and the probing depth to 92 nm. These figures will be of utmost importance in using the SPP waves as a biosensor (Chapter 5) or in designing opto-electronic plasmonic circuitry (Chapter 6).

4.2.5.2 *Interband transitions in plasmonics*

When dealing with complex values of the dielectric function ε_1, we know from Chapter 2 that a strong contribution to the losses stems from the *interband transitions* of the metal. These interband transitions play an important role in the optical properties of gold and drastically affect the SPP wave at the gold surface. The deviation in the resulting dispersion curve from the idealistic case of the Drude model is shown in Fig. 4.6. The figure shows how a visible wave can interact with an SPP wave. A wave in the visible range will have a circular frequency ω_0 in the range 2.4×10^{15} – 4.3×10^{15}rad/s. Figure 4.7 shows that the corresponding SPP wave vector will be very large, compared to the free propagating k_0. Let's consider a wave at 4×10^{15}rad/s in free space, as shown by the horizontal dashed line in Fig. 4.7. It would propagate with a wave vector $k_0 = 1 \times 10^6 \text{m}^{-1}$ ($\lambda_0 = 483$ nm). If we consider

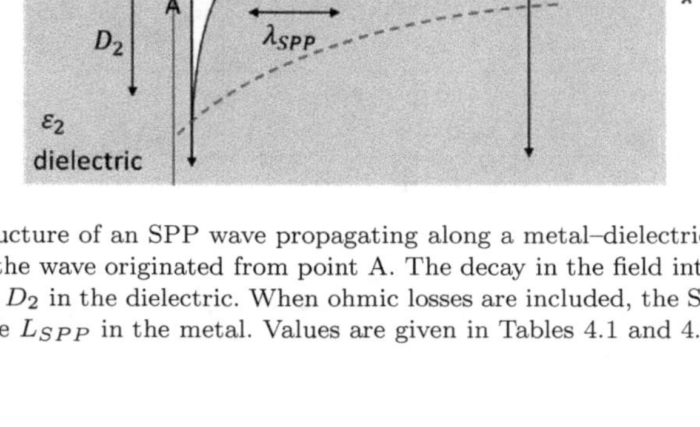

Fig. 4.6. Structure of an SPP wave propagating along a metal–dielectric interface. We suppose that the wave originated from point A. The decay in the field intensity is D_1 in the metal and D_2 in the dielectric. When ohmic losses are included, the SPP propagates over a distance L_{SPP} in the metal. Values are given in Tables 4.1 and 4.2.

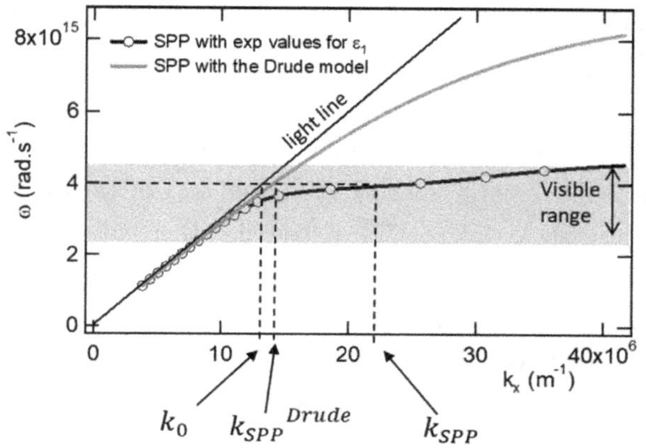

Fig. 4.7. Dispersion relation for the air–gold interface. Comparison of the calculation using the Drude model with the actual values of the dielectric function that also includes the interband transitions of gold. The discrepancy is very strong in the visible range.

the corresponding SPP wave predicted by the Drude model, the wave vector should be $k_{SPP}^{Drude} = 14\times10^6\text{m}^{-1}$, which is $k_{SPP}^{Drude} = 1.05k_0$. With the experimental values, the SPP wave vector is expected at $k_{SPP} = 24 \times 10^6\text{m}^{-1}$, which is $k_{SPP} = 1.86k_0$. This example shows that the wave vector of the plasmon wave is always greater than k_0 in free space. This is an essential property to remember when we need to couple a plasmon with an external excitation. This is actually what we will consider in the next section, where we discuss how to launch a plasmon wave.

4.3 Excitation of an SPP Wave

4.3.1 *The philosophy behind exciting an SPP wave*

So far, we have described the behavior of the SPP wave originating at a metal–dielectric interface. The question to be considered now is how to generate such a wave, or as it is usually said, how to *launch a plasmon wave.* Since the plasmon wave is a coupled oscillation between electrons and an electromagnetic field, there are two handles that can trigger this oscillation: either by injecting electrons into the interface or by coupling with an external optical wave. The first approach is carried out by bombarding a metallic surface with low-energy electrons and measuring the energy of the scattered electrons that exhibits the plasmon resonance. We will not develop this method here, although it has experienced a revival in recent years due to the improved performance of electron detectors (see Chapters 8 and 10). The second approach is based on optically exciting an SPP wave. More precisely, the goal is to transfer the energy and momentum of an external photon to a plasmon wave. Due to the conservation laws in physics, this external photon should have exactly the same energy and momentum as the plasmon we intend to launch. This is why the dispersion relation, where energy in plotted as a function of momentum, is so important. We seek the common parameters for which the dispersion relations of the external photon and of the SPP cross each other.

Let's consider a wave propagating in air at a frequency ω. Its wave vector is simply given by $\omega = kc$, which is the *light line* in Fig. 4.7. This light line does not cross the SPP dispersion relation. If now this wave impinges on the metal interface, it will generate a component $\omega = k_x c/\sin\theta$. Whatever the value of the incidence angle θ, this line is always steeper than the light line and has even less chance to cross the SPP dispersion relation. Therefore, by simply illuminating a "plasmonic" interface, there is no chance of launching an SPP wave. From Fig. 4.7, we understand that we need to find a way to generate values of k_x larger than k_0 in order to couple the photon with the plasmon. Finding ways of adding supplemental momentum to k_0 is precisely the challenge in launching surface plasmons. We will now discuss three traditional ways of achieving it.

4.3.2 *Launching SPPs in the Kretschmann configuration*

One method to achieve a larger k_x is to launch a wave inside a medium of optical index n_3 at an incidence angle greater than the critical angle. Let's make it clear: In addition to the metal and dielectric, we will now consider a third medium in the form of a glass prism. Let's consider a wave directed along $\boldsymbol{k}_3$, impinging inside the prism with an angle θ_{int}, as shown in Fig. 4.8. Then, $k_3 = n_3 k_0$, and the projection k_{3x} on the x-axis is $k_{3x} = n_3 \sin\theta_{int} k_0$. In order to produce wave vectors with k_x larger than k_0, it is obvious that

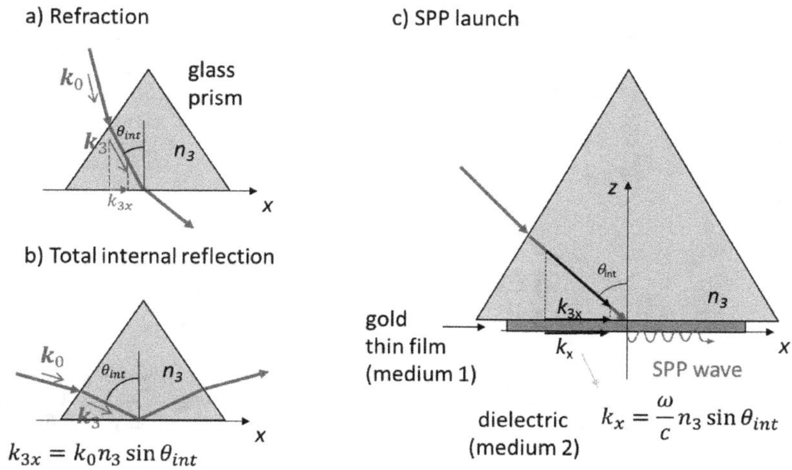

Fig. 4.8. Kretschmann configuration for exciting an SPP wave. In case (a), the incidence angle is smaller than the critical angle of the glass–air interface and the incident light beam is refracted. In case (b), the incident wave is totally reflected and this yields a value for k_x larger than k_0. In case (c), the incident wave is totally coupled with the SPP wave and there is no longer any reflected beam.

one needs to achieve $n_3 \sin\theta_{int} > 1$. This condition simply expresses that θ_{int} should be greater than the critical angle defined by $\theta_{int} = \operatorname{asin} 1/n_3$. This is the situation shown in Fig. 4.8(b). The wave is totally reflected and generates an evanescent wave on the other side of the interface.

The idea of the Kretschmann configuration is to place the gold surface on this interface to benefit from the large value of k_x. This is depicted in Fig. 4.8(c). The wave vector should be transmitted through the gold layer in order for it to reach the dielectric–metal interface.

The wave vector generated by this evanescent wave is written as

$$k_x = n_3 \sin\theta_{int} k_0 = n_3 \sin\theta_{int} \frac{\omega}{c}, \tag{4.30}$$

which can be expressed in a form adapted for plotting in the dispersion relation:

$$\omega = \frac{k_x c}{n_3 \sin\theta_{int}}. \tag{4.31}$$

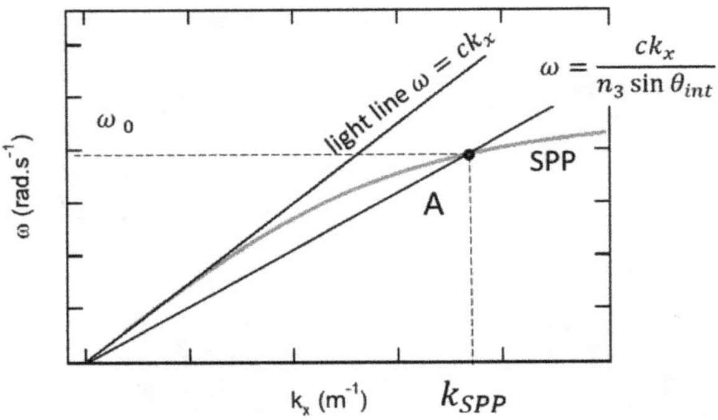

Fig. 4.9. In the Kretschmann configuration, the excitation beam impinges on the gold film according to a dispersion relation that crosses the SPP dispersion curve at point A. This makes possible the coupling of the beam with the SPP wave at a certain angle, θ_{int}.

Since $n_3 \sin\theta_{int} > 1$, the slope of this light line in the graph is lowered and the coupling between this wave and the SPP wave is possible for a given angle θ_{int}, as indicated by point A in Fig. 4.9. Therefore, the adjusting parameter for coupling a given laser of frequency ω is the angle θ_{int}.

Since the transfer of the momentum $\hbar k_x$ from the glass–gold interface to the gold–air interface occurs through the gold thin film, its thickness is critical. If the gold film is too thick, the energy of this evanescent wave is absorbed by it and the subsequent coupling is poor. Inside the gold thin film, the electromagnetic wave undergoes multiple reflections, and this can maximize the transfer of the momentum $\hbar k_x$ through the film thanks to constructive interferences. Therefore, if the film is too thin, no constructive interference takes place and the SP wave is weakly excited. An optimal thickness, which depends on the exciting wavelength, is required to maximize the momentum transfer. This will be discussed shortly.

4.3.3 *Simple setup for observing an SPP in the Kretschmann configuration*

An experimental setup for exciting an SPP wave consists of a glass prism with a thin metallic layer on one side. This prism should be illuminated by a monochromatic beam from a laser and mounted on a rotation stage in order

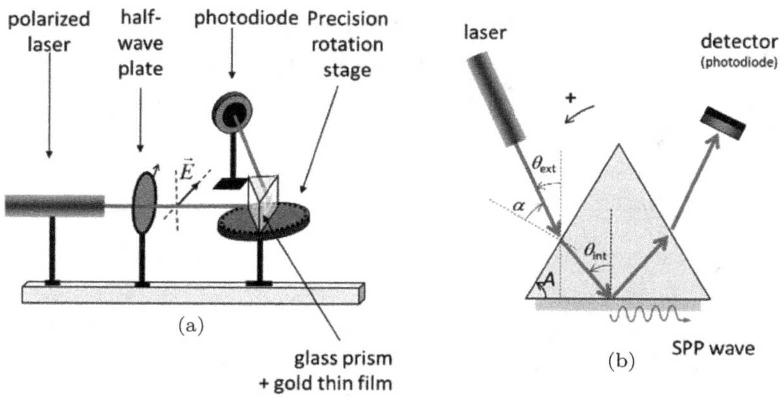

Fig. 4.10. Simple setup for detecting an SPP wave in the Kretschmann configuration.

to control the incidence angle of the laser. The light should be *p*-polarized. A simple setup is depicted in Fig. 4.10 and discussed in Ref. [2].

The SPP wave is a near-field wave that cannot easily be detected directly. In the setup presented above, a photodiode is constantly monitoring the reflected beam. If the laser beam couples with the SPP wave, the reflected intensity drops because its energy is transferred to the plasmon wave. This is the easiest way to indirectly detect the plasmon. When this coupling is achieved, the SPP dispersion relation (4.23) and the geometrical condition (4.30) should be equated so that

$$\frac{\omega}{c} n_3 \sin\theta_{int} = \frac{\omega}{\mathrm{c}} \cdot \sqrt{\frac{\varepsilon_1 \varepsilon_2}{\varepsilon_1 + \varepsilon_2}}. \tag{4.32}$$

In this relation, θ_{int} corresponds to the incidence angle inside the prism. However, in the practical experiment, this angle is not accessible, so θ_{ext} is measured (see Fig. 4.10(b) for notation). θ_{ext} is the angle between the incident laser beam and the face of the prism that bears the plasmonic thin metal film. The relation between θ_{int} and θ_{ext} can be obtained by using the Snell law:

$$\theta_{int} = \arcsin \frac{\sin(\theta_{ext} - \mathrm{A})}{n_3} + A, \tag{4.33}$$

where A is the angle of the prism indicated in Fig. 4.10(b). The angles are oriented: angles turning anticlockwise are positive and angle turning clockwise are negative.

4.3.4 *The extinction curve of the SPP*

By using the setup described above, a plot of the normalized intensity as a function of the external angle is obtained, as shown in Fig. 4.11. The prism is equilateral and made of flint glass with an index of refraction of 1.60 at 633 nm. The pronounced dip at 27.5° is the mark of the plasmon coupling: The intensity of the reflected beam drops to zero since all the energy of the incident light in transferred to the plasmon wave. Relation (4.33) allows calculating the internal angle θ_{int}, which is used in the dispersion relation (4.32). The bump at 23° corresponds to the critical angle of total internal reflection, as depicted in Fig. 4.8(b).

This shape of the extinction curve exhibits a very sharp minimum, typical of the SPP excitation. The position of this minimum is highly sensitive to the values of the local refraction index n_2, which is related to ε_2 since $\varepsilon_2 = n_2^2$. A minute change in the medium close to the metallic interface, such as the adsorption of a molecular layer, induces a change that can be

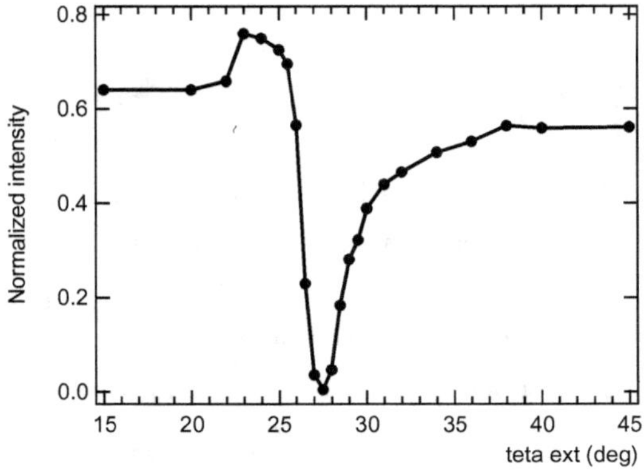

Fig. 4.11. Experimental plasmon extinction curve, where the coupling angle is measured at 27.5° (external angle).

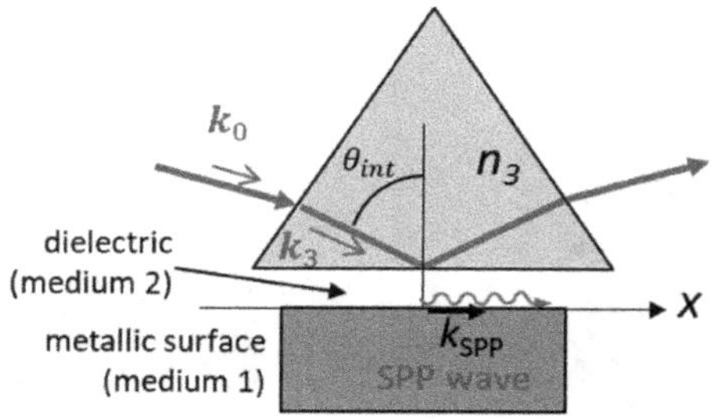

Fig. 4.12. Otto configuration for launching an SPP wave.

detected in the curve in Fig. 4.11. This is the basis of the surface plasmon sensors that will be described in the following chapter.

4.3.5 *Launching SPPs in the Otto configuration*

The Otto configuration is very similar to the Kretschmann configuration since the method for producing large values of k_x at the interface is still based on a glass prism. However, the metallic interface is slightly away from the glass prism, as shown in Fig. 4.12.

The detection of the SPP wave is carried out in the same way as in the Kretschmann configuration. The main difference is that the thickness of the metallic layer do not play a role anymore, which makes this approach more versatile for studying all kind of interfaces. However, the challenge is to control the distance between this metallic surface and the prism so that the evanescent wave generated at the prism can reach the metal. In practice, this requires the distance to be smaller than 1 μm, and the coupling efficiency between the exciting light and the SPP wave is very sensitive to any surface irregularities or roughness.

4.3.6 *Launching SPPs using the grating method*

A third method uses another approach for producing large k_x. The idea is that light of wavelength λ_0 when diffracted by a grating of period a acquires an additional wave vector contribution of value $p\lambda_0/a$, with p being an integer. Let's make it clear.

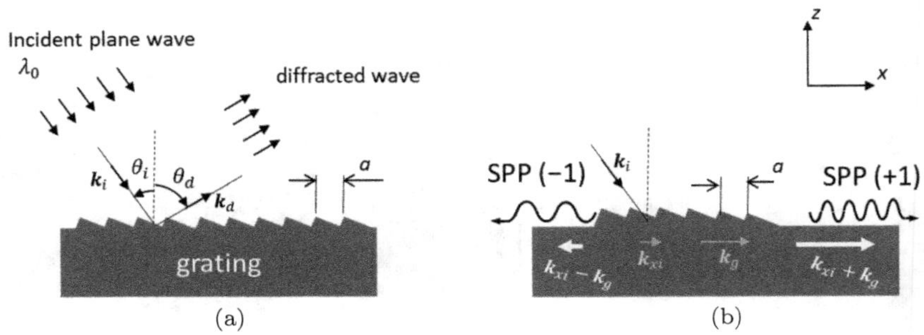

Fig. 4.13. (a) Diffraction of a plane wave by a reflective grating of period a. (b) The diffracted wave acquires a supplemental wave vector contribution, which is a multiple of $k_g = 2\pi/a$. Depending on the order of diffraction, various SPP waves can be generated: Here, two SPP waves with wave vectors $k_{SPP}(+1) = k_{xi} + k_g$ and $k_{SPP}(-1) = k_{xi} - k_g$ are represented. These two surface waves propagate in opposite directions.

We consider a reflective grating of period a lying in a medium of optical index n. A plane wave (wavelength λ_0) with a wave vector $\boldsymbol{k_i}$ impinges on the grating at an angle θ_i. This wave is diffracted by the grating and the direction of diffraction θ_d obeys the law of grating diffraction (see Fig. 4.13):

$$\sin\theta_d - \sin\theta_i = p\frac{\lambda_0}{na}, \tag{4.34}$$

with $p = \ldots, -2, -1, 0, +1, +2, +3, \ldots, p$ is an integer and is called the order of diffraction of the grating. Relation (4.34) can be transformed since the projection of $\boldsymbol{k_i}$ on the x-axis is $k_{xi} = 2\pi n/\lambda_0 \sin\theta_i$ and $k_{xd} = 2\pi n/\lambda_0 \sin\theta_d$. Moreover, we can define the grating wave vector as $k_g = 2\pi/a$. Finally, (4.34) is equivalent to

$$k_{xd} = k_{xi} + pk_g. \tag{4.35}$$

This shows that the wave vector of the diffracted wave acquires a supplemental contribution, which is a multiple of k_g. As discussed above, this supplemental contribution is necessary for coupling the incident wave with the plasmon wave at the grating–dielectric interface.

The coupling occurs when the "augmented" wave vector generated at the grating surface matches the wave vector of the SPP at a given frequency ω_0. There are several intersection points, as can be seen on the dispersion curve in Fig. 4.14. In order to visualize these conditions, we slightly modify

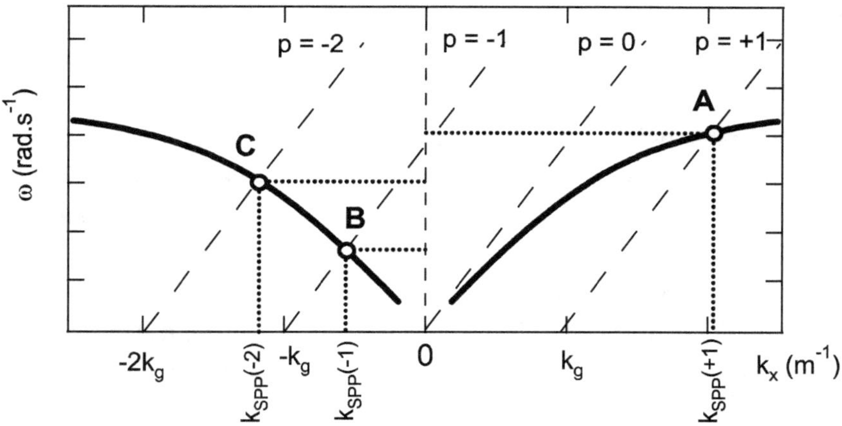

Fig. 4.14. The grating coupling of surface plasmons is illustrated in this dispersion curve. The x component of the incident light (k_x) is shown as dashed lines for the various diffraction orders of the grating (values of the integer p). For simplicity, only a few orders of the diffracted light are shown here. The plasmons are excited for the frequencies at which these lines intersect with the plasmon dispersion lines. Note that the plasmon dispersion lines are drawn for the two propagation directions ($k_x > 0$ and $k_x < 0$).

relation (4.34) by replacing k_{xi} with $n\omega \sin\theta_i/c$. Then, we obtain the dispersion relation of the grating wave:

$$\omega = \frac{c}{\sin\theta_i}(k_x - pk_g). \tag{4.36}$$

When $p = 0$, this condition is simply the light line and there is no intersection with the plasmon line. When $p = 1$, the two dispersion lines intersect at point **A**, as shown in Fig. 4.14. The wave vector takes the value $k_{\rm SPP}(+1)$ and the plasmon wave propagate in the $x > 0$ direction. When $p = -1$, the plasmon wave has the characteristics defined by point **B** and propagates with $k_{\rm SPP}(-1)$ in the $x < 0$ direction. Its frequency is different from that of SPP wave **A**.

In practice, the excitation of SPP using a grating requires cutting grooves equally spaced (with period a) on the metal surface where the plasmon needs to be launched. Since the excitation is usually carried out using a laser with a given frequency ω_0, the matching between the grating dispersion line and the plasmon line should be sought by adjusting the incidence angle θ_i.

4.3.7 *Relation between the SPP extinction curve and the Fresnel equations*

Let us now revisit the Kretschmann configuration. It can be viewed from a different perspective, as described above. Figure 4.8(c) shows an optical wave interacting with two interfaces between three media: glass, gold, and air (or another dielectric). This can be reduced to a one-dimensional problem of thin films. With the tool introduced in Chapter 3, namely the generalized Fresnel equations, it is possible to calculate the intensity reflected by this thin-film arrangement. The configuration is depicted in Fig. 4.15. We consider the wave after it has entered the coupling Kretschmann prism and is traveling in the glass medium of optical index n_1. This wave impinges on the interface at an angle θ_1 and interacts with the metallic thin film of thickness d_2. The metallic layer is in contact with a dielectric medium of optical index n_3, which can be either air or a liquid.

The intensity of the reflected beam can be computed according to the formula mentioned in Chapter 3:

$$\mathcal{R} = \frac{r_{12} + r_{23}e^{2i\delta_2}}{1 + r_{12} \cdot r_{23}e^{2i\delta_2}}. \tag{4.37}$$

The results for the interface discussed above are shown in Fig. 4.16. This approach is mostly numerical and does not allow the analysis of the nature of the SPP wave as precisely as carried out above. However, this has other uses, such as optimizing the thickness of the gold film and evaluating the sensitivity of the plasmon angle to the nature of the dielectric (see Figs. 4.17 and 4.18).

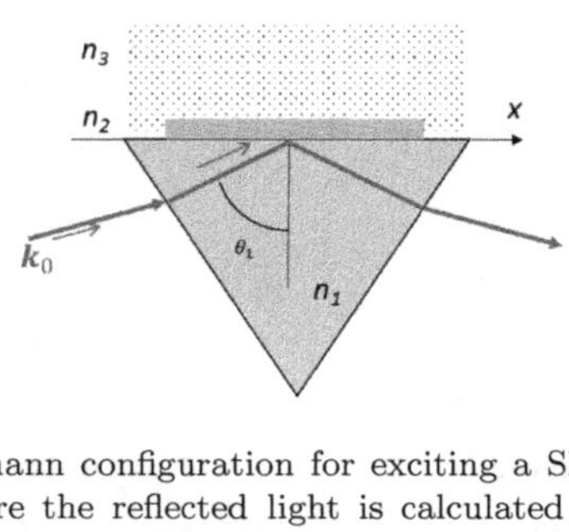

Fig. 4.15. The Kretschmann configuration for exciting a SPP wave can be viewed as a three-layer model, where the reflected light is calculated thanks to the generalized Fresnel equations. Medium 1 is glass, medium 2 is gold and medium 3 is air or another dielectrics.

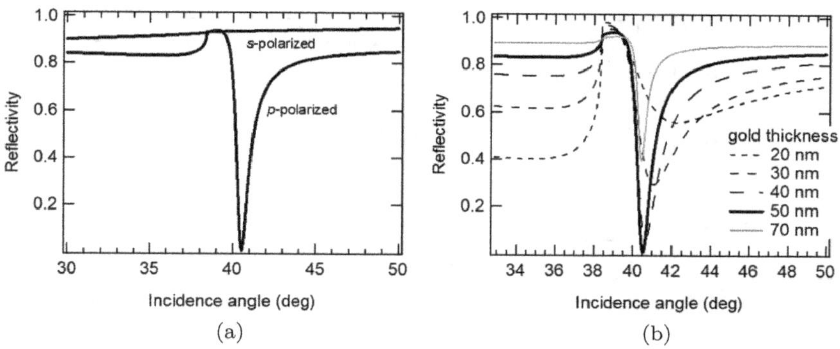

Fig. 4.16. SPP extinction curve with a 633 nm illumination in the Kretschmann configuration when the prism is placed in air. The prism is made of flint glass ($n = 1.61$). The index of Au is $n = 0.18 + 3.49i$. Note that the angle is the internal angle of the prism denoted by θ_1. The graph in (a) compares the case of the p-polarized light, where the SPP extinction is clearly visible at 40.5°, with s-polarization, which exhibits no plasmon line. The width at half minimum is 1.1°. The graph in (b) shows the evolution of the SPP extinction as a function of the gold layer thickness.

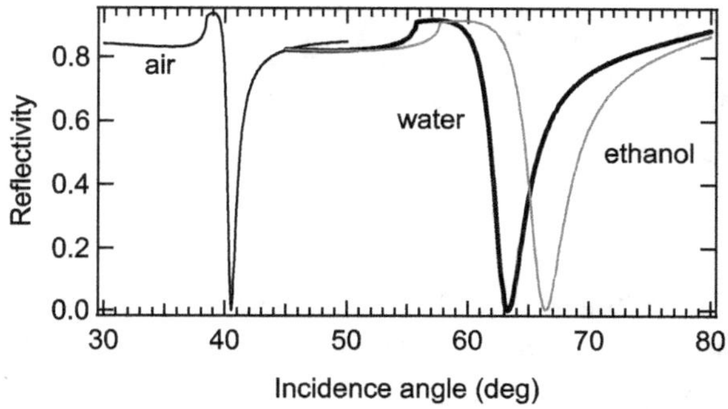

Fig. 4.17. Variation of the SPP extinction curve ($\lambda_0 = 633\,\text{nm}$) when the dielectric medium is changed from air ($n = 1$) to water ($n = 1.33$) or ethanol ($n = 1.36$). The extinction angle shifts from 40.5° to 63.3° or 66.4°, respectively.

4.4 Optical Detection of SPP Waves

4.4.1 *Indirect detection*

The indirect detection method is by far the easiest and most widely used approach. Its principle was explained in the previous section. Several variations of this method are available.

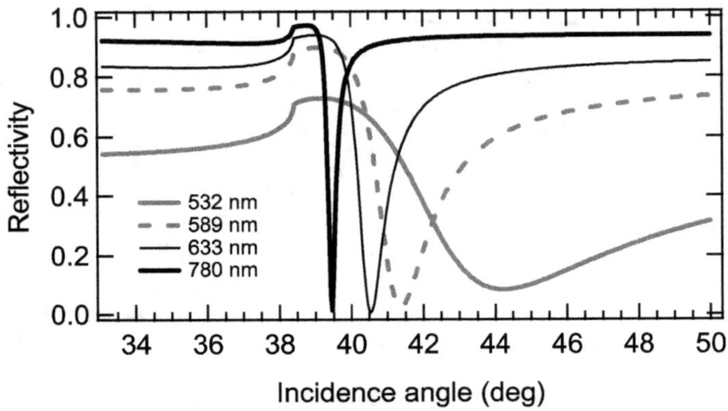

Fig. 4.18. Variation of the SPP extinction curve with the incident wavelength in the case of a 50 nm thick gold layer deposited on a glass prism ($n = 1.61$) in air. The extinction angle increases with the wavelength: 44.2° (532 nm, green light); 41.3° (589 nm, orange light); 40.5° (633 nm, red); 39.5° (780 nm, far red). More striking is the decrease in the width of the resonance when the wavelength increases toward infrared.

4.4.1.1 *Fixed angle detection*

First, an extinction plasmon curve, such as the one shown in Fig. 4.11, must be recorded. By measuring the reflected intensity at an angle at which the variation is the steepest (e.g. at 26° in Fig. 4.11), an extreme sensitivity can be obtained. This approach also allows fast detection and time-resolved acquisitions. This approach lays the foundation for biosensing with SPP waves since it takes advantage of the confined structure of this wave and its high sensitivity. This allows obtaining a signal from a 20 nm thin layer close the gold surface with a macroscopic laser beam. This approach will be discussed in depth in Chapter 5.

4.4.1.2 *Optical array detection*

Instead of using a parallel optical beam for illumination, it is a good idea to use a convergent beam on the plasmonic interface. This enables displaying the extinction curve in one shot. Various types of technology of detectors may be used depending on the application: linear arrays of detectors, CCD camera, or even a simple photo camera.

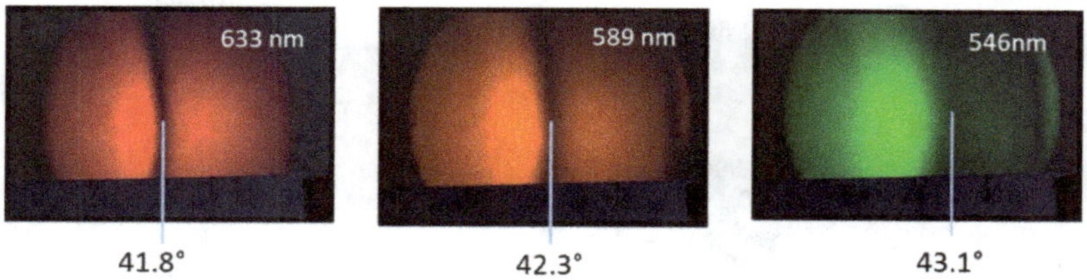

Fig. 4.19. Photographs of a screen on which the extinction curve was projected. The excitation beam is produced by a converging beam from a white lamp that is treated with a narrow-band pass filter centered at 633, 589, and 546 nm. The black strip is the plasmonic extinction angle. The values are the corresponding angles inside the glass prism.

This has been used for visualizing the effect of changing the excitation wavelength. The setup used for this is depicted in Fig. 4.10, where the light source is replaced by a white lamp whose beam is selected with a narrow-band pass filter centered at 633, 589, and 546 nm (see Fig. 4.19).

4.4.2 *Direct detection of the SPP wave in the near field*

As mentioned in Section 4.2.5, the SPP wave is confined to a very small volume close to the interface: It crawls at a maximum distance of 200 nm of the interface and over distances not longer than 10 μm. This is the near-field optic region. Obviously, it is very challenging to place a detector in such a small volume and detect the optical near field. Nevertheless, the development of scanning near-field optical microscopes (SNOMs) allows measuring directly the intensity of the SPP wave. Briefly, a SNOM is based on an optical fiber that is coated with a thin metal layer, leaving only a small aperture at the apex of the fiber. This aperture is smaller than the diffraction limit and acts as a near-field transducer that transforms the near field into far field, which is eventually collected by the optical fiber and recorded.

As an example, let's look at some results of research by Sierant *et al.* [3]. They fabricated a gold grating on glass with a period $d = 550$ nm (see Fig. 4.20(a)). The grating was milled over a surface of size $100 \times 100\,\mu\text{m}^2$ in the middle of the gold film (55 nm thickness). When the grating is illuminated from below with a 780 nm laser, it reflects the incoming beam

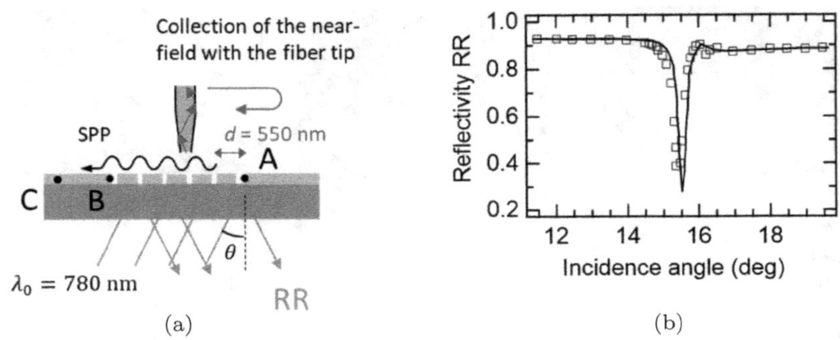

Fig. 4.20. (a) Schematic representation of the glass–gold grating used for launching an SPP wave at 780 nm. A collimated laser light is directed toward the back of the grating. The zeroth-order reflection *RR* is detected and exhibits a clear minimum when the plasmon wave is generated. The metal-coated optical fiber of SNOM scans the grating surface, which is able to measure the near field of the SPP wave. (b) Intensity of the zeroth-order reflection *RR* measured in the far-field regime. The minimum at 15.4° is the indication that the incoming energy is sent to the SPP wave.

Source: Reproduced with permission from Ref. [3]. Copyright © 2021 American Physical Society.

in the zeroth order. The reflected intensity *RR* exhibits a sharp dip when the incidence angle is swept from 10 to 20°, as shown in Fig. 4.20(b). This extinction (detected for $\theta = 15.4°$) corresponds to the excitation of the SPP wave at the gold–air interface, as explained in Section 4.3.6. In the present case, the mode SPP(−1) is launched, which means that the propagation occurs in the reverse direction (toward the left) compared to the direction of the incoming light (toward the right). The angle was therefore maintained at $\theta = 15.4°$ to ensure that an SPP wave can be collected by the fiber tip.

Then, the fiber tip was scanned over the gold grating of area $20 \times 20\,\mu\text{m}^2$ at different locations marked with the letters A, B, and C in Fig. 4.20(a). The intensity of the SNOM signal is shown in Fig. 4.21 after smoothing. Note that the raw signal exhibits pronounced oscillations due to a regular interference pattern resulting from the interference between three waves: incident wave, reflected light, and leakage radiation of the SPP. Some of this raw signal is shown in Fig. 4.21, but for more details, see discussion in the article by Sierant *et al.* [3]. The data in Fig. 4.21 clearly show that the SPP wave builds up progressively between point A and point B. The buildup follows an exponential growth with a constant $\xi_{bu} = 31\,\mu\text{m}$. Point

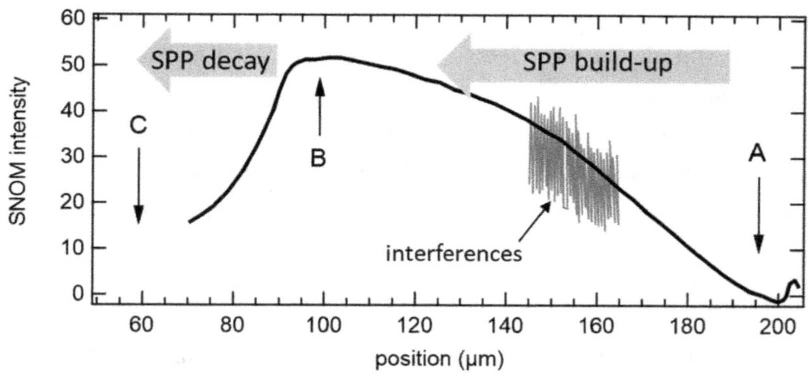

Fig. 4.21. Intensity of the SNOM signal collected by the optic fiber tip when scanning the SPP wave. Point A corresponds to one edge of the grating, where the plasmon originates. Between A and B, the SPP progressively builds up until the other end of the grating (B). Between B and C, the SPP propagation quickly decays due to the losses in the metal.
Source: Reproduced with permission from Ref. [3]. Copyright © 2021 American Physical Society.

B is the edge of the grating, and the SPP propagates at the gold–air interface. The rapid decay, which was already discussed in Section 4.2.5, is due to the ohmic losses in gold. This exponential decay is characterized by the SPP propagation length measured here at $\xi_p = 12\,\mu\text{m}$. Note that this experimental value is much smaller than the expected value ($21\,\mu\text{m}$), which is ascribed to irregularities in the top surface of the gold film that act as scatterers and waste away the SPP energy.

4.5 Conclusion

In this chapter, we have described the SPP wave and showed how this surface wave emerges at a metal–dielectric interface under special conditions. The plot of the dispersion relation proved to be crucial in determining these conditions. In particular, we have detailed the three main strategies for launching an SPP wave: with a prism in the Kretschmann configuration and in the Otto configuration, or with a grating. We have also shown that at a gold–air interface, the SPP wave was confined to a thickness of 90 nm or so, which is seven times thinner than the optical wavelength.

Therefore, producing an SPP wave is a first step into nano-optics, which aims at manipulating the electric fields of an optical wave within dimensions much smaller than the wavelength. It opens the door to two very exciting domains: detecting biomolecules with a rather simple optical setup and driving an optical signal for high-speed signal processing. These two application domains are discussed respectively in Chapters 5 and 6.

References

[1] Novotny L. and Hecht B. 2012. *Principles of Nano-Optics (2nd Ed.)* (Cambridge: Cambridge University Press).

[2] Pluchery O., Vayron R. and Van K.-M. 2011. Laboratory experiments for exploring the surface plasmon resonance. *European Journal of Physics* **32**, 585.

[3] Sierant A., Jany B. R. and Kawalec T. 2021. Near-field characterization of surface plasmon polaritons on a nanofabricated transmission structure. *Physical Review B* **103**, 165433.

Exercises

(Exercises marked with an asterisk (*) are difficult.)

(1) **Attenuation length of an optical wave inside a metal (skin depth):** The dielectric function of a metal is a complex number and is written as $\tilde{\varepsilon} = \varepsilon' + \varepsilon''$, with ε' being negative and ε'' much smaller than ε'. For silver, $\tilde{\varepsilon} = -15.05 + 0.73i$ at $\lambda_0 = 633\,\text{nm}$ (wavelength in vacuum).

Calculate the numerical value of the complex optical index in silver: $\tilde{n} = n' + n''$. Write down the wave vector k_m of a planar wave that would travel inside silver. Imagine a planar electromagnetic wave of amplitude E_0 propagating along the x direction inside a piece of silver. Write the electric field of this wave $E(x,t)$. Deduce the attenuation distance of this wave.

(2) **Conservation of k_x at an interface:** Consider an interface in the xOy plane between a dielectric medium and a metal characterized by their dielectric functions ε_1 and $\tilde{\varepsilon}_2$, respectively. $\tilde{\varepsilon}_2$ is a complex function.

Imagine a planar electric wave (of circular frequency ω) impinging on this interface at an incidence angle θ from the dielectric side.

Demonstrate that the parallel component k_x of the wave vector is preserved when the wave crosses the interface. Give the expression for k_x. Is it a real or an imaginary function?

(3) **Why is there no SPP wave at dielectric interfaces?** Consider the following interfaces and write down the corresponding dispersion relations of an SPP wave at an excitation wavelength of 633 nm. Discuss the existence of such an SPP wave. Interfaces: (i) gold–air, (ii) aluminum–air, and (iii) glass–air. For the values of the dielectric function of metals ε_2, use the simplified Drude relationship (2.7), and for glass, consider $\varepsilon_2 = 2.25$. (*) Explain, in particular, why there is no SPP wave at the air–glass interface.

(4) **SPP in the Kretschmann configuration for measuring the dielectric function of gold:** An SPP wave is excited with an equilateral prism in the Kretschmann configuration. The three faces of the prism are denoted by (a), (b), and (c), and its optical index is $n = 1.60$ (see figure below). A thin gold film was evaporated on face (b), and its dielectric function is denoted by ε_2. We first consider the case where the prism is placed in air and an He–Ne laser ($\lambda_0 = 633\,\text{nm}$) is used for the experiment.

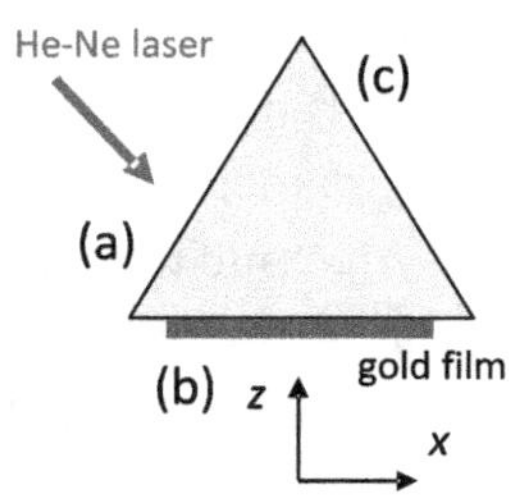

(i) The incident He–Ne beam is directed toward face (a) with an incidence angle $\theta_i = 10°$. Draw the optical path of the incident beam inside the prism.

(ii) We now consider the case where θ_i is adjusted so that the He–Ne beam launches an SPP wave. The beam inside the prism impinges on face (b) at an angle that is denoted by θ_{SPP}^{air}, which is called the *extinction angle*. Indicate θ_{SPP}^{air} in the drawing and wave vector $\boldsymbol{k}'$ on the internal face (b) of the prism. In particular, $\boldsymbol{k}' = k'_x\boldsymbol{u}_x + k'_z\boldsymbol{u}_z$.

(iii) Give the expression for k'_x as a function of n and θ.

(iv) Write down the SPP dispersion relation at the gold–air interface. Deduce the relationship between θ_{SPP}^{air} and the dielectric functions of air ε_1 and the metal ε_2.

(v) Experimentally, $\theta_{SPP}^{air} = 41.8°$. Derive the expression for the dielectric function of gold. Calculate the value. Given that the measurement accuracy of the angles is 0.2°, evaluate the accuracy of this method in measuring the dielectric function of gold at 633 nm. Is it an interesting method?

(vi) Calculate the critical angle θ_{crit} of the glass–air interface. Compare this value with θ_{SPP}^{air}. What would happen if $\theta_{SPP}^{air} < \theta_{crit}$?

(5) **Kretschmann configuration at the glass–water interface:** Consider now the same prism as in Exercise (4) whose face (b) is immersed in water ($n_{wat} = 1.33$).

(i) Use the formula derived in Exercise (4)-(iii) and calculate the new value for the extinction angle. This angle is denoted by θ_{SPP}^{water}. For gold, use $\varepsilon_2 = -10$.

(ii) Calculate the critical angle θ_{crit} of the glass–water interface. Is it possible to launch an SPP wave at the glass–gold interface?

(iii) With the equilateral prism depicted above, is it possible to launch an SPP wave? Replace the equilateral prism with an isosceles triangle with an angle at the vertex that is denoted A. Give a condition on A, for which the launching of an SPP wave is possible.

(6) **Propagation length of an SPP wave:** Consider an air–metal interface in the xOy plane. An SPP wave is launched at this interface with an He–Ne laser at $\lambda_0 = 633\,\text{nm}$. Make a schematic drawing of the SPP wave showing the decaying structure in the z direction away from the interface. Also, show the attenuation of the SPP wave in the x direction (damping of the SPP wave). Use the proper formula from Chapter 4 to calculate the decay length in the energy of the SPP

wave. Perform the numerical calculations for the following three metals: (i) gold: $\varepsilon_1 = -11.7 + 1.26i$; (ii) silver: $\varepsilon_1 = -18.3 + 0.48i$; and (iii) aluminum: $\varepsilon_1 = -54.7 + 21.8i$ (for other values of the dielectric function, see www.refractiveindex.info.)

(7) **(*) Numerical calculation of the extinction curve in the Kretschmann configuration:**

(i) Use a calculation software, such as MATLAB, IGOR Pro, or Python to calculate the reflected intensity of light at a glass–gold–air interface. Use the formula based on the generalized Fresnel formula to program a macro (subroutine) that calculates the complex reflection coefficients $\tilde{r}_{12}$ at the glass–gold interface and $\tilde{r}_{23}$ at the gold–air interface for *p*-polarized light. Then, calculate the three-layer reflection coefficient R in the case of a metallic film of thickness d_2. Use the following values for the optical indices at 633 nm: glass: $n_1 = 1.61$; gold: $\tilde{n}_2 = 0.18 + 3.49i$; and air: $n_3 = 1.00$. In particular, obtain the curve shown in Fig. 4.16(a) when $d_2 = 50\,\text{nm}$. Evaluate the SPP extinction angle θ_{SPP}.

(ii) The SPP extinction efficiency is defined by $\eta = 100 \times (R_1 - R_{\min})/R_1$, where $R_{\min}$ is the reflectivity at the minimum (i.e. at θ_{SPP}) and R_1 the reflectivity far away from the SPP extinction, e.g. at $\theta = 50°$. Calculate η for $d_2 = 50\,\text{nm}$.

(iii) Use your program to calculate the extinction curves for various gold thicknesses: $d_2 = 30\,\text{nm}$, 80 nm, and 100 nm. Evaluate the SPP efficiency η for these values.

(8) **(*) Working principle of an electronic nose based on the SPP wave in the Kretschmann configuration:** Exercise reproduced from the work by Hou; see *Anal. Chem.* **2018**, 90, 9879–9887 ("Highly-Selective Optoelectronic Nose Based on Surface Plasmon Resonance Imaging for Sensing Volatile Organic Compounds").

Consider an SPR system working in the Kretschmann configuration with the parameters described in Exercise (7)-(i). Here, the gold film is inserted inside a gas cell, which is connected to a gas line. Various gases can be flown through the cell. The initial atmosphere in the cell is air, whose optical index is $n_3 = 1.00027$ in normal conditions of temperature and pressure. A photodiode measures the reflected beam and the reflection coefficient in energy, $\%R$, is monitored (expressed in %).

(i) Explain how a change in refractive index n_3 can be monitored. Determine the best incidence angle that needs to be set for optimizing this sensor. This working angle θ_w will be kept at this value for sensing. Use either the program written for Exercise (7) or the curves in Section 4.3.7 to answer this question.

(ii) The following figure shows a zoomed-in section of a set of extinction SPR curves for increasing values of n_3. The presence of volatile organic compounds (VOCs) in air results in a slight increase in the optical index of air by approximately 10^{-4}RIU. Evaluate the sensitivity factor $\partial\%R/\partial n$ of this sensor for the VOCs.

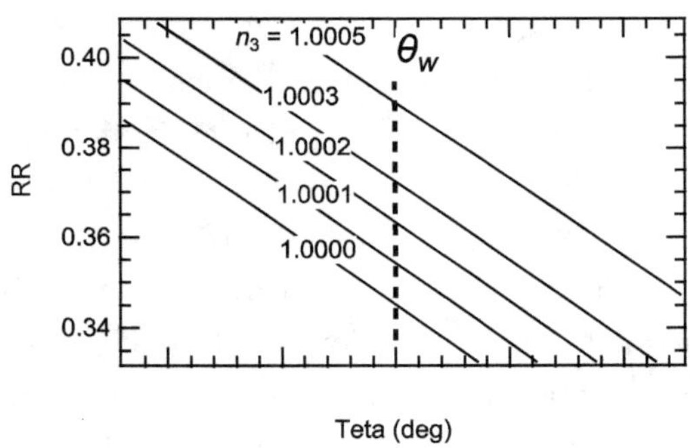

(iii) An amount of 100 ppm of toluene in air induces an increase in the optical index of $\Delta n_3 = 0.7 \times 10^{-4}$. Evaluate the change in reflectivity $\Delta\%R$.

(iv) Same as (iii) but with 1 ppm of toluene: $\Delta n_3 = 7 \times 10^{-7}$.

(v) Same as (iii) but with phenol, where 600 ppb in air leads to $\Delta n_3 = 2 \times 10^{-4}$.

(vi) The detection limit of this biosensor is set by various sources of noise of the detector that results in an accuracy of 0.05 % for R. Check if the three cases above correspond to a detectable amount of VOCs with this plasmonic nose.

(9) **SPP excitation with a grating:** We want to launch an SPP wave on a gold plane surface by imprinting a periodic grating on this surface. Let a be the period of this grating and θ the incidence angle of the excitation laser beam (He–Ne laser at 633 nm).

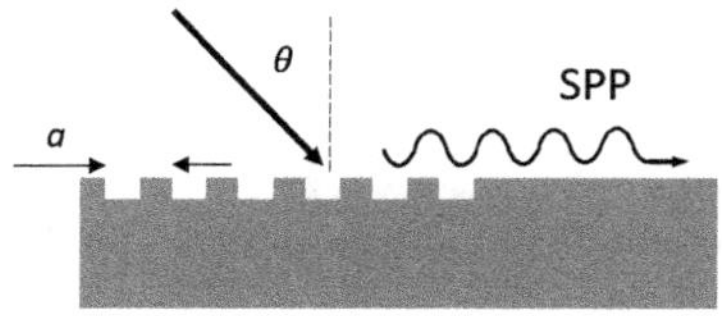

(i) Determine the possible values of a for launching the SPP, as indicated in the adjacent figure. Perform the calculations by using $\varepsilon_2 = -12$ for gold and $\theta = 45°$.

(ii) We now want to launch the SPP in the opposite direction (toward the left side in the figure above). Determine the values of a.

5

Propagative Plasmonics for Biosensing

Plasmonic biosensors are optical imaging systems that became popular due to their ability to characterize biomolecular interactions and their associated kinetics without any prior molecular labeling (targets). Under appropriate conditions, a **surface plasmon mode** can propagate at the metal–dielectric interface with near-optimal bulk and biofilm sensitivities. In this case, the evanescent electromagnetic field extends a few hundred nanometers in the surrounding medium where the molecules of interest have to be detected. In this chapter, the working principle of **plasmonic biosensors** is described with an explanation of how SPP is made sensitive to the presence of molecules. Several experimental configurations are explored and compared in terms of resolution and sensitivity. The experimental protocol on how to perform SPR measurements is also discussed, highlighting the performance and limits of SPR biosensors. Finally, the SPP measurements are linked to actual applications, with an emphasis on the best strategies to be used for **surface coating and surface functionalization.**

5.1 What Is a Practical Biosensor?

Being able to detect elements (such as proteins, molecules and ions) present in our environment (air, liquid) at low concentrations is a fundamental aspect of meeting environmental, health, or safety standards. It is also crucial for providing early diagnosis. For instance, ochratoxins are a group of mycotoxins produced by fungi. One of them, *ochratoxin A*, is a food contaminant found in cereal grains, coffee, beer, and wine [1, 2]. Due to their high toxicity, ochratoxins present a severe hazard to human health. Their concentrations have to be very low (under 10 ng/mL depending on the food [2], according to the EU commission), and hence, a sensitive device to identify, detect, and monitor them is required. Thus, the development of practical sensors, especially biosensors, is one of the main challenges of this century. So, what is a biosensor?

A biosensor [3] is defined by a transducer that detects and then converts the interaction between a biological recognition element (receptor) and a target molecule into a measurable signal (as shown in Fig. 5.1).

In this chapter, we focus on the optical transducer based on plasmonic properties. First, plasmonic biosensors can be separated into different categories depending on the nature of the studied plasmon mode: propagative plasmon (Chapter 4) or localized plasmon (Chapter 7). SPR biosensors are named after propagative surface plasmon resonance, while LSPR biosensors refer to sensors based on localized surface plasmon resonance.

SPR biosensors [4] are one of the two major families of plasmonic biosensors. They use propagative surface plasmons to measure the variation in the refractive index at a dielectric–metal interface. As discussed in Chapter 4, on the x-axis, the parallel component (projection in the mathematical sense) of the wave vector k_x at the interface is highly dependent on the dielectric constants (see the dispersion relation in Equation (4.17)). It consists of exciting the surface plasmon and then analyzing the optical properties of the reflected or transmitted light (depending on the experimental configuration). In addition, the SPR measurements are

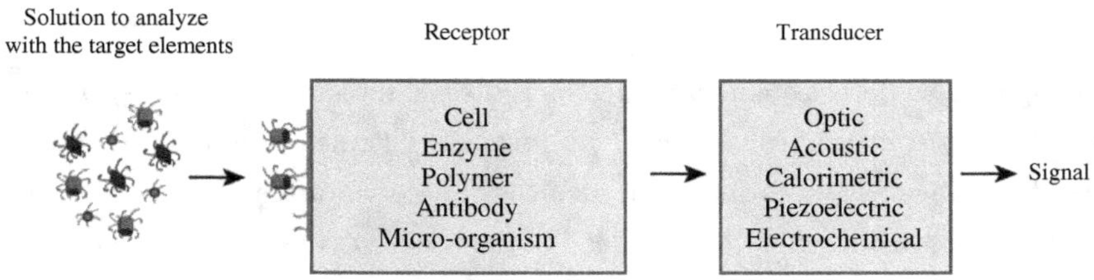

Fig. 5.1. Schematic diagram of a biosensor with the analyte, receptor, and transducer.

preferentially recorded with a CCD camera, allowing all the information from the entire biochip surface to be recorded at once.

5.2 Plasmonic Biosensors

5.2.1 *Principles*

The surrounding permittivity (ε_2) close to the metal surface can be modified either through a change in environment (air, water, or saline solution), temperature, or even the adhesion of a molecule. Thus, this variation (or disturbance) leads to a modification of ε_2 ($\varepsilon_2 \rightarrow \varepsilon_2 + \Delta\varepsilon$) as well as the resonance condition of the surface plasmon. This variation leads to a change in the optical properties of the reflected and transmitted waves. As described in the previous chapter, the wave vector of propagative surface plasmon in the case of two semi-infinite media follows Equation (4.17) and becomes

$$k_{SPP} = \frac{\omega}{c} \cdot \sqrt{\frac{\varepsilon_1 \varepsilon_2}{\varepsilon_1 + \varepsilon_2}} \rightarrow k_{SPP\_new} = \frac{\omega}{c} \cdot \sqrt{\frac{\varepsilon_1(\varepsilon_2 + \Delta\varepsilon)}{\varepsilon_1 + (\varepsilon_2 + \Delta\varepsilon)}}. \tag{5.1}$$

If the disturbance corresponds to the presence of a molecule (Fig. 5.2) on the surface, its adsorption will modify ε_2 to a new value ($\varepsilon_2 + \Delta\varepsilon$) depending on the concentration, density of the new layer, size of the molecule, and so on. This new ($\varepsilon_2 + \Delta\varepsilon$) will change the value of the local refraction index n_2 (since $\varepsilon = n^2$).

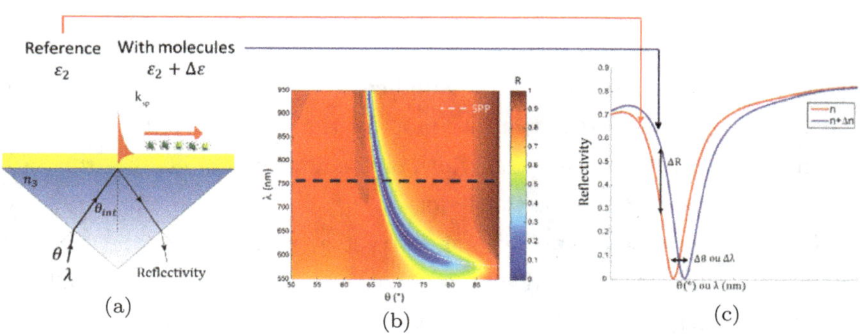

Fig. 5.2. (a) Schematic of the principle of biosensor (reference and measurement with molecules); (b) angulo-spectral reflectivity map; and (c) profile extracted at one angle or one wavelength. Copyright © J. F. Bryche.

Parameters such as the polarization, intensity, angle, and wavelength of resonance will be affected and can be measured to account for it. The phase that is correlated to the previous parameters is affected too. In general, the variation in the resonance condition of the surface plasmon is detected through a variation in the reflectivity, wavelength, or angle. These three methods of measurement are described in the following with concluding comments on the sensing method based on the phase measurement. The most common configuration of SPR is the *spectral interrogation.* To explain each method, we rely on Fig. 5.2(a) that shows the measurement of the reflectivity of a biochip as a function of two parameters in the Kretschmann configuration: excitation wavelength and incidence angle.

Several configurations described in the following allow SPR measurements either at a fixed angle or at a fixed wavelength. Nevertheless, it is possible to implement SPR devices that collect angular and spectral information sequentially to obtain an angulo-spectral SPR reflectivity map (Fig. 5.2(b)), where the intensity is plotted as a function of (λ, θ). The condition of the resonance is expressed by Equation (5.2) and corresponds to the equality between the excitation wave vector and the SPP wave vector k_{SPP}:

$$n_3 \sin \theta_{int} = \frac{\omega}{c} \cdot \sqrt{\frac{\varepsilon_1 \cdot \varepsilon_2}{\varepsilon_1 + \varepsilon_2}} = k_{SPP}. \tag{5.2}$$

When this equality is satisfied, a dip in the reflectivity is observed (blue area). The plot of k_{SPP} evolution is indicated by the dashed white line in the angulo-spectral SPR reflectivity map for the case of gold layer on BK7 prism (Fig. 5.2(b)). The spectral width of the blue area in comparison with the wave vector corresponds to the losses.

From the complete map, we can extract a profile at a given angle or wavelength (black curve) and get a conventional measurement (Fig. 5.2(c)). A complete map extends the study to all the properties of a substrate, especially when it is nanostructured (see the end of the chapter).

Since the position of the resonance is determined by the previous equation, the new value $\varepsilon_2 + \Delta\varepsilon$ of ε_2 caused by the presence of molecules will induce a shift in the angle or wavelength resonance, as depicted between the red and blue curves in Fig. 5.2(c). Most of the time, this shift will be a red shift (higher wavelength) or higher angle. From the angulo-spectral map and the profiles obtained, we can identify strategies to optimize the sensitivity of measurement (working in the highest variation of the slope for instance) or during signal processing.

5.2.2 *Refractive index unit*

To compare several different biosensors, we focus on two parameters: first, the *dynamics of the system*, corresponding to the variation of the amplitude of the signal, and second, the *resolution*, which is the threshold below which no signal can be recorded. Both are expressed in refractive index unit (RIU), a dimensionless number that is simply the refractive index (since $\varepsilon = n^2$) when it varies due to local changes in concentration, temperature, or other transformations. The idea is to be independent of the choice of the setup, of the configuration, or of the analyte, which allows universal comparisons.

For instance, we can compare two liquids at the same wavelength (589 nm), such as pure water (refractive index of 1.333) and pure ethanol (refractive index of 1.361), the variation will be 0.028 RIU, which is a huge value. Typical values that sensors can measure are close to $10^{-5} - 10^{-7}$ RIU (depending on the signal-to-noise ratio of the setup, the configuration, and the working wavelength). It shows that the biosensors must be calibrated to know their response to the variation in RIU.

5.2.3 *Configurations and setups*

5.2.3.1 *The reflectivity query or amplitude modulation*

This configuration was the first and simplest method to complete SPR experiments. It consists in acquiring the reflectivity for a given angle and wavelength. A variation of the refractive index will lead to a variation in the reflectivity ΔR (light intensity), as shown in Fig. 5.2. Experimentally, reflectivity is measured by a charge-coupled device (CCD) camera. This method eliminates the need for an angle or wavelength sweep and allows kinetic tracking (explained later in this chapter). The main drawback of this characterization method is the disparity in the intensity of the biochip linked to the inhomogeneity of materials, deposits, or surface functionalization previously carried out. In the same way, the light source must be extremely stable or a real-time normalization must be carried out by introducing a beam splitter in the setup.

5.2.3.2 *Spectral interrogation*

Spectral interrogation is based on the same principle. A change in the refractive index of the medium at the metal–dielectric interface leads to a modification of the coupling strength of light with the surface plasmon of the gold layer. It leads to a spectral shift $\Delta\lambda$ of the minimum reflectivity at a given angle. The tracking of the spectral position of the minimum reflectivity can be performed by making a Lorentzian interpolation of the response of the biochip. This method is very effective for thin layers. It is also more robust since the spectral shift will not depend on a variation in the intensity of the source or the inhomogeneity of the substrate. However, this method requires a more complex treatment when the biochip has a surface which is micro- or nanostructured. One last element concerns the impossibility of extracting this spectral offset in real time. The data processing is done after the experiment, which complicates the kinetic monitoring (combination of a spectrophotometer and a camera at a given point on the biochip only).

5.2.3.3 *Angular interrogation*

Compared to the previous method, the measurement is carried out at a given wavelength and an angular sweep is performed. Like before, the modification of the refractive index will be perceived by an angular offset $\Delta\theta$ induced by this variation. It is interesting to note that the shift may be higher for the same concentration in angular interrogation than in the spectral one. This can be understood from the map in Fig. 5.2(b) with the shape of the plasmonic resonance. At high wavelengths, the configuration in $\Delta\lambda$ is preferable since the dip is sharper following a horizontal line in this graph. In contrast, at low wavelengths, it is more efficient to work within $\Delta\theta$.

5.2.3.4 *Interrogation of the phase*

The use of phase interrogation for a biosensor relies on the steep phase jump experienced by the TM-polarized wave in the vicinity of surface plasmon resonance. In general, the phase is measured by means of an experimental device in Mach–Zehnder configuration (Fig. 5.3). The sensitivity is excellent and evaluated theoretically at 4×10^{-8} RIU [5]. The explanation originates from the Fresnel formulas, where the reflection of light from a metal surface is followed by a change in the phase of the light beam. In the configuration shown in Fig. 5.3, the initial beam is separated into two paths by a beam splitter. Each beam travels along the optical distance *d*. Given the refractive index n of the surrounding medium and the wavelength λ, it leads to a phase of $\varphi = 2\pi nd/\lambda$. A phase variation can be attributed to a variation in the refractive index at the metallic surface itself, which is linked to the presence of molecules. Experimentally, the two beams are combined, and

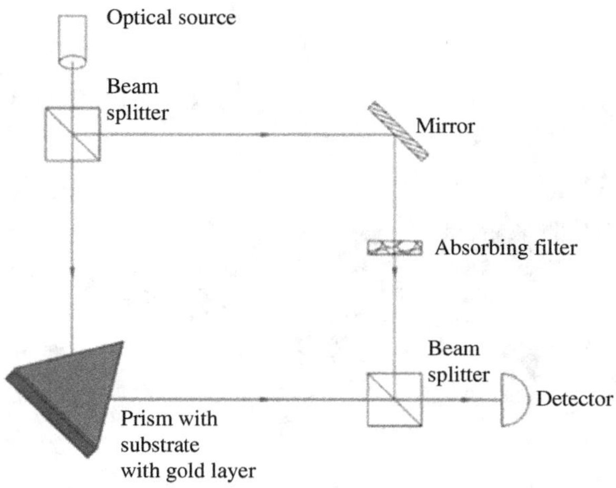

Fig. 5.3. Schematic of SPR interferometer with a beam split in two: one in the reference arm, the second one in the arm with the prism and the sample. The absorbing filter allows intensity adjustments of the first arm to maximize sensitivity. This is carried out in the initial adjustment state ($\Delta\varphi = \pi/2$).

their interferences can be measured by a photodetector, which results in a current variation: $i \propto 1/2\,(1 + \cos \Delta\varphi)$. However, this strong sensitivity also has drawbacks and leads to an unstable system caused by temperature or environmental index fluctuations. Due to these limitations, this configuration is less used experimentally for routine measurements. Nevertheless, its high accuracy makes it an appealing configuration for research.

5.3 How to Carry Out an SPR Measurement?

In this section, we guide the reader on how to perform an SPR measurement and process the data. To illustrate the experiment, a photograph of an experimental setup used by the group of Canva and Moreau at Laboratory Charles Fabry in France is shown in Fig. 5.4. Small adjustments to this protocol can be made to accommodate it to your instrumental bench. In this

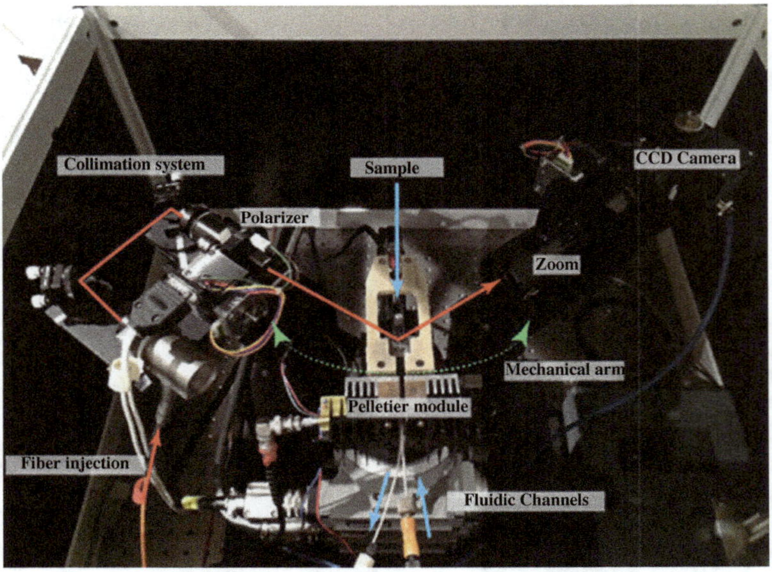

Fig. 5.4. Experimental setup allowing both angle and spectral measurements of the reflectivity of a biochip. Copyright © J. Moreau and J. F. Bryche.

section, we use the notations TE and TM, also called s and p polarizations respectively, depending on the community.

The setup is used in the Kretschmann configuration to excite a SPP on a gold film layer:

(1) An optical fiber conveys the light to the setup from a thermal light source combined with a monochromator to select the wavelength. The optical fiber is appropriate when a motorized arm is used.
(2) The light goes through a collimated system to increase the beam size without any deformation. Moreover, it allows to have only one angle of incidence, which is very important in such experiments.
(3) A polarizer allows choosing either TE or TM excitation before the prism and the sample disposed in the Kretschmann configuration.
(4) A microfluidic system allows working with liquid solutions.
(5) To measure the reflectivity, the signal from each pixel of the sample is recorded using a CCD camera. Each pixel yields a reflectivity curve similar to that shown in Fig. 5.2(c).
(6) If necessary, the angle can be modified thanks to the two mechanical arms.

To process the data and quantify the adhesion of molecules, the measurement must follow several steps whose important features are as follows:

- Initialize the setup (mechanical position, temperature, etc.) and record a dark picture (at 0° for instance) to allow normalization of the TE or TM pictures. This picture corresponds to the light noise of the environment.
- Take a series of $\text{TE}_{\text{1st serie}}$ pictures (at each wavelength and angle) to normalize $\text{TM}_{\text{1st serie}}$ pictures. $\text{TE}_{\text{1st serie}}$ pictures where no plasmon is generated allow subtracting any inhomogeneity in the sample. It is also useful to suppress the presence of dust on the CCD or optical lens.
- Take a series of $\text{TM}_{\text{1st serie}}$ pictures.
- Inject the active solution and wait for stabilization (1–20 min).

- Replace the active solution with the initial buffer solution.
- Record a second series of $\mathrm{TE}_{\mathrm{2nd\,serie}}$ and $\mathrm{TM}_{\mathrm{2nd\,serie}}$ pictures.
- Process the data as follows ($\mathrm{TM}_{\mathrm{2nd\,serie}}/\mathrm{TE}_{\mathrm{2nd\,serie}}$ and $\mathrm{TM}_{\mathrm{1st\,serie}}/\mathrm{TE}_{\mathrm{1st\,serie}}$) for each angle and wavelength. You will obtain a normalized series of pictures before (reference) and after the injection.
- Interpolation can be performed on the data if necessary.

Note: For kinetic measurement, the experiment will be done at one wavelength and one angle, and you will acquire the normalized reflectivity (TM/TE).

Figure 5.5 shows all the pictures, from the dark reference to the normalized pictures and associated profiles reported above. In order to have visible elements in the picture, several nanostructure arrays with different periods are introduced. You can note that the normalized curve for gold film does

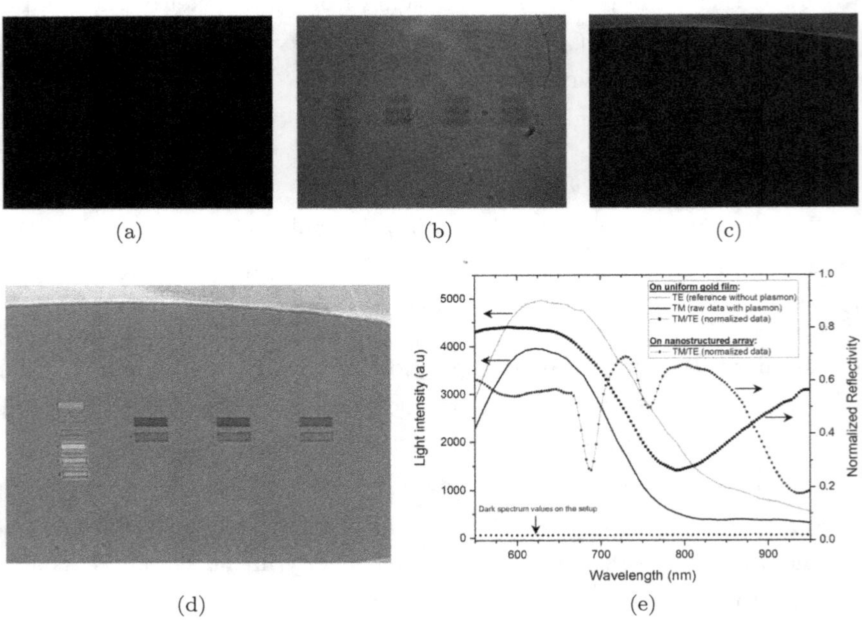

Fig. 5.5. Collection of pictures where an SPR profile is obtained (a) from a dark spectrum, (b) reference spectrum in TE polarization, (c) raw spectrum in TM, (d) normalized spectrum, (e) and associated profile.

not go down to 0 at the resonance (around 785 nm) due to the thickness of the film (30 nm), the adhesive layer (Ti) and the measurement angle of 16°. The surrounding media is water. The curve for the nanostructure array shows several resonances associated with different plasmonic modes.

5.3.1 *Performance and sensitivity of SPR sensors*

5.3.1.1 *How to evaluate the performance of an SPR sensor?*

A non-exhaustive list of several characteristics is given in the following, which provides a better understanding of the key elements of biosensor performance:

- **Limit of detection (LOD) or resolution:** the concentration or minimum number of the quantity to be measured that can be clearly detected over the experimental noise (extraction method data and instrument).
- **Sensitivity:** ratio between the increase in the sensor response and the corresponding variation of the quantity to be measured. In our case, this quantity corresponds to the refractive index.
- **Selectivity:** the ability to distinguish between two different targeted entities.
- **Reproducibility:** the correspondence of responses and their repeated measurements on the same set of biosensors.
- **Accuracy/Absolute error:** the agreement between the measurement and the true value.

This list highlights the limits related not only to the instruments and manufacturing processes but also to the very nature of the physical process used.

In the previous chapter, an important element has been described, the decay length of the SPP in the dielectric media for intensity ($D_2 = d_2/2 = 1/2k''_{2z}$). This decay length corresponds to the theoretical probing depth. Any molecules or proteins at this distance or closer to the interface will be detected by the sensor, as the condition to induce a sufficient modification of the refractive index. In the visible range (400–800 nm), D_2 varies from 30 to 170 nm in water solution, with intermediate values of 42, 92, and

103 nm at 532, 633, and 660 nm (usual working wavelength), respectively. These distances do not describe an abrupt LOD (only a reduction of 1/e), but they allow us to probe at specific distances away from the interface. Especially, in the case of functionalized surfaces, the thickness of the layer could be several tens of nanometers, which could push the analyte far from the vicinity of the metal–dielectric interface. Therefore, we need to select with precaution the wavelength and the functionalization before working.

5.3.1.2 *Sensitivity of an SPR biosensor*

The sensitivity of the sensor is defined by

$$S_{Sensors} = \frac{\partial Y}{\partial C} = \frac{\partial Y}{\partial \Delta n}\frac{\partial \Delta n(C)}{\partial C} = S_{RI} S_{analyte}. \tag{5.3}$$

Δn corresponds to the refractive index change when the analyte of concentration C binds to the surface of the sensors.

$$S_{RI} = \frac{\partial \lambda}{\partial n_{ef}}\frac{\partial n_{ef}}{\partial n}, \tag{5.4}$$

where the quantities $\partial\lambda/(\partial n_{ef})$ and $(\partial n_{ef})/\partial n$ correspond to the setup contribution sensitivity (link to the camera for instance) and the sensitivity of the effective index of a surface plasmon to refractive index, respectively. $(\partial n_{ef})/\partial n$ is independent of the method of excitation of the propagative surface plasmon. S_{RI} can be estimated by a calibration curve without an analyte in solution.

5.3.2 *Limits*

In the past decade, classical SPR sensors based on uniform metallic layers have reached the ultimate resolution predicted by Homola *et al.* [7] and is illustrated in Fig. 5.6. The figure shows the ultimate theoretical resolution (line) and the results from various groups around the world. As reported, the resolution is mainly limited by the properties of the surface plasmon and the optical components (detectors with low noise).

One way of tackling this limit is by optimizing the plasmonic properties of the substrate. For instance, a long-range surface plasmon has shown a higher resolution (up to five times) with narrower resonances. These hybrid

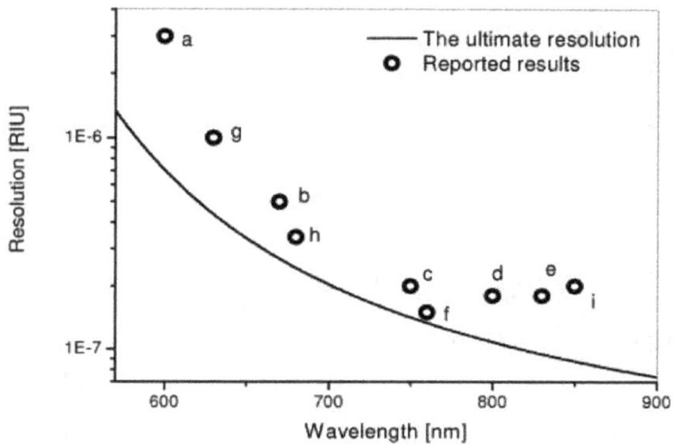

Fig. 5.6. The ultimate resolution of an SPR sensor (equation and parameters given in Ref. [7]) compared with the best experimental results reported: (a) Stemmler *et al.*, (b) Thirstrup *et al.*, (c) Piliarik *et al.*, (d) Nenninger *et al.*, (e) Chinowsky *et al.*, (f) Biacore 3000 (GE Healthcare, USA), (g) Wu *et al.*, (h) Bardin *et al.*, and (i) Piliarik *et al.*

Source: Reproduced from Ref. [7].

surface plasmons take place along a symmetric waveguide or in a metal–insulator–metal (MIM) configuration (see Chapter 6). Another way is to combine the propagative surface plasmon with a localized surface plasmon [8,9]. The properties of the localized surface plasmon will be described in Chapter 7. One of them is to confine the electromagnetic field close to the vicinity of the nanoparticles or nanostructures. In some specific configurations (periodicity or size), the SPP and LSPR modes can coexist. Each time their wave vectors become equal, a phenomenon of the band gap is observed, which gives birth to this hybrid mode described by two branches or resonances [9,10].

5.3.3 *Numerical aspect*

Based on the Fresnel reflection and transmission coefficients calculation for the two polarizations (TM and TE) with each layer crossed (see Chapter 4), the Rouard method [11] based on Yeh's matrix formulation [12] describes the behavior of a stack of thin layers as a function of their refractive index (n) and their thickness (h). This method extends Fresnel's calculations to absorbing films and therefore to complex indices (metals type).

By recurrence, it is possible to determine the reflection coefficient between the indexed layer $p-1$ and the layer $p+1$ while considering the intermediate layer p. This reflection coefficient is written as

$$r_{p-1,p+1} = \frac{r_{p-1,p} + r_{p,p+1}r^{-2j\phi_p}}{1 + r_{p-1,p}r_{p,p+1}e^{-2j\phi_p}}, \tag{5.5}$$

where $\phi_p = n_p h_p \cos(\theta_p)\frac{2\pi}{\lambda}$ is the phase shift during the propagation through the layer p. By recurrence, we can then describe the behavior of the biochip for a multiple stack of layer (see details in Chapter 3).

This method allows to predict the response of a biochip theoretically. Therefore, assuming that a biological layer can be approximated as a film of average effective index, it is possible to predict the angular or spectral offset that will be measured. It is also a good method to estimate the best thickness of the metallic layer and the position of the plasmonic resonance.

5.4 Surface Coating, Kinetics, and Applications

As described in the beginning of this chapter, SPR measurement can be carried out with spectral or angular interrogation, and the shift observed corresponds to the adsorption of a molecule (Fig. 5.7(a)). However, this adhesion can also be measured through a reflectivity variation (Fig. 5.7(b)), and it is easy to move from the first to the second graph by data

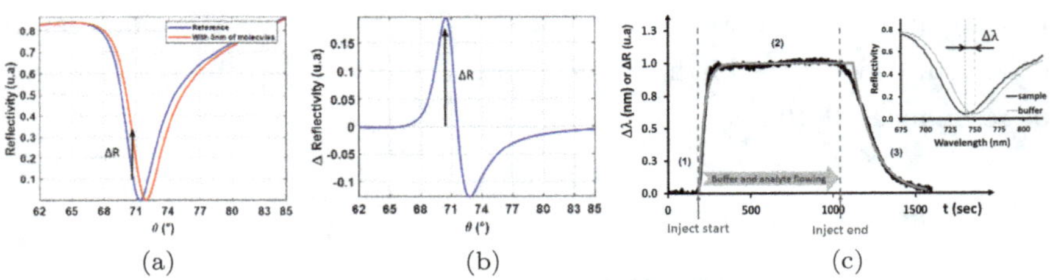

Fig. 5.7. (a) SPR measurement of a binding molecule in angular interrogation. (b) Correspondence in the variation of reflectivity. (c) Typical sensorgram or kinetics of the binding molecules on the surface (black rough curve) and the kinetic-model-fitted association and dissociation phases (gray smooth curve), with the inset showing the determination of $\Delta\lambda$ on y-axis.

Source: Reproduced with permission from Ref. [13]. Copyright © 2016 American Chemical Society. (Variations can also be expressed in ΔR.)

post-treatment. The ΔR is proportional to the molecule concentration and varies with time until stabilization. This point is important because it makes it possible to follow the adsorption kinetics of molecules onto a surface (Fig. 5.7(c)) and to explore the chemical reaction.

Until now, the SPR measurement was described only for the physical adsorption of a molecule directly on the metal layer. This approach allows us to describe the principle of SPR biosensors and the angular or wavelength shift observed when the molecules stick to the surface. However, in actual applications, a solution may contain many proteins, molecules, or elements, and we need to prepare the surface by coating a ligand (interaction partner) in order to increase the specificity and selectivity of the measurement.

If we consider a model with two entities A and B that can form a complex AB (such as DNA–DNA, biotin–streptavidin, and antigen–antibody), the following equation describes the kinetic reaction:

$$A + B \underset{k_d}{\overset{k_a}{\leftrightharpoons}} AB, \tag{5.6}$$

where k_a and k_d correspond to the rate constants of association (complex formation) and dissociation (breakdown) for the complex AB, respectively. Every step, from the association step of the two entities on the biochip until the regeneration step, is explained in the following and illustrated in Fig. 5.8:

(1) **Association:** entities A and B bind to each other (k_a).
(2) **Equilibrium/steady state:** the number of binding molecules is equal (compensate) the number of broken bonds.
(3) **Dissociation:** the bonds between A and B are broken.
(4) **Regeneration:** the sample is washed with a buffer solution (with a specific pH for instance) to obtain a clean sample with only the molecule A bound to the surface. The solvent breaks the interaction between A and B.

Note that small differences are observed between Figs. 5.7(c) and 5.8. In the sensorgram shown in Fig. 5.7(c), at the end of the dissociation stage, the

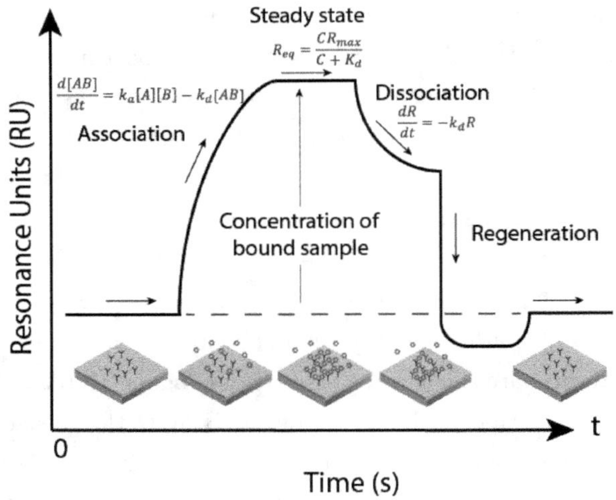

Fig. 5.8. Evolution of the reflectivity of biochip as a function of time. The curve is associated with different steps involved in the binding and dissociation of the molecules onto the surface.

signal returns to the initial threshold (buffer flow). However, the sensorgram shown in Fig. 5.8 illustrates the general case, where some molecules remain on the surface and the signal remains at higher values than before the interaction. A rinsing solution is required to clean the biochip properly.

Regarding Fig. 5.7(c), we follow the detailed study given in Ref. [14] and determine k_a and k_d as follows:

$$\frac{d[AB]}{dt} = k_a\,[A]\,[B] - k_d[AB]. \tag{5.7}$$

After the beginning of the reaction, we have $[B] = [B]_0 - [AB]$, with $[B]_0$ the initial concentration $[B]$ at $t = 0$ stuck on the surface. Thus, Equation (5.7) becomes

$$\frac{d[AB]}{dt} = k_a\,[A]\,([B]_0 - [AB]) - k_d[AB]. \tag{5.8}$$

Usually, one of the elements is immobilized onto the surface of the biochip $[B]_0$, while the other is continuously introduced in the fluidic chamber $[A]$. The signal measured, R, is proportional to the formation of AB complexes, and R_{max}, the maximum signal, corresponds to the capacity of the immobilized ligand surface expressed in resonance units. Equation 5.8 becomes

$$\frac{dR}{dt} = k_a C\left(R_{max} - R\right) - k_d R, \tag{5.9}$$

where dR/dt correspond to the rate of formation of surface-associated complexes (i.e. the derivative of the response curve). C is defined as the constant concentration of the ligate ($[A]_0$). $R_{max} - R$ is equivalent to the number of unoccupied surface binding sites at time t. By rearranging the previous equation, we obtain

$$\frac{dR}{dt} = k_a C R_{max} - \left(k_a C + k_d\right) R. \tag{5.10}$$

The first term on the right-hand side of the equation is constant, and we can define k_{obs} for the second part as $k_{obs} = k_a C + k_d$. The value k_{obs} corresponds to the slope of the graph, dR/dt, versus R. With several experiments with different concentrations C (or $[A]_0$), we can determine the value of k_a (the slope) and k_d (the y-intercept value) in the plot (C vs. k_{obs}), as shown in Fig. 5.9.

The values of k_a and k_d are 1.210 ± 0.151 $\mathrm{M^{-1}s^{-1}}$ and 0.0234 ± 0.0004 $\mathrm{s^{-1}}$. One other method to determine the two constants is to use a global fit from Equation 5.11 [14], which is not explained in detail here but shown in Fig. 5.7(c) as a grey curve. This second method allows us to determine the two constants for every binding experiment where two species are considered:

$$R\left(t\right) = \frac{k_a C R_{max}\left(1 - e^{-\left(k_a C + k_d\right)t}\right)}{k_a C + k_d} + R_{baseline}. \tag{5.11}$$

Finally, the dissociation equation ($dR/dt = k_a\, C(R_{max} - R) - k_d\, R$) or, in the integrated form, ($R(t) = Ae^{-k_d\, t}$) can be used to determine the value k_d, with A being the amplitude of the dissociation process.

Note that from Equation 5.10, we can find the equation in Fig. 5.8 for the steady state by setting $dR/dt = 0$, $R = R_{eq}$, and $K_D = k_d/k_a$.

Figure 5.10 shows a concrete example of an SPR experiment with a negative control (migG, corresponding to a constant chain of immunoglobulin)

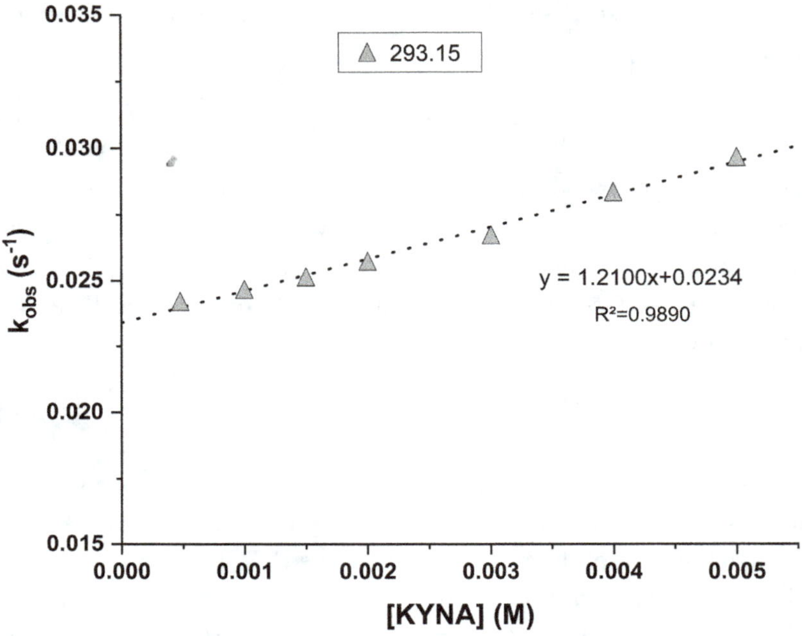

Fig. 5.9. Value of k_{obs} as a function of the concentration [KYNA], which is equivalent to C in this book. The data are related to the experiment shown in Fig. 5.7(c).

Source: Reproduced from Ref. 13. Copyright © 2016 American Chemical Society.

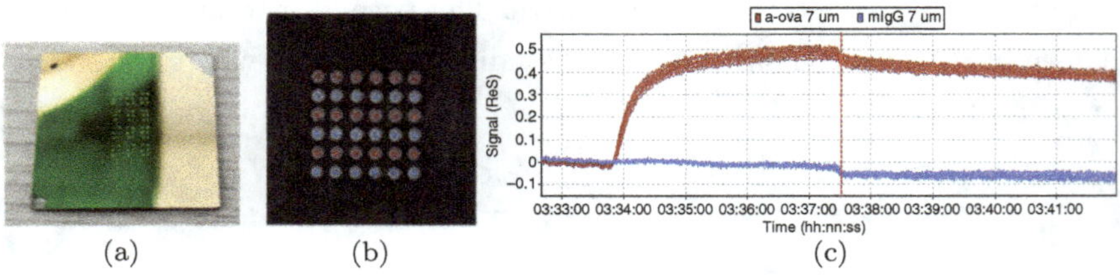

Fig. 5.10. Example of SPR experiment with gold biochip: (a) photograph provided by Beltrami; (b) an example of a functionalization spot; and (c) the sensorgram associated with two molecules (a-OVA and mIgG) ((b) and (c) provided by Horiba).

that does not bind to the surface and the molecules of interest (a-OVA, ovalbumin). This experiment is very interesting as it illustrates the selectivity of biosensors. Another element to keep in mind is the dispersion of the measurement. As we work with biological elements, the signal will vary from one surface of a biochip to the other due to small variation in the binding or functionalization. Here, the variation is very small, which shows the high quality of the biochip in clearly determining the presence of the molecules a-OVA. The multiple red spots allow us to avoid a false positive. In this example, the binding kinetics is fast (few minutes). If we use low concentrations, it would be interesting to consider the transport time of the analyte through the solution to the interaction surface. This is called mass transport limitation (MTL). This causes a modification of the association constant k_a, which is smaller than the true value. In general, the MTL should be minimized as much as possible, and the strategy of microfluidic systems, functionalization coverage, or active capture induced by external forces can be employed.

5.5 Conclusion

Plasmonic biosensors are very versatile setups enabling precise and real-time measurements of the presence of molecules near the metallic surface thanks to the propagative surface plasmon. Since the propagative surface plasmons are very sensitive to index variations, it is possible to exploit this property for biosensing. This type of biosensor is widely used today and its use has increased significantly in the past 15 years for food and medical applications. To go beyond the detection limits, the propagating surface plasmons are now associated with other plasmonic or photonic modes (some of which will be detailed in the following chapters).

In this chapter, we have shown how to use the properties of propagative surface plasmons to detect the presence of molecules. In particular, the molecules attached at the metal–dielectric interface modify the effective index. This index will change the behavior (the position of the resonance) of the propagative surface plasmon, which can be observed through an experimental setup (angular, wavelength, reflectivity, or phase interrogation). Following this aspect, we have described how to do SPR measurements correctly with data post-treatment. Finally, we have shown that SPR experiments can be used to study kinetics and chemical reactions and allow us to determine the association and dissociation constants for a first-order reaction.

References

[1] J. Yuan, *et al.* 2009. Surface plasmon resonance biosensor for the detection of ochratoxin A in cereals and beverages. *Analytica Chimica Acta* **656**(1–2), 63–71.

[2] A. Karczmarczyk, *et al.* 2016. Fast and sensitive detection of ochratoxin A in red wine by nanoparticle-enhanced SPR. *Analytica Chimica Acta* **937**, 143–150.

[3] D. R. Thévenot, *et al.* 2001. Electrochemical biosensors: Recommended definitions and classification. *Biosensors and Bioelectronics* **16**(1–2), 121–131.

[4] M. Li, *et al.* 2015. Plasmon-enhanced optical sensors: A review. *The Analyst* **140**(2), 386–406.

[5] P. I. Nikitin, *et al.* 1999. Surface plasmon resonance interferometry for biological and chemical sensing. *Sensors and Actuators B: Chemical* **54**(1–2), 43–50.

[6] M. Nakkach, *et al.* 2010. Angulo-spectral surface plasmon resonance imaging of nanofabricated grating surfaces. *Optics Letters* **35**(13), 2209.

[7] M. Piliarik, *et al.* 2009. Surface plasmon resonance (SPR) sensors: Approaching their limits? *Optics Express* **17**(19), 16505.

[8] C. J. Alleyne, *et al.* 2007. Enhanced SPR sensitivity using periodic metallic structures. *Optics Express* **15**(13), 8163.

[9] J.-F. Bryche, *et al.* 2019. Experimental and numerical investigation of biosensors plasmonic substrates induced differences by e-beam, soft and hard UV-NIL fabrication techniques. *Micro and Nano Engineering*. Elsevier BV, **2**, 122–130.

[10] M. Sarkar, *et al.* 2015. Generalized analytical model based on harmonic coupling for hybrid plasmonic modes: Comparison with numerical and experimental results. *Optics Express* **23**(21), 27376.

[11] P. Lecaruyer, *et al.* 2006. Generalization of the Rouard method to an absorbing thin-film stack and application to surface plasmon resonance. *Applied Optics* **45**(33), 8419.

[12] P. Yeh. 2005. *Optical Waves in Layered Media* (Wiley & Sons).

[13] Á. Juhász, *et al.* 2016. Kinetic and thermodynamic evaluation of kynurenic acid binding to GluR1270–300 polypeptide by surface plasmon resonance experiments. *The Journal of Physical Chemistry B* **120**(32), 7844–7850.

[14] D. J. Oshannessy, *et al.* 1993. Determination of rate and equilibrium binding constants for macromolecular interactions using surface plasmon resonance: Use of nonlinear least squares analysis methods. *Analytical Biochemistry* **212**(2), 457–468.

[15] M. Dolci, *et al.* 2018. Robust clicked assembly based on iron oxide nanoparticles for a new type of SPR biosensor. *Journal of Materials Chemistry C. Royal Society of Chemistry (RSC)* **6**(34), 9102–9110.

[16] L. S. Jung, C. T. Campbell, T. M. Chinowsky, M. N. Mar and S. S. Yee. 1998. Quantitative interpretation of the response of surface plasmon resonance sensors to adsorbed films. *Langmuir* **14**(19), 5636–5648.

[17] Willets K. A. and Van Duyne R. P. 2007. Localized Surface Plasmon Resonance Spectroscopy and Sensing. *Annual Review of Physical Chemistry* **58**, 267–297.

Exercises

(1) **Sensing length:** In the case of two semi-infinite media (gold and water), in what range can a molecule be detected above the gold surface? The experiment parameters are a wavelength of 660 nm and an internal angle of 62°.

Consider a pre-functionalization step of the gold surface, with molecule A to be selective to the binding molecule B. Molecules A form a 25 nm thick monolayer on the gold surface with full coverage. Can you still detect the binding of B molecules?

(2) **Estimation of sensitivity:** Consider the following table extracted from measured values for a 30 nm gold film at three wavelengths for a 71.15° internal angle. The measurements were done with the SPR setup shown in Fig. 5.4 on a testing sample. To simplify the calculations, we consider a 1 nm layer of molecules stuck to the surface and a variation of 10^{-2} in the refractive index of the solution. What are the sensitivities (S_{sensor}) associated with each wavelength?

Wavelength	$\Delta\lambda_{\text{molecule}}$	$\Delta\lambda_{\text{solution}}$
622 nm	5.4 nm	24 nm
707 nm	5 nm	42 nm
727 nm	4.2 nm	54 nm

Why it is interesting to have an SPR instrument with several wavelengths (or angles)?

(3) **Estimation of sensitivity:** Comment on the effect of temperature on the SPR measurements. You can use the dependency of the dielectric

permittivity on the temperature. Why is it important to monitor the temperature?

(4) **Open question on catching strategies:** Why is it important to think about your catching strategies when working with low concentrations? Can you provide at least two methods to increase the performance of molecule capture? What do you need to consider in regard to your plasmonic system before using combined strategies?

(5) **Determination of kinetic constants:** From the following figure, determine the kinetic constants k_a and k_d using Equation (5.10).

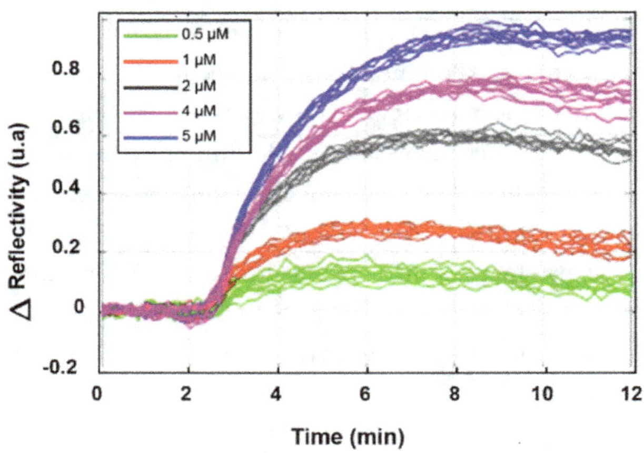

Evolution of the reflectivity of a biochip as a function of time. Each curve corresponds to one concentration.

(6) **More complex biochips:** In real studies, several capture sites are usually present on the biochip to provide a reference, information redundancy, and negative controls. With a commercial device, you can quickly identify the spot areas and perform an analysis of the curves obtained. In this example, determine the sensitivity of the microarray to antigens.

In view of the spectrogram, have we reached saturation for the strongest concentration?

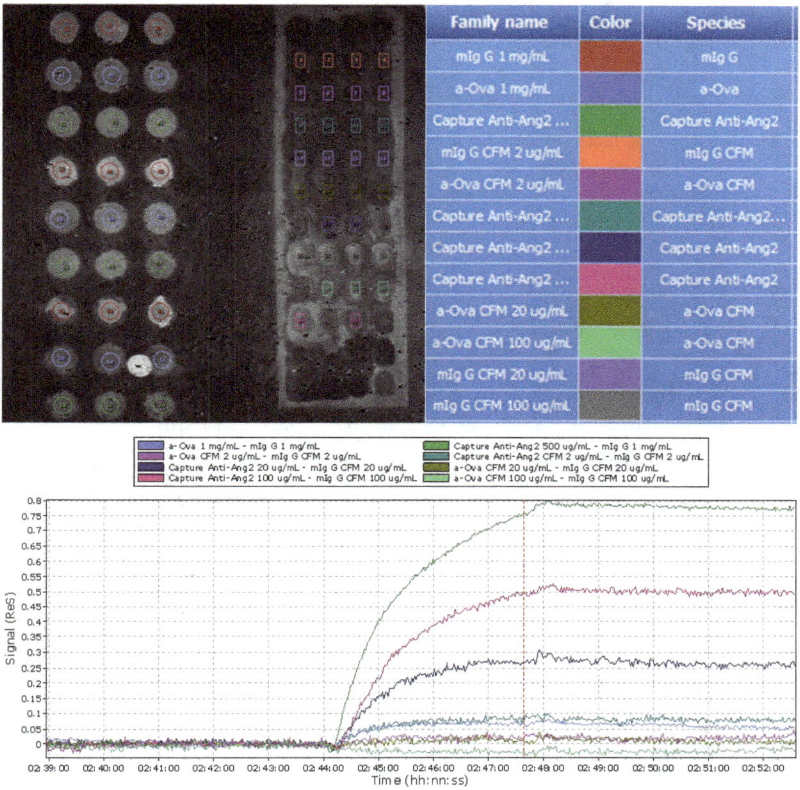

Evolution of the reflectivity of biochip as a function of time. The curve is associated with different steps involved in the binding and dissociation of the molecules onto the surface. (data provided by Horiba).

Finally, determine the k_a and k_d values for one species of your choice.

(7) **New type of SPR biosensor (combining gold film with nanoparticles):** The bottom-up approach to the growth of nanoparticles was mentioned at the beginning of the chapter. To increase the SPR sensitivity factor, Dolci *et al.* [15] have used iron oxide particles with different surface coverages on a gold film. The nanoparticles were functionalized with biotin to induce selective capture of streptavidin. Here, we propose to calculate the sensitivity factor m defined in Chapter 7 with the Campbell model [16] (Equation (7.37)), which was reused and confirmed experimentally by Van Duyne *et al.* [17]:

$$\Delta\lambda_{LSPR} = m \cdot \Delta n \left[1 - \exp\left(-\frac{2d}{l_d}\right)\right].$$

Some parameters are given in the following table. Also, $n_{absorbate} = 1.57$ and the work is done in water.

Sample	Effective thickness d	Decay length l_d
0% Fe_3O_4	0.8 nm	196 nm
15% Fe_3O_4	1.6 nm	218 nm
100% Fe_3O_4	4.6 nm	313 nm

- From the different values of m, what can you deduce about the benefits of combining nanoparticles with a flat gold film?
- From a chemical point of view, is there any benefit to using larger or smaller particles? If yes, what would be the impact on the SPR measurement afterward.

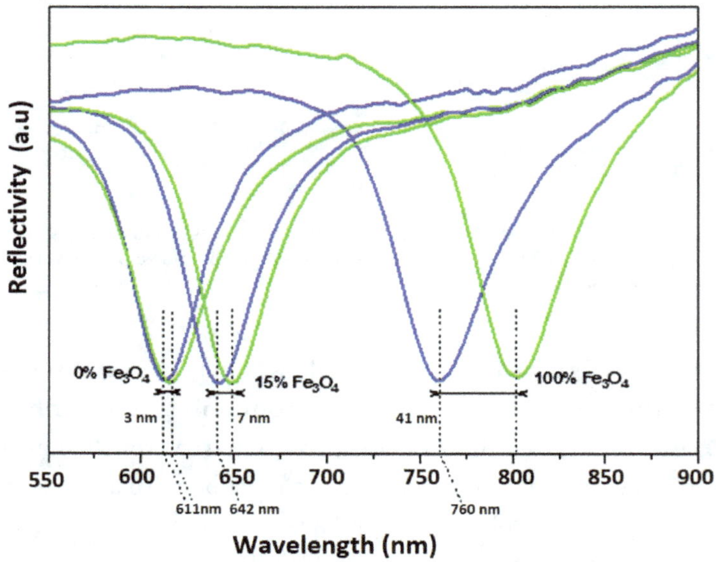

Variation of reflectivity for different surface coverages of gold film by Fe_3O_4 nanoparticles. The blue lines correspond to the reference measures, and the green lines correspond to the detection of binding streptavidin.

Source: Reproduced with permission from Ref. [15].

6

Propagative Plasmons in Waveguides

The propagation of the surface plasmon polariton is reviewed for various waveguides: **2D waveguides**, which accommodate the gap plasmon or the thin-film plasmon, and **3D waveguides**, such as the V-groove waveguides. The **propagation lengths** of these waveguides are systematically discussed, as well as their ability to confine the electromagnetic waves more or less efficiently. Finally, some devices are also explained, such as the **plasmonic laser** (SPASER), and some plasmonic-based electro-active devices.

6.1 Introduction

We have shown in Chapter 4 that surface plasmon polaritons (SPPs) are surface waves that can propagate at a metal–dielectric interface. They are made of the oscillation of an electromagnetic field coupled with the oscillation of free electrons of the metal. This wave is confined in an ultrathin volume at the interface with a subwavelength compactness. For example, at an air–gold interface, a 633 nm SPP wave extends over 28 nm on the gold side and 340 nm on the air side. The electric field exhibits an exponential decay, as shown in Fig. 6.1(a). In other words, a simple metallic film of a few nm thickness is able to convey an optical wave. If this wave can be propagated, routed, multiplexed, amplified, and detected, it signifies that it can be used as a carrier wave for signal processing. This opens up very exciting

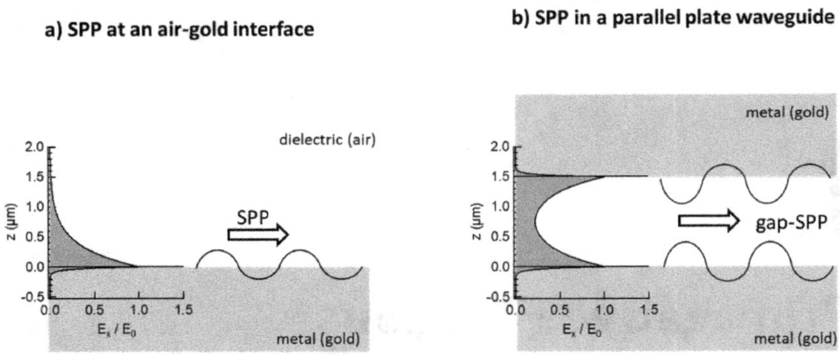

Fig. 6.1. (a) SPP wave at the gold–air interface in the case of a 633 nm wavelength. The electric field component parallel to the interface decays exponentially on both sides of the interface. The decay constants are 28 nm in gold and 340 nm in air. (b) When two gold plates are placed in close vicinity (1.5 μm in this example), they form a parallel plate waveguide. The two SPPs are coupled, and new plasmon modes are generated, called gap plasmons.

perspectives because signal processing based on optical waves leads to much larger bandwidth and better performance than electrical signals. A dream in terms of the speed of signal processing would be to fabricate an all-optical computer. Hopes are that plasmonics will contribute to turning this dream into a reality.

6.2 Basic Figures of Signal Processing

6.2.1 *Signal processing with SPP*

Plasmonic propagation allows squeezing optical waves over distances smaller than the wavelength and breaking, in some ways, the diffraction limit. Sub-micrometer confinement of optical signals is necessary for building all-optical microchips, and plasmonics offer a real opportunity for mixing the best of both worlds: the nanometer compacity of the electronic circuitry and the ultra-high speed of optical signal processing. Therefore, plasmonic signal processing should address some key basic functions, such as transmission, routing, multiplexing, commutation, and amplification (see Fig. 6.2). This will be explained in this chapter.

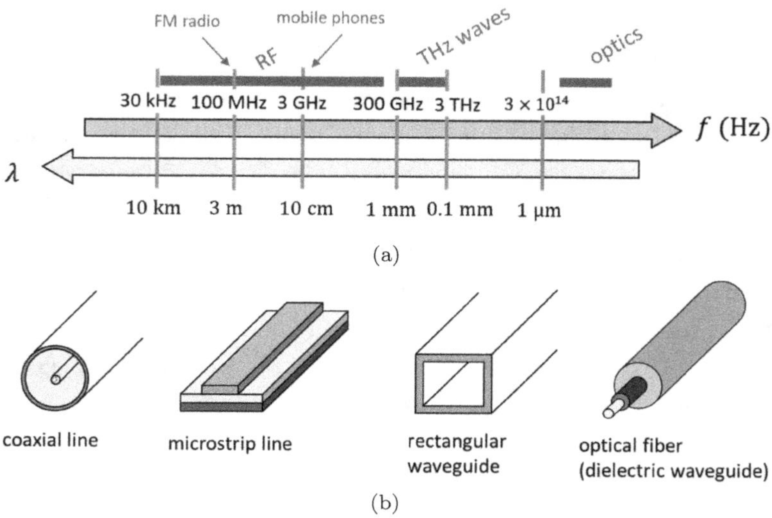

Fig. 6.2. (a) The frequency range of the electromagnetic waves. (b) Examples of different electromagnetic waveguides optimized for transporting different electromagnetic waves. Coaxial cables are used for RF electric signal, microstrips for microwaves in integrated circuits. Rectangular waveguides are used for transferring large amounts of the power of the wave of frequencies higher than 1 GHz. Optical fibers also convey electromagnetic waves but at optical frequencies and are made of nonconductive materials, as opposed to the other waveguides, which are all metallic.

6.2.2 *Waveguides*

A waveguide is a structure that guides electromagnetic waves with minimal loss of energy by restricting the transmission of energy to one direction. The most common waveguides are used for high-frequency radio waves, particularly microwaves, and are a hollow conductive metal pipe. The idea is to transfer this concept to plasmonic waves, given that these SPP waves follow the metal–dielectric interface. However, one main obstacle is the short propagation length of the SPP waves: At a wavelength of 633 nm, the propagation length L_{SPP} is just 11 μm for an air–gold interface (see Chapter 4). At 1550 nm, L_{SPP} increases to 335 μm. This short propagation length is due to the very strong ohmic losses undergone by the wave in the metal. At optical frequencies (10^{15} Hz–300,000 GHz), the metal is far from being a *perfect conductor*, and it is mainly absorbing. Propagation over such short distances for SPP makes it difficult to manipulate the signal. This is not the case at radio frequencies (RFs). A visualization of the most common frequency ranges considered for electromagnetic wave propagation is shown in Fig. 6.2(a). One way of improving the propagation length is to confine the wave inside a waveguide. Due to the boundary conditions imposed by the waveguide, the wave might be strongly modified, and eventually, its propagation length is increased. Ohmic losses occur for the portion of the wave that propagates inside the metallic layer. If the waveguide can favor the field propagation in the dielectric medium and de-favor the propagation in the metal, one can expect an increase in the propagation length. For example, the simplest waveguide is represented in Fig. 6.1(b), where two parallel metallic plates are placed sufficiently close to each other so that the two SPP waves interact. As a result, the electric field increases in the air section between the two plates and L_{SPP} will slightly increase. This case is addressed and discussed in the following.

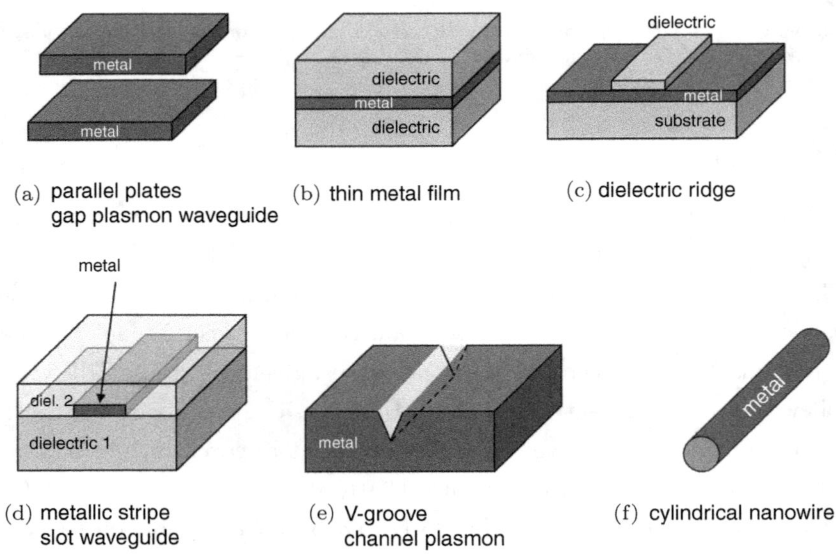

Fig. 6.3. Most important plasmonic waveguides used for guiding SPP waves. They differ by the way they confine the electromagnetic field, giving rise to various SPP modes and SPP propagation lengths. The metal is often gold.

Waveguides have been developed for decades for RF components in order to transport electromagnetic signals or electromagnetic power. Their shape depends mainly on the frequency range of the wave to be conveyed (see Fig. 6.2). Coaxial cables are widely used for RF electric signals up to 3 GHz. Microstrips are used for microwaves in integrated circuits. Rectangular waveguides are used for transferring large amounts of the power of the wave at frequencies higher than 1 GHz. Optical fibers are also excellent waveguides for electromagnetic waves at optical frequencies with extremely low losses.

Propagation of SPP can also be controlled by using waveguides. Although propagation of RF waves and plasmon waves are fully described by the same Maxwell equations, they are not in the same frequency ranges. Therefore, the waveguides optimized for RF waves cannot be directly transposed to SPP waves. This is due to the very different behavior of the dielectric permittivity ε at RF frequencies and at optical frequencies. In the microwave and RF regimes, the metal can be approximated to a perfectly electric conductor (PEC), which is absolutely not the case at optical frequencies. Therefore, the waveguide for SPP waves adopt very different geometries, as shown in Fig. 6.3.

6.2.3 *Some figures for understanding signal processing in waveguides*

A digital signal is made of individual *bits* than can take the value 0 or 1. In an electrical wire, *bits* are typically coded as 0 or +5 V signals. Assembling these *bits* together gives rise to more complex signals coded in the binary format. A set of 8 *bits* is called a *byte* and can take $2^8 = 256$ values. For example, in the ASCII standard, the letter "*e*" is coded as 01100101, and the corresponding chronogram is shown in Fig. 6.4, where τ is the time delay between two bits. The fundamental frequency of this signal is given by $1/\tau$. For a typical coaxial cable, τ is 100 ns, and the corresponding fundamental frequency is 10 MHz. The bits are produced by modulating a carrier frequency whose frequency is typically 10 times the fundamental frequency (see Table 6.1).

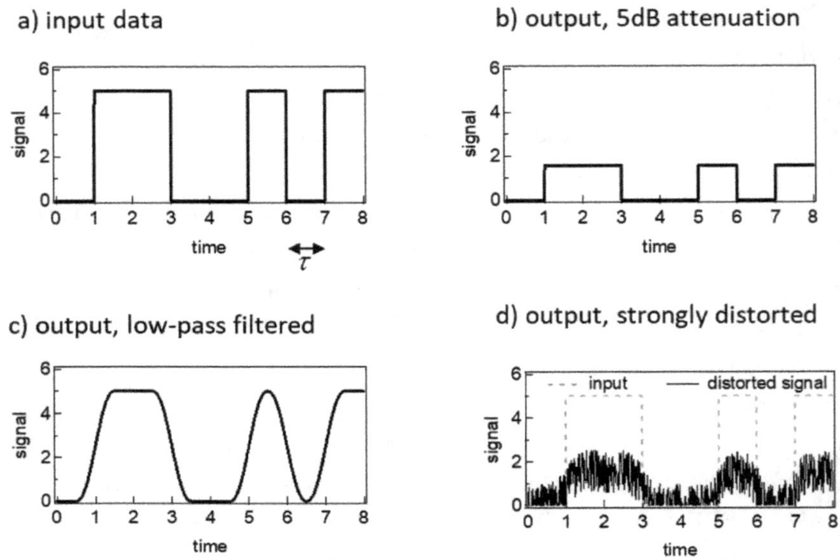

Fig. 6.4. Illustration of the transmission of a digital signal: (a) a binary signal (here, the chronogram is the binary code for the letter "e": 01100101), and the time delay between two successive bits is τ; (b) this input signal after having undergone an attenuation of 5 dB (division by 3.2) is shown; (c) representation of the signal after transmission through a line with a too low cutoff frequency, and the signal is smeared out; (d) typical transmission when the lines exhibit attenuation and low-pass defects and noise is added. Such a signal will hardly be recoverable.

Table 6.1. Key figures of traditional transmission lines.

	Attenuation (dB/km)	Carrier frequency	Bandwidth	Two-way voice channels
Coaxial cable @ 1 MHz	20	1 MHz	100 kHz	<2,000
Coaxial cable @ 100 MHz	220	100 MHz	10 MHz	13,000
Optical fiber @ 1550 nm	0.2	1000 THz	40 THz	3,000,000

Signals needs to be transported, and this can be done either with electric cables or with optical fibers. Two main properties of a transmission line are its *attenuation* and its *bandwidth*. Attenuation is measured in decibel per unit distance and is defined as follows: If P_i is the input power into the transmission line and P_o is the output power measured after a distance of say 1 km, the attenuation constant is defined as

$$\alpha = 10 \log P_i/P_o \text{ in dB/km.} \tag{6.1}$$

A typical optical fiber has an attenuation coefficient $\alpha = 0.2$ dB/km at a wavelength of 1550 nm (input signal divided by 1.05 after 1 km). This should be compared to coaxial cables, where the attenuation of the electric signal is around 20 dB/km. This means that after 1 km, the power of the electric signal is divided by 100! In Fig. 6.4(b), the input signal was attenuated by 5 dB, which is an attenuation factor of 3.2. Such attenuation occurs after 25 km of transmission through an optical fiber and 250 m through a coaxial cable. This shows that optical fibers are the most efficient channel for data transmission.

Another limitation of the transmission lines is their limited *bandwidth*. It corresponds to the highest frequency that the line can propagate. It will limit the capacity of the line for transmitting a large load of data. It is related to the cutoff frequency of the line f_c and is illustrated in Fig. 6.4(c). For example, a coaxial cable can only transmit 2,000 simultaneous conversations (see Table 6.1).

6.2.4 *Why are metallic waveguides with optical frequencies more complex than in microwave or THz ranges?*

When dealing with waveguides, we can distinguish three ranges of frequencies: microwaves, terahertz (THz) waves, and optical frequencies (see Fig. 6.2). Within the RFs, **microwaves** correspond to electromagnetic waves of frequencies below 300 GHz (wavelengths longer than 1 mm). These waves and those with lower frequencies that do not penetrate into metallic surfaces, or more precisely, the corresponding skin depth is much smaller than their wavelength. For this range of frequencies, metals are considered as PECs. In practice, this means that the electric field is null inside the metal. The physics of waveguides has been developed mostly for these kinds of waves [1]. **Optical frequencies** are situated at above 3 THz (3×10^{12} Hz) and correspond to wavelengths shorter than 100 μm. They include infrared, visible, and UV light. As explained in Chapter 3, the skin depth of such waves is of the same order of magnitude as their wavelength, and they can propagate inside a metallic material over very small distances. Surface plasmon waves originate from these oscillations close to the surface. Applying the physics of waveguides to plasmonic waves will be strenuous since it not only requires taking into account the geometrical boundary conditions of the waveguides but also the boundary conditions where the electric field is non-zero at the metallic walls. These will be explained for the most simple cases in the following.

Finally, **THz waves** are electromagnetic waves of intermediate frequencies between 0.3 and 3 THz (wavelengths between 0.1 and 1 mm). They can penetrate into very thin metallic objects. This is a recent topic of research, with applications in biomedical imaging, security screening, and high bandwidth data transmission.

6.3 Parallel Plate Waveguide or Gap Plasmon Waveguide

Let's now focus on the most elementary waveguide and derive its main optical properties. We consider two infinite and parallel metallic planes separated by a distance b. A dielectric lies in between. This is a structure of type metal–insulator–metal (MIM). Each of the interface can accommodate an SPP wave, and when the plates come close enough, the SPP waves overlap and give rise to coupled plasmonic modes: This is the **gap plasmon**. For deriving the main properties of this gap SPP, we refer Leal-Sevillano [2] and Bozhevolnyi [3, 4].

6.3.1 *Derivation of the dispersion relation of the gap SPP*

Set of equations: The plates are parallel to the (x, y) plane, and we seek a wave that would propagate parallel to these plates. We set the $x-$axis along this propagation direction. Therefore, the real part of the wave vector will be along x and is written as k_x. $\boldsymbol{k}$ is a complex number, and the wave vector is expected to exhibit x and z components, $\tilde{k}_x$ and $\tilde{k}_z$, respectively, with real and imaginary parts (damping and evanescence of the wave, respectively). Moreover, $\tilde{k}_x$ is uniform when crossing the interfaces (see Chapter 4).

Therefore, the electromagnetic fields in the dielectric at a circular frequency ω can be written as

$$E_d = \begin{bmatrix} E_{dx} \\ E_{dy} \\ E_{dz} \end{bmatrix} \exp\left(i\left(\tilde{k}_x.x - \omega t\right)\right) \cdot \exp\left(i\tilde{k}_{zd}.z\right), \tag{6.2}$$

$$H_d = \begin{bmatrix} H_{dx} \\ H_{dy} \\ H_{dz} \end{bmatrix} \exp\left(i\left(\tilde{k}_x.x - \omega t\right)\right) \cdot \exp\left(i\tilde{k}_{zd}.z\right). \tag{6.3}$$

Note that, *a priori*, the amplitudes E_{dx}, E_{dy}, and E_{dz} are functions of the three space coordinates x, y, and z, respectively. But this problem is invariant in the (x, y) directions so that the functions E_{dx}, E_{dy}, and E_{dz} depend on z only and not on the x and y coordinates. And this is the same for $\boldsymbol{H}_d$ as well as in the metal (fields $\boldsymbol{H}_m$ and $\boldsymbol{E}_m$).

The wave equation can be written for each of these six components in the three media (dielectric and two plates: 18 wave equations). We follow the approach adopted in the waveguide physics and separate the propagation into *longitudinal* fields and *transverse* fields. If the longitudinal fields are known (six fields), the others can be deduced [1]. The wave equation for the longitudinal fields in the dielectric is written as

$$\frac{\partial^2 E_{dx}(z)}{\partial z^2} - k_x^2 E_{dx}(z) + \frac{\varepsilon_{\mathrm{d}}}{c^2}\omega^2 E_{dx}(z) = 0, \tag{6.4}$$

$$\text{which simplifies into } \frac{\partial^2 E_{dx}(z)}{\partial z^2} - k_{zd}^2 E_{dx}(z) = 0, \tag{6.5}$$

$$\text{where } k_{zd}^2 = k_x^2 - \varepsilon_{\mathrm{d}}\omega^2/c^2, \tag{6.6}$$

$$\text{and for the magnetic field, } \frac{\partial^2 H_{dx}(z)}{\partial z^2} - k_{zd}^2 H_{dx}(z) = 0. \tag{6.7}$$

Similar equations can be written for media **1** and **3**: $k_{zm}^2 = k_x^2 - \varepsilon_m\, \omega^2/c^2$.

Equations (6.5) and (6.7) are simple second-order differential equations with easy solutions:

$$\begin{cases} E_{xd} = A_d \exp(-k_{zd}z) + B_d \exp(k_{zd}z) \\ H_{xd} = C_d \exp(-k_{zd}z) + D_d \exp(k_{zd}z) \end{cases}, \tag{6.8}$$

$$\begin{cases} E_{xm} = A_m \exp(-k_{zm}z) + B_m \exp(k_{zm}z) \\ H_{xm} = C_m \exp(-k_{zm}z) + D_m \exp(k_{zm}z) \end{cases}. \tag{6.9}$$

Symmetries: Many symmetries help in simplifying the set of equations. The (x, y) plane is the plane of symmetry for the waveguide; therefore, the electromagnetic fields will respect this symmetry. Two sets of solutions will arise: either the electric field is symmetric with respect to the (x, y) plane and the magnetic field is antisymmetric (case of the *perfect electric wall*, PEW) or the reverse (case of the *perfect magnetic wall*, PMW). This symmetry allows us to solve only for the upper half-space and deduce the other half by symmetry. Moreover, as demonstrated in Chapter 4, the plasmon wave is TM ($H_{xd} = 0$). We limit our calculations to the PEW case, for which $C_d = D_d$ and $A_d = -B_d$. The transverse components of the field can be derived from the longitudinal ones expressed by Equations (6.8) and (6.9).

Boundary conditions: The continuity of the tangential components of the fields and the discontinuity of the normal components (see Chapter 3) of the electric field at the interface $z = b/2$ lead to a system of equations that can be split into two uncoupled systems [2]:

$$\begin{bmatrix} -e^{-k_{zm}b/2} & 2\sinh(k_{zd}b/2) \\ \frac{\varepsilon_m}{k_{zm}} e^{-k_{zm}b/2} & 2\frac{\varepsilon_d}{\gamma_d}\cosh(k_{zd}b/2) \end{bmatrix} \begin{bmatrix} A_m \\ B_d \end{bmatrix} = \begin{bmatrix} 0 \\ 0 \end{bmatrix}, \text{ and } C_m = D_d = 0, \quad (6.10)$$

$$\begin{bmatrix} \frac{\mu_m}{k_{zm}} e^{-k_{zm}b/2} & 2\frac{\mu_d}{\gamma_d}\sinh(k_{zd}b/2) \\ -e^{-k_{zm}b/2} & 2\cosh(k_{zd}b/2) \end{bmatrix} \begin{bmatrix} C_m \\ D_d \end{bmatrix} = \begin{bmatrix} 0 \\ 0 \end{bmatrix}, \text{ and } A_m = B_d = 0. \quad (6.11)$$

These two sets of equations correspond to the TM and TE solutions, respectively. Solutions are possible only if the corresponding determinants are zero (same principle as in Chapter 4), and this leads to the dispersion equations. A similar derivation is carried out for the PMW case, where the magnetic field is antisymmetric and the electric field symmetric. The following systems are obtained (TM and TE modes):

$$\begin{bmatrix} -e^{-k_{zm}b/2} & 2\cosh(k_{zd}b/2) \\ \frac{\varepsilon_m}{k_{zm}} e^{-k_{zm}b/2} & 2\frac{\varepsilon_d}{\gamma_d}\sinh(k_{zd}b/2) \end{bmatrix} \begin{bmatrix} A_m \\ B_d \end{bmatrix} = \begin{bmatrix} 0 \\ 0 \end{bmatrix}, \text{ and } C_m = D_d = 0, \quad (6.12)$$

$$\begin{bmatrix} \frac{\mu_m}{k_{zm}} e^{-k_{zm}b/2} & 2\frac{\mu_d}{\gamma_d}\cosh(k_{zd}b/2) \\ -e^{-k_{zm}b/2} & 2\sinh(k_{zd}b/2) \end{bmatrix} \begin{bmatrix} C_m \\ D_d \end{bmatrix} = \begin{bmatrix} 0 \\ 0 \end{bmatrix}, \text{ and } A_m = B_d = 0. \quad (6.13)$$

As discussed in Chapter 4, surfaces waves are possible only in TM modes, and we end up with the two dispersion relations obtained from Equations (6.10) and (6.12) for the gap SPP:

$$\tanh k_{zd}b/2 = -\frac{\varepsilon_d k_{zm}}{\varepsilon_m k_{zd}} \quad \text{symmetric/LR-SPP}, \quad (6.14)$$

$$\tanh k_{zd}b/2 = -\frac{\varepsilon_m k_{zd}}{\varepsilon_d k_{zm}} \quad \text{antisymmetric/SR-SPP}. \quad (6.15)$$

In these relations, $k_{zd}^2 = k_x^2 - \varepsilon_d\ \omega^2/c^2$ and $k_{zm}^2 = k_x^2 - \varepsilon_m\ \omega^2/c^2$. All these values are complex functions. The entire problem is reduced to finding values for the propagation constant k_x that verifies the dispersion relation (6.14) or (6.15). Analytic solutions are not possible in the general case [2, 3]. But we discuss some general properties of the gap SPP in the following section.

6.3.2 *Propagation of the gap SPP*

As depicted in Figs. 6.5(b) and 6.5(c), when the two SPPs of each metal–dielectric interface overlap, the plasmon modes split into a symmetric and an antisymmetric mode. It lifts the degeneracy and gives birth to three different modes that are able to propagate: (i) a forward symmetric,

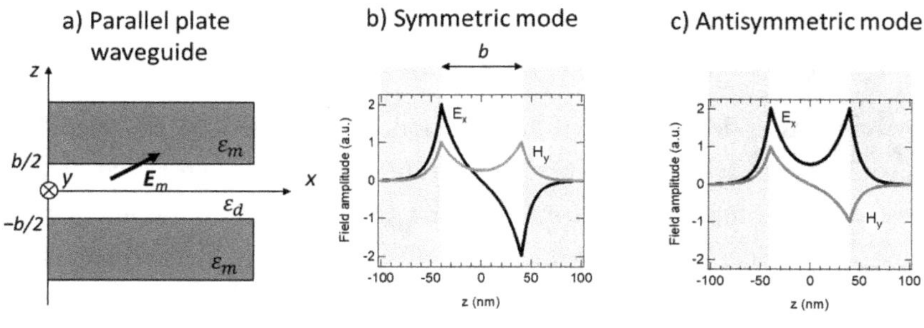

Fig. 6.5. (a) Geometry of the plasmonic parallel plate waveguide. (b) Magnetic and electric fields calculated for a 40 nm wide waveguide in the symmetric case. (c) *Idem* for the asymmetric mode.

Source: Reproduced from Ref. [3].

(ii) a forward antisymmetric, and a (iii) backward antisymmetric [3]. The existence of these different modes depends on the wavelength (in free space) and on the gap thickness b. Mode (i) is the only mode that exists for any values of b and is the most studied one. This is the *gap SPP*. Mode (ii) is detected for ultrathin gaps. And mode (iii) corresponds to surface waves propagating in the reverse direction and with high losses. A detailed discussion of these plasmonic modes is given by Davoyan and Bozelvolnyi [3]. In the following, we restrict ourselves to the symmetric mode (i), the gap SPP.

By solving Equation (6.14), $k_x = k_{SPP}$ is obtained. From this, k_{zm} and k_{zd} are calculated, and the propagative factor is fully determined (see relations (6.2) and (6.3)). The imaginary part of k_{SPP} is the inverse of the propagation length of the SPP (see Chapter 4), which yields

$$L = \frac{1}{2} \cdot \frac{1}{\mathrm{Im}\,(k_{SPP})}. \tag{6.16}$$

This propagation length L is plotted in Fig. 6.6 as a function of the gap width (parameter b) [5]. At 1.55 μm and for a gap width of 3 μm, the propagation length is 260 μm, and at a wavelength of 653 nm, it is reduced to 8.5 μm (see point B in Fig 6.6). A very interesting feature at 653 nm is that, by squeezing the fields (decreasing the gap width), the propagation wavelength slightly increases up to 9.1 μm when the gap passes the 800 nm point (point A).

6.3.3 *The effective refractive index in the waveguide*

For any waveguide, the wave propagation is controlled by the dispersion relation, where k_{SPP} is expressed as a function of the dielectric permittivities (or the complex optical indices) and the geometric parameters of the waveguide. For a given waveguide, the effective index is defined as

$$N_{SPP} = \frac{k_{SPP}}{k_0}, \tag{6.17}$$

where $k_0 = \omega/c$ is the free-space wave vector. Of course, N_{SPP} is a complex effective index. The imaginary part of the index determines the SPP power

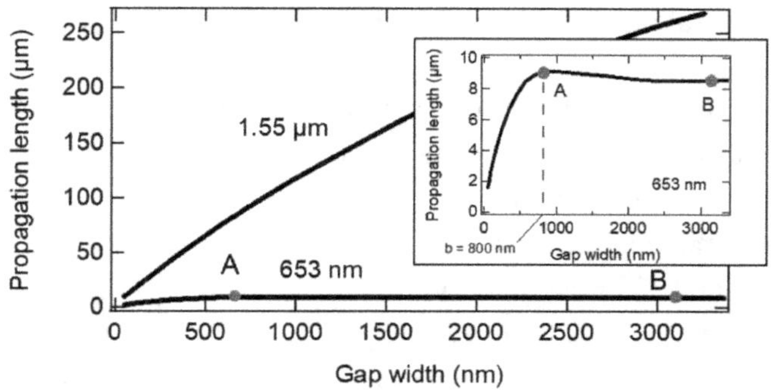

Fig. 6.6. Evolution of the propagation length of the symmetric gap SPP as a function of the width of the gap at two wavelengths (653 nm and 1.55 μm). The metal is gold, and the dielectric is air. Inset shows the case of the 653 nm free-space wavelength, with two particular points: A is the maximum of the SPP propagation length obtained for a width of 800 nm, and B the propagation length that matches that of a single-interface plasmon.

Source: Reproduced from Ref. [5].

loss, which is denoted by l and expressed in dB/mm:

$$l = 3.52 \times 10^{-4} \times \mathrm{Im}(N_{SPP}). \tag{6.18}$$

This formula is very similar to (6.16) above.

With one simple complex number at a given wavelength, the effective index describes the propagation of the SPP wave. The propagation can be conceptually considered as a simple propagation in a homogeneous medium of optical index N_{SPP}. This property is useful for deriving the dispersion relation of 2D waveguides, following the effective index method (EIM) [6].

6.3.4 *Conclusion on the gap SPP*

The gap SPP has a more general significance for plasmonic waveguides since the behavior and some simplified forms of the dispersion relation can be applied to close geometries, such as the V-groove and trench waveguides (see Fig. 6.3(e)).

Although the parallel-plate waveguide is probably the simplest plasmonic waveguide, the previous section has shown that the calculations become very quickly difficult and tedious. We will not go into the details of how to derive the dispersion relations.

6.4 Thin Metal Film SPP Waveguide

We now consider an IMI structure with a metallic thin film of thickness b embedded in a dielectric, as depicted in Figs. 6.3(b) and 6.7 [7,8]. Each metal–dielectric interface at $z = +b/2$ and $-b/2$ support an SPP wave. When the metallic film is thin enough, these waves interact and give rise to a symmetric and an antisymmetric mode. The symmetric SPP mode occurs when the transverse fields E_z and H_y are symmetric with respect to the plane $z = 0$. In this case, the longitudinal field E_x is antisymmetric. For the antisymmetric mode, the situation is reversed. These definitions have important consequences. Since the damping of the SPP mode is mainly set by the longitudinal component E_x, this damping is minimized when E_x remains as small as possible. In the case of antisymmetric E_x, the plane $z = 0$ is a node where $E_x = 0$. Therefore, the damping will be much smaller for the symmetric SPP mode than for the antisymmetric one. The symmetric SPP mode is called **long-range SPP mode** (LR-SPP), whereas the antisymmetric one is called **short-range SPP** (SR-SPP).

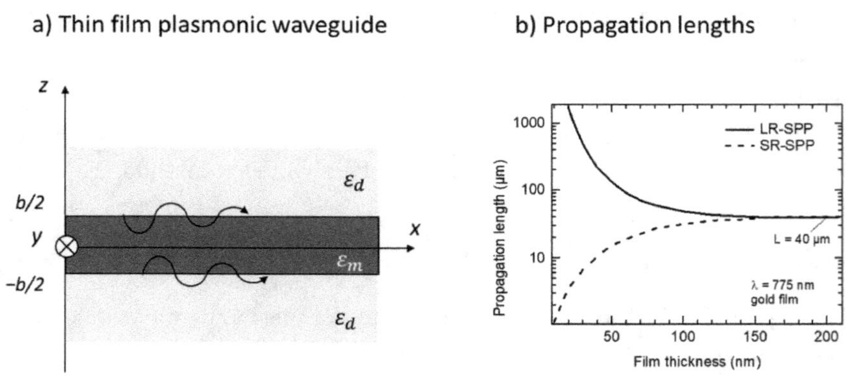

Fig. 6.7. (a) A thin-film plasmonic waveguide is made of a thin metallic plate embedded in a dielectric. Two plasmonic modes can be launched, with either symmetric or antisymmetric behavior, and are called long-range SPP (LR-SPP) or short-range SPP (SR-SPP) waves. (b) The corresponding propagation lengths strongly differ when the metal thickness b decreases below 150 nm in the case of a gold film in air.

Source: Reproduced from Ref. [8].

6.4.1 *Dispersion relations of LR-SPP and SR-SPP*

LR-SPP: The approach is similar to the derivation explained above in the case of the gap-plasmon mode. It is based on the boundary conditions and the use of the symmetries of the fields. The results are very similar. When E_z and H_y are symmetric, this is the LR-SPP, and the dispersion relation is written as

$$\tanh k_{zm}b/2 = -\frac{\varepsilon_m k_{zd}}{\varepsilon_d k_{zm}} \text{ (LR – SPP)}, \tag{6.19}$$

with $k_{zd}^2 = k_{SPP}^2 - \varepsilon_d \omega^2/c^2$ and $k_{zm}^2 = k_{SPP}^2 - \varepsilon_m\, \omega^2/c^2$.

Extracting a solution from this dispersion relation means being able to provide the value(s) of k_{SPP} for a given ω. This is far from easy since k_{SPP} appears with a square root in k_{zd} and k_{zm}. Then, the hyperbolic tangent function needs to be inversed. An analytical solution is not possible in the general case. However, the relation simplifies for very thin metal films ($b \to 0$) since $\tanh x \approx x$. This case can be written as ($k_0 = \omega/c$)

$$k_{SPP} \approx k_0 \sqrt{\varepsilon_d + (bk_0\varepsilon_d/2)^2 \cdot (1 - \varepsilon_d/\varepsilon_m)^2}. \tag{6.20}$$

This simplified relation is valid only for films thinner than 40 nm. Note that for ultrathin metal films, $k_{SPP} \to k_0\sqrt{\varepsilon_d}$, which is the light line in the dielectric. This case represents the normal light propagation inside a medium of index $n = \sqrt{\varepsilon_d}$, and this is no longer a plasmon wave, which is not surprising since the metal is thinned down to zero. In this extreme limit case, the wave is not confined anymore.

SR-SPP: The SR-SPP exhibits an asymmetric behavior with respect to the plane $z = 0$ for the fields E_z and H_y. The dispersion relation for the SR-SPP is written as

$$\tanh k_{zm}b/2 = -\frac{\varepsilon_d k_{zm}}{\varepsilon_m k_{zd}} \quad \text{(SR-SPP)}. \tag{6.21}$$

It can also be simplified for ultrathin metallic plates ($b \to 0$) into

$$k_{SPP} \approx k_0 \sqrt{\varepsilon_d + [2\varepsilon_d/(bk_0\varepsilon_m)]^2}. \tag{6.22}$$

6.4.2 *Discussion of the SPP propagation in a thin-film plasmonic waveguide*

We use the expression of the dispersion relations for computing the propagation lengths of SPP waves launched in a thin-film waveguide depicted in Fig. 6.7. The propagation length is given by Equation (6.16) above and plotted as a function of the film thickness in Fig. 6.7(b) [8]. We now focus our discussion on the case of a gold film in air at a wavelength of 775 nm. The figure shows that both plasmons merge into the usual gold–dielectric SPP, with a propagation length of 40 μm when $b \to 0$. The degeneracy is lifted when the metal becomes thinner than 150 nm, and both SR-SPP and LR-SPP can be launched. The plasmon waves at each interface are coupled. The propagation length of the LR-SPP diverges when $b \to 0$ because the wave is progressively expelled out of the metal and tends to a regular propagation in the dielectric, as mentioned above. There is no damping in the dielectrics in this case, and the propagation is limitless. For a 20 nm thin metal film, the LR-SPP propagates over 1700 μm and the SR-SPP over 3.4 μm. The large propagation length of 1700 μm for the LR-SPP should be compared with that of the simple gold–air interface, where it is only 40 μm. However, this strong increase results from a trade-off with wave confinement. The LR-SPP is poorly confined and not really guided anymore.

6.4.3 *Mode confinement vs. propagation length in a waveguide*

The wave confinement is a measure of the extent to which energy is squeezed inside the waveguide.

A waveguide should guide and confine an optical wave. An SPP wave is always a coupled propagation of a wave in the metal and a wave in the dielectric. The maximum confinement occurs on the metal side but at the expense of wave damping and reduced propagation length. When the energy of the SPP wave is expelled out of the metal, the propagation length increases but not the confinement, and the guiding property tends to be lost. Therefore, the conception of a waveguide is a trade-off between these two figures: propagation length and confinement.

6.5 Other Plasmonic Waveguides: Stripes, Slots, V-grooves, and Nanowires

In the previous sections, we have considered 1D confinement only when dealing with the gap SPP and thin-film SPP. Now, we discuss more realistic waveguides, where waves are confined to two dimensions (2D confinement), as shown in Figs. 6.3(c)–6.3(e).

6.5.1 *About the stripe waveguide*

A stripe waveguide, shown in Fig. 6.3(d), consists of a metallic stripe of width w, thickness h, and an infinite length. It is deposited on a dielectric substrate (medium 1) and embedded in a superstrate (medium 2). Such a metal stripe supports an important LR-SPP mode that has attracted considerable attention for developing optical processing since 1981, even before the strong focus on plasmonics [9]. The SPP propagation in a stripe waveguide is briefly discussed in Ref. [10] and reviewed by Berini in 2009 [11]. A stripe waveguide of silver of width $w = 1$ μm and thickness $h = 10$ nm allows the propagation of a symmetric mode at 633 nm over a distance of 400 μm when the two dielectric media are identical ($\varepsilon_1 = \varepsilon_2 = 4$). However, this large propagation length is linked to the low confinement of the SPP wave to the metal, and the wave spans mostly in the dielectric over distances of a few wavelengths. This propagation length drops to 40 μm when h increases to 30 nm [10]. Therefore, the long-range propagation for the stripe waveguide is obtained at the price of poor confinement. Therefore, such stripe plasmonic waveguides will be hardly useful for subwavelength photonic integration. However, they might find interesting applications for sensors.

6.5.2 *Effective index method in the case of the slot waveguide*

There is no simple analytical derivation for obtaining the wave vector of any of the waveguides with 2D confinement. Usually, theoretical simulations are employed based on finite-element modeling (FDTD,

COMSOL Multiphysics®). However, an efficient approximation was established that can handle some rectangular waveguides: this is the effective index method (EIM).

The EIM is based on the fact that the most important feature of SPP propagation is captured by the plasmon wave vector. Let's consider, as an illustrative example, the case of a plasmonic slot waveguide shown in Fig. 6.8(a) [10]. It consists of a rectangular trench cut into a metallic thin film (dielectric constant ε_m and thickness b) that was deposited on a dielectric substrate (dielectric constant ε_{d1}). The trench has a thickness h and a width w. The structure is embedded in a second dielectric (dielectric constant ε_{d2}). For simplicity, we consider $\varepsilon_{d2} = \varepsilon_{d1}$ here. This problem is solved in two steps. **In the first step**, we restrict the problem to a 1D confinement and analyze an SPP wave guide between the two metallic plates spaced at w (see Fig. 6.8(b)). The propagation in the x direction is a gap SPP, as explained in Section 6.3 and described by an effective index $N_{eff\,0}$. **In the second step**, the slot waveguide is represented by an effective waveguide made of a medium of optical index $N_{eff\,0}$ and thickness h between the two dielectrics (see Fig. 6.8(c)). This last waveguide can be solved with the approach of the thin-film waveguide (see Section 1.4). Note that a rigorous analysis of the wave vector in both cases involves solving transcendent equations; therefore, no exact analytical solution is possible unless approximations similar to those considered in Equations (6.20) and (6.22) are used.

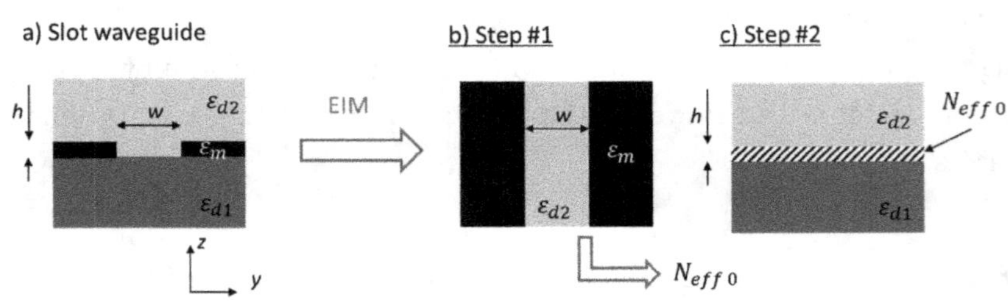

Fig. 6.8. (a) A slot waveguide where the plasmon wave is confined to two directions (directions y and z). The SPP waves propagates in the x direction. (a) and (b) The effective index method (EIM) allows decomposing the problem into two 1D confined waveguides: (b) a parallel-plate waveguide and (c) a thin-film waveguide.

Let's look at an example where the SPP is excited at 1550 nm, the metal is gold ($\varepsilon_m = -116 + 11i$), and the dielectrics 1 and 2 are composed of the same material with a refractive index of 1.45 (such as PMMA). Here, we follow Ref. [10]. We fix the gold layer thickness at $h = 100$ nm and the width of the trench is varied from 20 to 800 nm. The corresponding propagation length is plotted in Fig. 6.9, along with the propagation lengths of a parallel-plate of identical thickness (waveguide depicted in Fig. 6.8(b)) and the edge waveguide. It shows that when the width is small compared to the height, it behaves like a parallel-plate waveguide with propagation in the gap SPP mode. When the width is larger than the height, the waveguide approaches the behavior of a metallic edge [10].

The most interesting feature of the slot waveguide is illustrated in Fig. 6.10, where the component of the electric field E_x is plotted in false colors in the case of an excitation wavelength of 1550 nm. E_x is the major component and can be considered as a good indicator of the energy confinement in the waveguide. This field was calculated using COMSOL [10]. In the case of $w = 50$ nm, the field is much more intense (red color) than in the case of $w = 100$ nm (white color). This is an illustration of the

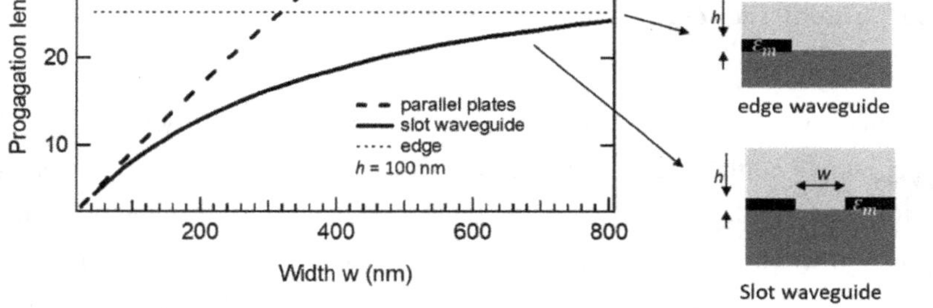

Fig. 6.9. The propagation length of a slot waveguide with a gold film (thickness $h =$ 100 nm) as a function of the width w of the trench for $\lambda_0 = 1550$ nm. This propagation is compared to the parallel-plate and edge waveguides.

Source: Reproduced from Ref. 10.

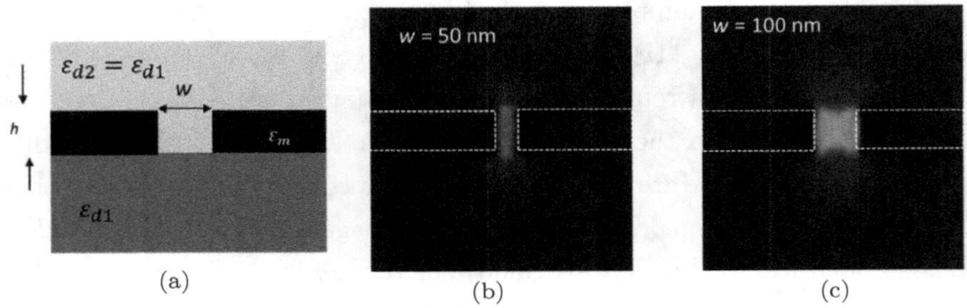

Fig. 6.10. (a) Schematic of a slot waveguide made of a gold layer of height $h = 100$ nm deposited on a dielectric substrate. The component of the electric field E_x is represented in false color in the case of a wavelength of 1550 nm and the two values of the width: (b) $w = 50$ nm and (c) $w = 100$ nm. It shows a high confinement of the electric field between the two metallic layers.
Source: Reproduced with permission from Ref. [10]. Copyright © 2012 American Institute of Physics.

high confinement inside the waveguide. Moreover, since the field penetrates very little into the metallic layers, it is squeezed into the gap and exhibits a very high confinement, with little spillover into the dielectric. The 50 nm waveguide is therefore very favorable for guiding parallel optical waves when several waveguides are designed on a chip. It allows high integration, avoiding cross talks and signal interferences. Moreover, this waveguide supports a fundamental mode with low bending losses, which is an important feature for fabricating circuits not limited to straight connections [12].

6.5.3 *Channel waveguides and V-grooves*

The most interesting and promising plasmonic waveguide is probably the V-groove: A triangular groove is cut into a metallic layer with a depth h and a groove angle θ. The resulting width at the top of the groove is w (see Fig. 6.11). Usually, the surrounding dielectric in simply air. The eigenmode supported by this V-groove is termed the channel plasmon-polariton (CPP) mode. The CPP mode has received quite a lot of attention in the past few years, and some plasmonic devices based on this mode have been experimentally realized [13]. The derivation of the properties of the CPP mode could be approached by the EIM model, considering a set of slot waveguides piled up with decreasing gap width [5]. But the most effective way is

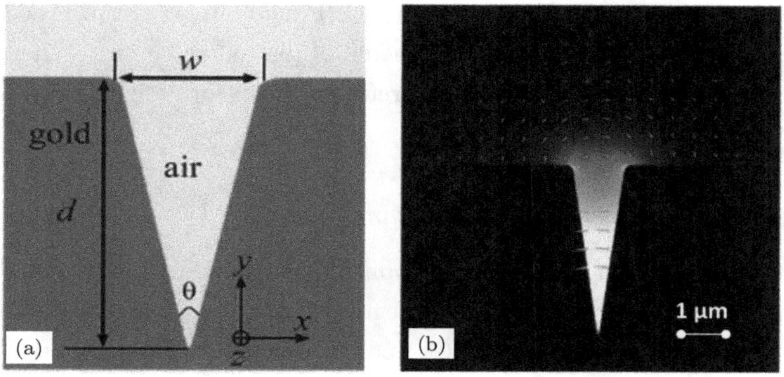

Fig. 6.11. (a) Schematic of a V-groove with an angle $\theta = 20°$ cut into a gold substrate with a depth $d = 3$ µm. The width is 1.06 µm. The dielectric is in the groove, and above it is air. (b) A V-groove can support the propagation of a channel plate plasmon (CPP), and the figure displays in false colors the intensity of the electric field. The plasmon mode is confined to the bottom of the groove.

Source: Reproduced with permission from Ref. 10. Copyright © 2012 American Institute of Physics.

to use finite-element model in COMSOL, where a geometric structure, as depicted in Fig. 6.11(a), is created. The metallic substrate is gold, and the dielectric is air. Here, we discuss the case of a groove of depth $d = 3$ µm and of opening $\theta = 20°$. The corresponding dispersion relation is discussed in Ref. [14] and will not be detailed here.

Several CPP modes can propagate in the groove at $\lambda = 1550$ nm. Fig. 6.11(b) shows the field amplitude of the fundamental mode in false colors. The field is maximum at the bottom of the groove. It shows that this mode travels at the bottom of the groove. Actually, if the wavelength is increased, the mode is slowly expelled out of the groove. At a wavelength of 1550 nm, the fundamental mode shown in Fig. 6.11(b) is efficiently confined to the bottom of the groove. Also note that at the edges the field is also a bit more intense due to a coupling with edge plasmon modes. In this example, the edges of the grooves are smoothed (radius of curvature is 10 nm here) to minimize these edge plasmon modes, since in actual fabrication processes the edges are never infinitely sharp. The propagation of the plasmon modes strongly depends on the groove angle (see Ref. [14] for a more detailed discussion).

The typical propagation length of a CPP mode in a V-groove is 62 μm, which is sufficient to design the elements of plasmonic circuits. Therefore, these plasmonic waveguides are the preferred choice for designing plasmonic circuitry.

6.5.4 *SPP guided into cylindrical nanowires*

We finish this section on the plasmonic waveguides with the cylindrical waveguide. This is certainly the waveguide that most closely resembles an electric wire or an optical fiber. Compared to an optical fiber whose radius is typically 50 μm, the aim is to confine light to waveguides that are 200 times smaller and made of conductive material instead of a dielectric, such as glass. We consider the metal to be a cylindrical nanowire of silver (ε_1) with a radius $a = 100$ nm placed in a dielectric (ε_2). We suppose its length to be infinite. Using the boundary conditions in cylindrical coordinates, the dispersion equation can be derived analytically [15]. The wire supports several modes, and the fundamental one is a pure TM mode that obeys the following dispersion equation:

$$\frac{\varepsilon_1 I_1(\xi_1)}{\xi_1 I_0(\xi_1)} + \frac{\varepsilon_2 K_1(\xi_2)}{\xi_2 K_0(\xi_2)} = 0, \tag{6.23}$$

where I_0 and I_1 are the first-kind modified Bessel functions of orders 0 and 1, respectively, and K_0 and K_1 the second-kind modified Bessel functions.

$\xi_1 = a\sqrt{k_x^2 - \varepsilon_1 k_0}$ and $\xi_2 = a\sqrt{k_x^2 - \varepsilon_2 k_0}$, where k_x is the propagation wave vector and k_0 is the free-space wave vector.

Figure 6.12(b) shows the distribution of the electric field in the fundamental mode (mode of order 0). The field is TM and normal to the nanowire (axial symmetry), with a rapid decay away from the wire. In the first-order mode, shown in Fig. 6.12(c), the electric field is no longer perpendicular to the wire and has also a longitudinal component. The properties of the different modes are given by the corresponding dispersion equations: mode effective index, direction of the electromagnetic field, and propagation length. For example, in the case of a 100 nm radius silver nanowire excited at a wavelength of 633 nm, the propagation length is 20 μm in the fundamental mode and 165 μm in the first-order mode. The strong

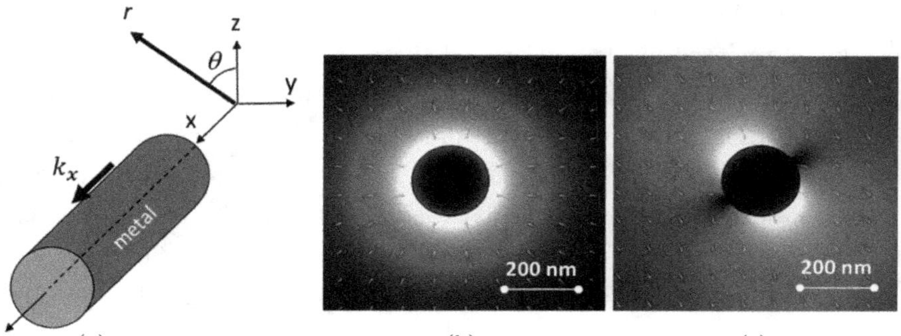

(a) (b) (c)

Fig. 6.12. (a) Schematic of a cylindrical waveguide in cylindrical coordinates used to derive the dispersion relation. Here, the wire is made of silver, with a radius of 100 nm. The propagation wave vector k_x is represented along the wire. Representation of the electric field of the (b) fundamental mode and (c) the first mode.

Source: Reproduced with permission from Ref. 10. Copyright © 2012 American Institute of Physics.

difference between the two modes originates from the fact that the field is expelled out of the wire in the case of the first-order mode and approaches the free-space propagation. Note that when the diameter of the nanowire is of the order of magnitude of the wavelength, the two modes behave similarly, with propagation lengths of 33 μm and 45 μm for the fundamental and first-order modes, respectively, when the nanowire diameter is 600 nm (a = 300 nm) [10].

6.6 Comparison of the Performance of the Principal Waveguides

6.6.1 *Propagation length and plasmonic mode confinement*

A wide range of plasmonic waveguides have been proposed and investigated since the beginning of 2000, and it is natural to compare their performances. The key characteristics are propagation length, mode confinement, cross talk, and bend losses.

The **propagation length** is defined as the distance over which the Poynting vector decays by a factor e, which is equivalent of the electric field decaying by e^2. According to Equation (6.16), it is directly related to the propagation wave vector k_{SPP}.

The **mode confinement** describes how much the electromagnetic energy is squeezed inside and around the metallic part of the waveguide. The goal is usually to produce plasmon waves with a lateral extension much smaller than the wavelength. This is described by the mode area A_e and the lateral mode width w_0. To make it clear, let us consider a stripe waveguide with a propagation along the x-axis, as shown in Fig. 6.13. The x component of the Poynting vector $P_x(y, z)$ describes the power flux along the waveguide. The **mode area** A_e is the surface in the (y, z) plane encompassing half of the total power. The **lateral mode width** w_0 is obtained by integrating $P_x(y, z)$ over the vertical component z, which corresponds to the average variation of power along the y direction (see Fig. 6.13). The lateral mode width tells us how close a waveguide can be positioned to another component on a plasmonic circuit. For distances smaller than w_0, the signals transported in the waveguide will interfere and may be strongly disturbed.

The **cross talk** between two plasmonic waveguides positioned parallel on a plasmonic circuit can be characterized by the coupling length L_c, which is the distance over which the optical power is completely transferred to

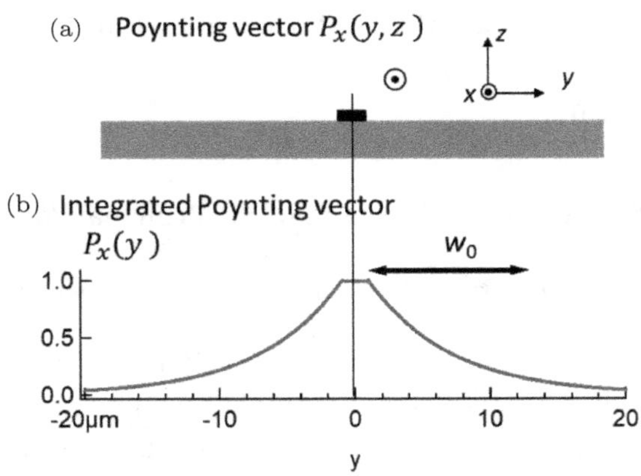

Fig. 6.13. (a) Plasmonic mode confinement in a stripe waveguide of 30 nm thickness and 2 µm width. The direction of the x component of the Poynting vector $P_x(y, z)$ is given. It describes the electromagnetic power transported by the plasmonic waveguide. (b) Plot of the integrated component $P_x(y)$ that shows the lateral expansion of the electromagnetic energy. The lateral mode width w_0 is 12 µm in this case. Energy is poorly confined to the plasmonic stripe waveguide.

the neighboring waveguides. This figure is important for designing compact circuits but will not be developed further here.

Finally, the **bend loss** is the smallest radius of a bend that the plasmonic wave can cross without too high a loss.

6.6.2 *Comparison of the five most important plasmonic waveguides*

The various plasmonic waveguides display very different behaviors in terms of propagation length and electromagnetic confinement. We present in the following the main characteristics of five representative waveguides based on a gold stripe or gold plate for an excitation wavelength at 1550 nm, which is the telecom wavelength. We compare their performance with the simple plasmon polaritons at a gold–air interface [10]. The five plasmonic waveguides are presented in Fig. 6.14, along with the simplest plasmonic device, which is the gold–air interface. All these waveguides have dimensions well below the free-space wavelength.

Table 6.2 illustrates the strong differences in terms of propagation lengths for the different waveguide geometries. Compared to the simple case of the gold–air interface, where the propagation length is 335 μm, the so-called "long-range" waveguides offer much longer propagation lengths: the LR-SPPW and the LR-DLSPPW, with a maximum of 7.45 mm for the first one. But there is a trade-off with the wave confinement. This long-range plasmon is poorly confined, exhibiting a mode area of 62 μm^2. Comparing this value to $\lambda^2 = 2.4$ μm^2 shows that this plasmon mode resembles free-space propagation. The best confined plasmon propagation occurs in the

Table 6.2. Comparison of the performances of plasmonic waveguides at 1.55 μm. Their major dimensions are indicated in Fig. 6.14. A_e is the mode area, w_0 is the lateral mode width, and L_p is the propagation length. In the case of gold–air interface, there is no lateral confinement and the values of A_e and w_0 are not relevant.

Waveguide	Gold–air SPP	LR-SPPW	DLSPPW	LR-DLSPPW	CPP	HSPPW
L_p(μm)	335	7450	49	3125	20	27
A_e(μm^2)	na	62	0.24	1.2	0.18	0.059
w_0(μm)	na	12	0.96	1.5	0.27	0.37

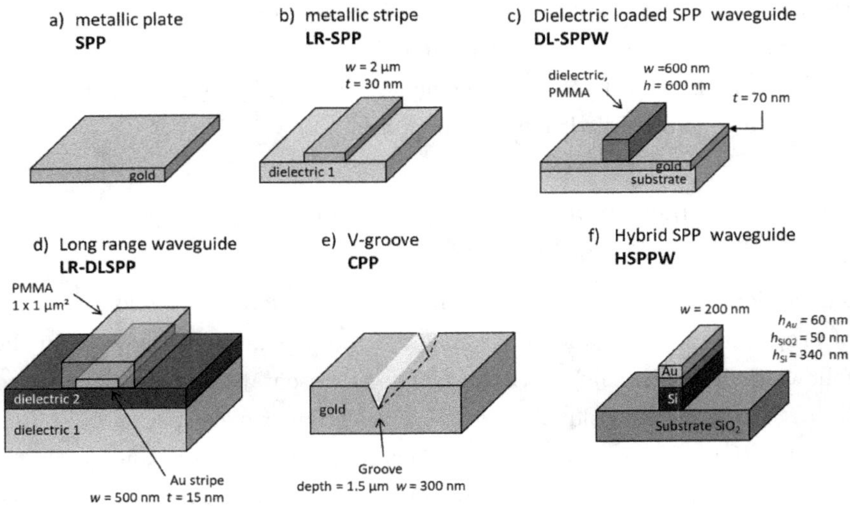

Fig. 6.14. Schematic of a set of SPP waveguides based on gold stripes or plates: (a) The gold–air interface can propagate an SPP wave that is not confined. (b) The metallic stripe gives rise to several modes, among which is an LR-SPP mode. Two configurations of dielectric-loaded SPP waveguides are presented in (c) and (d). (e) A V-shaped groove in a gold plate supports a channel plasmon polariton wave. A hybrid SPP waveguide is presented in (f).

HSPPW, with a mode area of just 0.059 μm^2, which is 40 times smaller than λ^2. In terms of component integration on an optical chip, the lateral mode width should be as small as possible: This is achieved for the V-groove, where the plasmon wave is squeezed down to 0.27 μm, which is six times smaller than the wavelength and well below the diffraction limit.

6.7 Amplification of the SPP Signals: The SPASER

6.7.1 *Working principles of a plasmonic laser*

We have stressed that in plasmonic waveguides, light can be squeezed well below the wavelength, but its propagation length is quite limited to a few millimeters at best. Therefore, it is crucial to be able to generate light directly on a chip. One could imagine using laser diodes, but their sizes are much greater than the size of the plasmonic guide, and the coupling of light with an SPP waveguide is often tricky. A good approach is to use surface plasmon lasers, known as SPASERs (surface plasmon amplification by stimulated emission of radiation), for generating a plasmon wave directly into a plasmonic waveguide [16].

SPASERs are not really a new class of lasers since they are based on nanowire lasers, which themselves are semiconductor lasers shaped as nanowires [17, 18]. The first stimulated emission in a nanowire laser was observed in 2001 by Yang at Berlekely [19]. A laser is a combination of two elements: an optical cavity and a gain medium. Lasing is obtained when a photon is amplified by stimulated emission in the gain medium inside the cavity. In the case of semiconductor lasers, the amplification takes place in direct-gap semiconductors, such as GaAs. The pumping can be achieved optically by illuminating the semiconductor with a high-power pump laser at an energy higher than the band gap. The excited electrons will decay quickly to the bottom of the conduction band and eventually undergo a transition to the valence band triggered by another photon, circulating in the cavity. This is the stimulated emission, and this process is shown in Fig. 6.15(a). Optical amplification is possible for a sufficiently high carrier density in the conduction band, and this requires a proper doping of the semiconductor. In the case of nanowire lasers, the two ends of the nanowire serve as mirrors of a Fabry–Perot cavity due to the sudden optical index change. Since the mode propagation inside the nanowire is very different from free-space propagation, the reflectivity of these mirrors should be calculated numerically. A typical value of 80% is expected for the first TE mode.

6.7.2 *Practical illustration of a SPASER*

An example of SPASER is described in the article by Bermúdez-Ureña *et al.* [20]. A nanowire laser made of a GaAs nanowire of 6.8 μm length and 590 nm diameter is placed at the bottom of a V-groove plasmonic waveguide, as shown in Fig. 6.15(b). The pumping is achieved with a pulsed laser (with a pulse duration of 200 fs and repetition rate of 80 MHz) at 730 nm. The V-groove waveguide is fabricated on a silicon substrate by lithography and well-controlled chemical etching. The groove itself is 30 μm

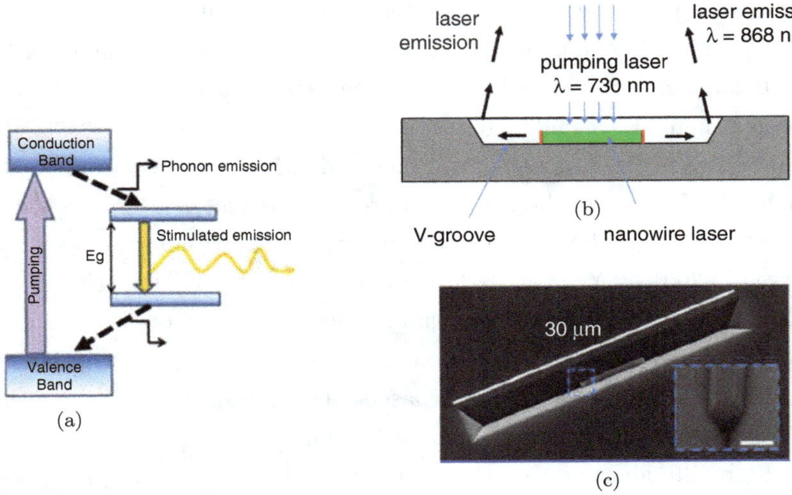

Fig. 6.15. Working principle of a plasmon polariton laser known as SPASER. Gain medium of the laser is identical to that of a semiconductor laser, (a) where the pumping is achieved by photons of energy above the bandgap of a direct-gap semiconductor, such as GaAs. Stimulated emission occurs at near the bandgap energy. (b) Schematic of a SPASER, where a nanowire laser is deposited at the bottom of a V-groove. The laser light emitted by the nanowire laser is coupled with the channel plasmon polariton (CPP) mode and detected at the two ends of the V-groove. In this case, the V-groove has a length of 30 μm and the nanowire (GaAs) is 6.8 μm long, with a diameter of 590 nm. (c) SEM image of the nanowire positioned inside the V-groove.

Source: Reproduced from Ref. [20] used under ACS AuthorChoice Open Access license.

long and 3.5 μm wide. Because of the etching process that is directed by the crystallographic orientation of silicon, the V-groove sidewalls and the termination mirrors have an angle of 55° from the surface plane. This is why the laser emission occurs in oblique directions, as shown in Fig. 6.15(b). A gold film of 70 nm is deposited on the walls of the V-groove. A scanning electronic microscope image of the device is shown in Fig. 6.15(c), with the nanowire being visible.

The operation of a SPASER requires pumping by a 730 nm laser. When the power of the pump is below the lasing threshold, the spontaneous emission dominates and the device produces a broadband light emission, represented by the blue curve in Fig. 6.16(a). However, above this threshold, the laser emits in two modes at 875 and 868 nm. The laser emission is tuned

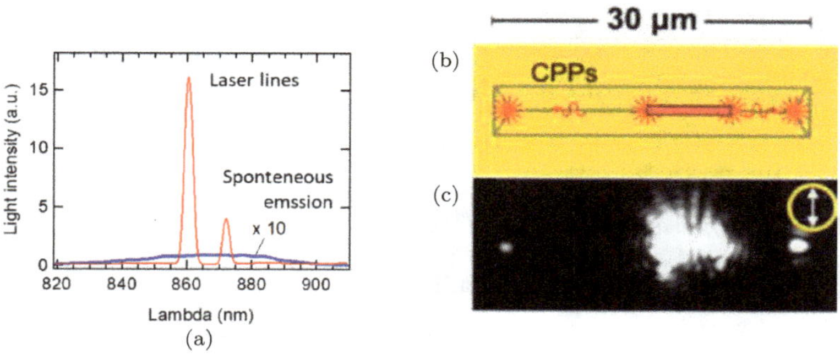

Fig. 6.16. (a) Spectral characterization of a SPASER when the pumping laser is below the lasing threshold (blue line) and above (red line). With a pumping laser at 730 nm, this laser emits two laser modes at 868 and 875 nm. (b) Schematic of the 30 µm long V-groove waveguide that can accommodate CPP modes. The nanowire laser is also shown. (c) Image detected with an electron-multiplying CCD camera of the light emission of the device. The nanowire laser leaks photons from its side (white features), and the two spots are the light guided by the V-groove.

Source: Reproduced from Ref. [20] used under ACS AuthorChoice Open Access license.

with the CPP mode of the V-groove and can travel in this waveguide (see Fig. 6.16(b)). In this setup, the authors are able to detect the photons diffused in the far field with an electron-multiplying CCD (EM-CCD) camera, as shown in Fig. 6.16(c). The two white spots at the end of the V-groove are the portion of the light coupled to the far field.

SPASERs are still in their infancy, but they could provide some interesting solutions to some of the weaknesses of SPP waveguiding. They could compensate for the loss of propagation in such waveguides. Above all, they offer a solution of integrating the laser source into plasmonic chips, and the high-sensitivity sensing platforms could prove to be the most accessible.

6.8 Plasmonic Circuits

If we make an analogy with microelectronic circuits, the plasmonic waveguides we have described until now are just the wires and the connections between the various active parts of an electronic circuits. There is already a wide set of plasmonic elements that has been demonstrated, such as routers, switches, electrically controlled modulators, detectors, splitters and Mach–Zender interferometers [21–24]. All these elements handle SPP waves in various plasmonic waveguides. In the following, we describe some of these elements.

6.8.1 *Basic operations with SPP waves*

Bozhevolnyi and Ebbesen have built a series of simple plasmonic circuits based on V-groove waveguides, such as a splitter that is able to separate the plasmonic signals into two identical signals, or a Mach–Zender interferometer [23]. Figures 6.17(a)–6.17(d) show these circuits. Atomic force microscopy (AFM) images are presented, as well as the SNOM images that are able to show the intensity of the SPP wave. A Mach–Zender interferometer becomes an interesting circuit when it is possible to insert a dephasing system on one of the arms. This makes the modulation of the transmitted signals and the building of a controllable gating system possible; if the dephasing is zero, the signals of the two arms yield constructive interferences, and the device allows the information to pass. If the dephasing is 180°, the interference becomes destructive, and the device blocks the transmission.

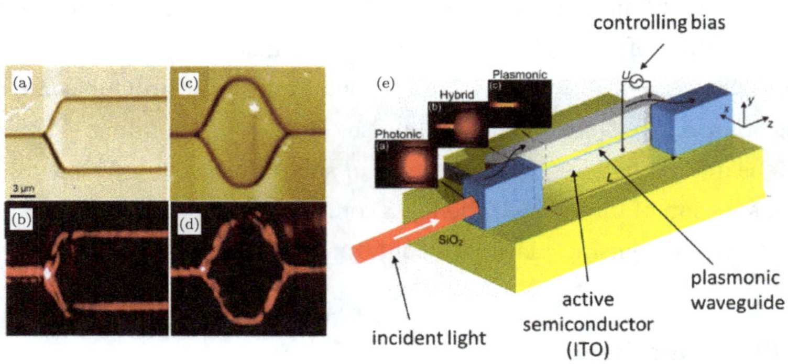

Fig. 6.17. Example of two basic plasmonic circuits: (a) a Y-splitter and (c) a Mach–Zehnder interferometers. The V-grooves are imaged with AFM. The propagation of the confined plasmon is imaged with an SNOM (images (b) and (d)). A plasmonic switch is presented in (e), where the SPP wave inside the plasmonic waveguide is controlled with an electrical bias.

Source: Reproduced from Refs. 23 and 25.

6.8.2 *Active electro-optic plasmonic device*

A more complex device is presented in Fig. 6.17. This is an electrically controlled switch where the stripe plasmonic waveguide can be opened or closed by applying an electrical bias [25]. The active region is based on a stack of layers with ITO, SiO_2, and Ag. The SiO_2–Ag interface serves as the stripe waveguide. When a bias is applied at the ITO–SiO_2–Ag junction, the carrier concentration of the semiconducting ITO is modified. It changes the value of the refractive index of ITO and alters the propagation of the SPP in the stripe waveguide. By adjusting the different thicknesses, Melikyan and his colleagues anticipate that their device can switch the SPP signal ON or OFF at a speed higher than 100 Gbits/s. This theoretical study has been tested with a slightly different configuration in 2015 by Berini and his colleagues [26].

6.8.3 *Logical circuits with SPP waves*

Based on the plasmonic waveguides presented above, a set of linear devices can be engineered. Using the wave nature of the plasmon, it is possible to rely on constructive or destructive interferences to perform addition or subtraction of signals. In Fig. 6.18, a set of simple logic operations is proposed. A Bragg grating serves as a filter that will only transmit the wavelength that matches with the period of the Bragg grating. A ring resonator also transmits a set of wavelengths. A Mach–Zehnder interferometer is a powerful element when it is coupled to a phase shift control because it allows controlling the transmitted energy through the device. If the phase shift is 0°, the signal is totally transmitted, and if it is 180°, the two arms interfere destructively and the signal is blocked. A Y-junction works similar to an OR gate in electronics. If the arm B undergoes a dephasing of 180°, it works like an XOR gate. The major challenge is to actually design such devices.

Some plasmonic device elements have been realized, and the age of plasmonic circuitry has thus begun. There are, however, a few major challenges: being able to integrate plasmonic circuitry on CMOS-compatible substrates [27], checking the maximum speed of commutation allowed by

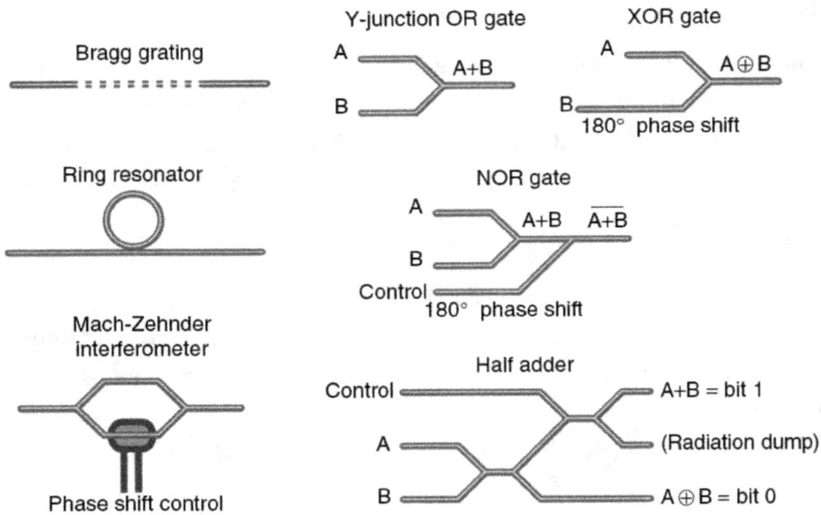

Fig. 6.18. Design of a set of linear operations with plasmonic waveguides.
Source: Reproduced from Ref. [24] used under Creative Commons CC-BY-NC-ND license.

plasmonics [28], and establishing strategies to limit the propagation losses in waveguides.

6.9 Conclusion

In this chapter, we have shown that the plasmon surface wave, which is intrinsically a 2D wave, can be squeezed into a 1D structure and can be guided along plasmonic waveguides. This is analogous to the electric wires in electronics. With the waveguide approach, light is guided in nontransparent media, such as gold, silver, and copper. But the major drawback is that the propagation length is limited to just 20 μm in the case of the most confined line (the V-groove) at the optical wavelength of 1.55 μm. In this case, this optical wire is just 5 μm wide. For a poorly confined waveguide, this length reaches 7.4 mm for the LR-SPPW. This makes this "optical wire" efficient for very compact devices but not for long-distant signal transport. However, the major key elements of a circuit have been designed in plasmonics, and it is only a matter of years before an all-optical plasmonic device for ultrahigh-rate communications is created.

References

[1] Orfanidis S. J. 2016. *Electromagnetic Waves and Antennas.* http://eceweb1.rutgers.edu/~orfanidi/ewa/.

[2] Leal-Sevillano C. A., Ruiz-Cruz J. A., Montejo-Garai J. R. and Rebollar J. M. 2012. Rigorous analysis of the parallel plate waveguide: From the transverse electromagnetic mode to the surface plasmon polariton. *Radio Science* **47**(6).

[3] Davoyan A. R., Shadrivov I. V., Bozhevolnyi S. I. and Kivshar Y. S. 2010. Backward and forward modes guided by metal-dielectric-metal plasmonic waveguides. *Journal of Nanophotonics* **4**, 043509.

[4] Han Z. and Bozhevolnyi S. I. 2014. *Modern Plasmonics* (Maradudin: Elsevier Science).

[5] Bozhevolnyi S. I. 2006. Effective-index modeling of channel plasmon polaritons. *Optics Express* **14**, 9467–9476.

[6] Bozhevolnyi S. 2009. *Plasmonic Nanoguides and Circuits* (Singapore: Pan Stanford).

[7] Maradudin A., Sambles J. R. and Barnes W. L. 2014. *Modern Plasmonics* (Amsterdam: Elsevier Science).

[8] Bozhevolnyi S. I. and Söndergaard T. 2007. General properties of slow-plasmon resonant nanostructures: Nano-antennas and resonators. *Optics Express* **15**, 10869–10877.

[9] Sarid D. 1981. Long-range surface-plasma waves on very thin metal films. *Physical Review Letters* **47**, 1927–1930.

[10] Han Z. and Bozhevolnyi S. I. 2012. Radiation guiding with surface plasmon polaritons. *Reports on Progress in Physics* **76**, 016402.

[11] Berini P. 2009. Long-range surface plasmon polaritons. *Advances in Optics and Photonics* **1**, 484–588.

[12] Maier S. A. and Atwater H. A. 2005. Plasmonics: Localization and guiding of electromagnetic energy in metal/dielectric structures. *Journal of Applied Physics* **98**, 011101.

[13] Bozhevolnyi S. I., Volkov V. S., Devaux E. S. and Ebbesen T. W. 2005. Channel plasmon-polariton guiding by subwavelength metal grooves. *Physical Review Letters* **95**, 046802.

[14] Moreno E., Garcia-Vidal F. J., Rodrigo S. G., Martin-Moreno L. and Bozhevolnyi S. I. 2006. Channel plasmon-polaritons: Modal shape, dispersion, and losses. *Optics Letters* **31**, 3447–3449.

[15] Takahara J., Yamagishi S., Taki H., Morimoto A. and Kobayashi T. 1997. Guiding of a one-dimensional optical beam with nanometer diameter. *Optics Letters* **22**, 475–477.

[16] Berini P. 2014. *Modern Plasmonics* (Maradudin: Elsevier Science).

[17] Eaton S. W., Fu A., Wong A. B., Ning C.-Z. and Yang P. 2016. Semiconductor nanowire lasers. *Nature Reviews Materials* **1**, 16028.

[18] Couteau C., Larrue A., Wilhelm C. and Soci C. 2015. Nanowire lasers. *Nanophotonics* **4**, 90.

[19] Huang M. H., Mao S., Feick H., Yan H., Wu Y., Kind H., Weber E., Russo R. and Yang P. 2001. Room-temperature ultraviolet nanowire nanolasers. *Science* **292**, 1897–1899.

[20] Bermúdez-Ureña E., Tutuncuoglu G., Cuerda J., Smith C. L. C., Bravo-Abad J., Bozhevolnyi S. I., Fontcuberta i Morral A., Garcia-Vidal F. J. and Quidant R. 2017. Plasmonic waveguide-integrated nanowire laser. *Nano Letters* **17**, 747–754.

[21] Fang Y. and Sun M. 2015. Nanoplasmonic waveguides: Towards applications in integrated nanophotonic circuits. *Light: Science &Amp; Applications* **4**, e294.

[22] Sorger V. J., Oulton R. F., Ma R.-M. and Zhang X. 2012. Toward integrated plasmonic circuits. *MRS Bulletin* **37**, 728–738.

[23] Bozhevolnyi S. I., Volkov V. S., Devaux E. S., Laluet J.-Y. and Ebbesen T. W. 2006. Channel plasmon subwavelength waveguide components including interferometers and ring resonators. *Nature* **440**, 508.

[24] Davis T. J., Gomez D. E. and Roberts A. 2016. Plasmonic circuits for manipulating optical information. *Nanophotonics* **6**, 543.

[25] Melikyan A., Lindenmann N., Walheim S., Leufke P. M., Ulrich S., Ye J., Vincze P., Hahn H., Schimmel T., Koos C., Freude W. and Leuthold J. 2011. Surface plasmon polariton absorption modulator. *Optics Express* **19**, 8855–8869.

[26] Olivieri A., Chen C., Hassan S. a., Lisicka-Skrzek E., Tait R. N. and Berini P. 2015. Plasmonic nanostructured metal-oxide-semiconductor reflection modulators. *Nano Letters* **15**, 2304–2311.

[27] Haffner C., Chelladurai D., Fedoryshyn Y., Josten A., Baeuerle B., Heni W., Watanabe T., Cui T., Cheng B., Saha S., Elder D. L., Dalton L. R., Boltasseva A., Shalaev V. M., Kinsey N. and Leuthold J. 2018. Low-loss plasmon-assisted electro-optic modulator. *Nature* **556**, 483–486.

[28] Liu K., Ye C. R., Khan S. and Sorger V. J. 2015. Review and perspective on ultrafast wavelength-size electro-optic modulators. *Laser & Photonics Reviews* **9**, 172–194.

[29] Shahzad M., Medhi G., Peale R. E., Buchwald W. R., Cleary J. W., Soref R., Boreman G. D. and Edwards O. 2011. Infrared surface plasmons on heavily doped silicon. *Journal of Applied Physics* **110**, 123105.

[30] Cleary J. W., Peale R. E., Shelton D. J., Boreman G. D., Smith C. W., Ishigami M., Soref R., Drehman A. and Buchwald W. R. 2010. IR permittivities for silicides and doped silicon. *Journal of the Optical Society of America B* **27**, 730–734.

Exercises

(Exercises with an asterisk (*) are difficult.)

(1) Waveguides are used for transporting an electromagnetic signal. They are made with metallic walls, and their shape and profiles are crucial for optimizing the signal transport. Compare how waves interact with copper walls in three important frequency domains: radio-frequency (VHF), terahertz waves, and optical waves. For these three frequency domains, calculate the skin depth β_2, ratio β_2/λ_0, and reflection coefficient. Discuss why waveguides behave differently in these three frequency domains. Use the value for the complex refractive index of copper $\tilde{n}_{Cu}$ given in the following table.

Complex refractive index of copper at different frequencies.

Frequency domain	VHF	Terahertz	Optical
ν_0	3 GHz	3 THz	6×10^{14} Hz
$\tilde{n}_{Cu}$	$7.32 \times 10^4\,(1+i)$	$156 + 186i$	$1.21 + 2.57i$

(2) **Figure of merit (FOM) of plasmonic metals:** An *FOM* is a dimensionless parameter defined to compare various situations or various materials. In the case of plasmonic propagation, we wish to evaluate the local field enhancement compared to the losses in the metal. We consider a simple metal–air interface, with the respective dielectric permittivities $\tilde{\varepsilon}_1$ and ε_2.

(a) Express the field enhancement for the perpendicular components, $\eta = E_{2z}/E_{1z}$. Express the SPP plasmon length L_{SPP} (see Chapter 4). Justify the FOM $\xi = -\varepsilon'_1/\varepsilon''_1$.

(b) Use the online database to determine the FOM of various metals, gold, silver, copper, and aluminum at the following wavelengths: 520, 750, and 1550 nm (https://refractiveindex.info/). The following table provides some values. Comment on the use of these metals for plasmonic waveguides.

Complex dielectric permittivities of gold and silver.

λ (nm)	Au	Ag
520 nm	$-3.89 + 2.63i$	$-11 + 0.33i$
750 nm	$-20.1 + 1.25i$	$-27 + 0.32i$
1550 nm	$-115 + 11.3i$	$-129 + 3.28i$

(3) **Is silicon a plasmonic material?** The complex dielectric permittivities of silicon depend on its dopant concentration. For undoped silicon, they can be found on the online database (https://refractiveindex.info/). For *p*-doped silicon, they have been measured in the range 2–40 μm and reported in *J. Appl. Phys.* **110**, 123105, (2011), as shown in the following figure [29]. We study the possibility of the propagation of an SPP wave at the interface between medium **1** (ε_1) and medium **2** (ε_2).

(a) Recall the conditions of ε_1 and ε_2 for the SPP waves to exist.

(b) Is intrinsic silicon (undoped silicon) a plasmonic material? What about *p*-doped silicon? Use the following figure to answer, and evaluate the wavelength range if an SPP wave is expected to be launched for both dopant concentrations.

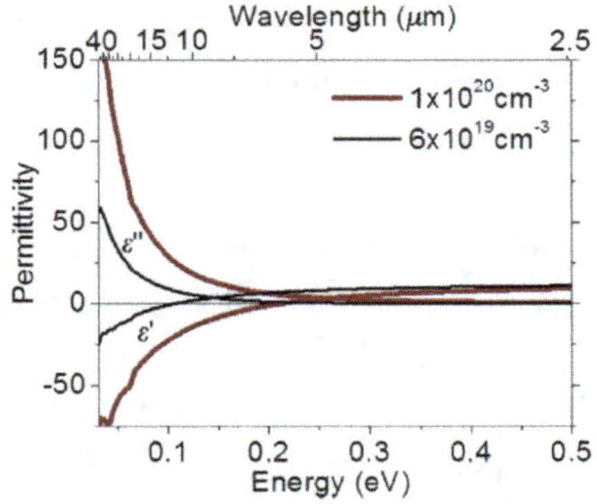

Real and imaginary parts of the permittivity for heavily doped *p*-type silicon of different carrier concentrations, as indicated in the legend.

Source: Reproduced with permission from Ref. [29].

(4) The dielectric permittivity of silicon can be described with the Drude model when it is heavily doped. The corresponding Drude parameters are given in the following table taken from Ref. [30]. From these values, explain on what wavelength range silicon is a plasmonic material that can accommodate an SPP wave. Calculate the FOM (see Exercise (2)) of *p*-doped silicon at 12 μm.

Drude parameters for two doped silicon substrates.

	Dopant concentration	$\hbar\omega_p$	$\hbar\gamma$
n-doped	$2 \times 10^{18}\,\text{cm}^{-3}$	0.036 eV	0.037 eV
p-doped	$4 \times 10^{19}\,\text{cm}^{-3}$	0.140 eV	0.070 eV

Source: Reproduced from Ref. [30].

(5) **Attenuation in a transmission line:** Imagine an optical fiber line with a loss of 0.4 dB/km, which is typical for transmitting signals at a wavelength of 1300 nm and at a bit rate of 2.5 Gbit/s.

(a) Calculate the maximum length of this optical fiber if a signal attenuation of 90% is acceptable.

(b) In practice, the signal can still be properly detected if only 0.02% remains after traveling through the fiber. Calculate the maximum length of this optical fiber.

(6) The **gap plasmon waveguide** accommodates two plasmonic modes: a symmetric and an antisymmetric mode. Their existence depends on the thickness of the dielectric layer, denoted by b. The two dispersion relations are given by relations (6.14) and (6.15). We assume that the metal is gold, and the dielectric is PMMA ($\varepsilon_d = 2.17$). Here, we neglect losses in the metal (imaginary part $\varepsilon''_m = 0$) in gold, and we set $\varepsilon_m = -115$ and $\lambda_0 = 1550\,\text{nm}$.

(a) We define $\alpha_x = k_x/k_0$ (reduced wave vector) and $\beta = b/\lambda_0$. Rewrite the two dispersions relations (6.14) and (6.15) with these reduced parameters.

(b) We are now interested in the gap waveguide with $b = 0.02\lambda_0$. Solve graphically the reduced version of Equation (6.14), and deduce for this symmetric mode the value for the SPP wave vector k_x, which is noted by k_{SPP}.

(c) Plot the electric field in the dielectric gap for this symmetric mode.

(7) **Thin-film waveguide in air:** Consider a thin-film plasmon waveguide like in Fig. 6.7 with an air gap of $b = 40$ nm.

(a) Using the simplified form of the dispersion relation, calculate the propagation length of the SPP wave at 775 nm and 1550 nm. Use the following values for ε_m: $\varepsilon_m = -24 + 1.7i$ and $\varepsilon_m = -116 + 11i$, respectively.

(b) Calculate the vertical confinement of $E(z)$.

(c) Evaluate the volume of the plasmonic mode.

(8) **Slot waveguides:** Imagine a plasmonic circuit that is designed to measure the losses induced by curved waveguides: see the following figure. The metal is gold. The dielectric is PMMA (index of 1.45), with $h = 100$ nm and $w = 200$ nm. The laser beam is directed toward the points marked A_1, A_2, A_3, and A_4. The SPP wave is then transported through the waveguide, and the positions marked B_1–B_4 are used for radiating the SPP away in the far field and collecting the remaining SPP energy.

(a) Explain why the areas A_1–A_4 are made of four regularly spaced stripes. What is the periodicity adapted to launch an SPP wave at 1550 nm? The effective waveguide index at this wavelength is $n_{\mathit{eff}} = 1.61$.

(b) Use the plot in Fig. 6.9 to predict the propagation length L_{SPP}. What is the power remaining in B_1?

(c) Calculate the loss in dB/mm.

(d) From the plot shown in the following fig. (b), predict the power remaining in B_2, B_3, and B_4.

(e) (*)Evaluate the decay of the electric field inside the metal. Deduce from this value the minimum distance between the stripes if we want to avoid cross talk between different neighboring slot waveguides.

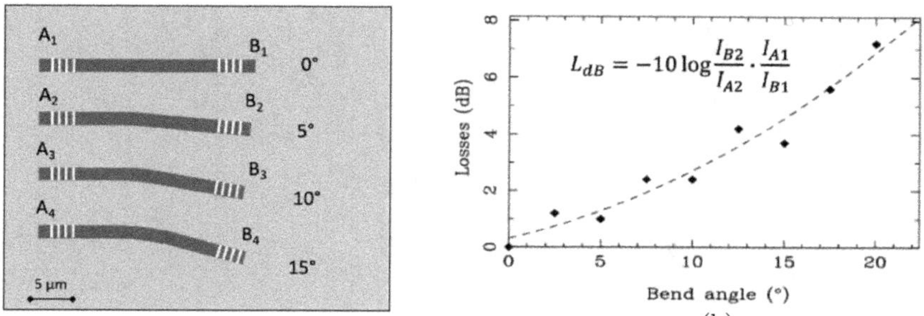

(a) (b)

(a) Schematic of a series of four slot waveguides used for measuring the plasmonic losses in curved waveguides. (b) Plot of the losses caused by the curve.

Source: Reproduced from the work of Bouhelier.

7

Fundamentals of Localized Plasmon in Nanospheres

The **localized surface plasmon resonance** (LSPR) takes place in metallic nanostructures when their size is much smaller than the wavelength of the incoming optical plane wave. Here, the principal properties of LSPR are derived, starting from the case of a spherical plasmonic nanoparticle within the so-called **electrostatic approach**. The **optical cross sections** are calculated for absorption, scattering, and extinction. The link between the extinction cross section and the **Beer–Lambert law** is established. This case of nanosphere also illustrates the efficiency of the **dipolar model** in describing plasmonics, and it is used to calculate how the electric field radiated by a nanoparticle evolves when going from the **near field** to the far field. This is a good introduction to **nano-optics**.

7.1 Introduction

We now address the *localized plasmon resonance*, which is the second type of plasmon and should be distinguished from the surface plasmon polaritons (SPPs) described in the three previous chapters. The *localized plasmon* occurs in metallic nanostructures, such as nanoparticles, where the electromagnetic field is confined within the three dimensions. This plasmon oscillation is usually termed localized surface plasmon resonance (LSPR), although surface phenomena are not dominant here.

7.1.1 *A brief overview of the localized surface plasmon resonance*

The LSPR corresponds to a resonance of the electromagnetic field inside a metallic nanostructure whose size is smaller than the wavelength of the excitation optical wave. This resonance is intrinsically coupled to an oscillation of charges and can be viewed as a global oscillation of the conduction electrons inside the metal. This oscillation of charges would not occur in a bulk material with large dimensions, and its existence is linked to the constraints set by the boundaries of a nanostructure. LSPR is not a propagation but an oscillation of charges that eventually is coupled to an optical wave. The most striking characteristic of the LSPR is the strong optical absorption at the resonance frequency, and when the latter is in the visible range, an assembly of nanoparticles acquire a very distinctive color. For example, gold nanoparticles appear red and silver nanoparticles appear yellow. We detail these properties in the following and explore how they can be used for a variety of applications, such as nanosensors, advanced therapies, coloring materials, and concentrating light with nano-antennas.

7.1.2 *Historical landmarks in the development of LSPR*

The Lycurgus cup (see Fig. 7.1)**:** The most renown example of the visual effect of LSPR is found in the Lycurgus cup, where nanoparticles made of an alloy of gold and silver were employed for giving glass an intriguing red color. This cup is a carefully designed jewelry artwork dated to the fourth century and attributed to a Roman glassmaker. This artwork is exhibited in the British Museum in London. On the cup, a mythological episode, the victory of the god Dionysus over Lycurgus, a king of the Thracians (*circa* 800 BC) is represented: One of Dionysus' maenads, Ambrosia, transformed into a vine by Mother Earth, retains Lycurgus, while Dionysus instructs to kill him [1, 2]. What is exceptional with the color of this glass is that it actually exhibits two colors: when observed in transparency, the cup appears red and when observed by diffusion, the color turns green [3]. Such an art piece is unique and was not well understood until recently [4] (see also Chapter 10). The process used by the Roman glassmakers is still unknown, and it is unclear whether the Lycurgus cup was obtained by chance or was

Fig. 7.1. The Lycurgus cup fabricated during the Late Roman Empire, dated to the fourth century AD: The cup appears (a) red when illuminated from inside and (b) green when illuminated from outside.

Source: Reproduced with permission from the British Museum free image service.

a controlled technology. The scientific explanations of the color of objects at that time were still very rudimentary. There are other examples in the medieval times of ruby-red glass, also called the *purple of Cassius*, whose color results from the LSPR, but we will not detail this aspect here. You may refer to [5] for a detailed explanation.

Michael Faraday wrote in 1857 that the color of the *purple of Cassius* was due to gold present in solution in a "finely divided metallic state" [6]. In the 19th century, many scientists tried to elucidate the origin of this purple coloration. Faraday's assumption was only demonstrated in 1905 by Richard Adolf Zsigmondy, who used a slit ultra-microscope, which is the ancestor of the dark-field microscope used for observing the light scattered by finely divided gold particles reduced by stannic acid. The complete story can be found in Carbert's article [7].

Gustav Mie, in 1908, confirmed these peculiar optical properties of small particles by providing a detailed analytical model of the absorption and scattering efficiency of metallic particles [8]. The Mie theory remains the reference for understanding the LSPR nowadays. But at that time, all the consequences had not been drawn, and the notion of the plasmon resonance had not been clarified. It was David Pines who first used the name "plasmon" in 1956 to describe the quantum of energy associated with the eigenfrequency of the collective oscillation of electrons in a metal [9]. This concept was applied to the SPP (see Chapter 4), and the interest in plasmonic oscillations in nanostructures came into focus with the need to explain the incredible sensitivity of surface-enhanced Raman scattering (SERS) in the late 1970s, thanks in particular to the work of Richard Van Duyne [10].

Uwe Kreibig was the one who carried out the complete derivation of the localized plasmon resonance in nanoparticles in 1985 [11,12]. Since then, the topic of localized plasmonic has been taken over by numerous scientists and applied to many fields. It benefited a lot from the huge progress in nano-optics and instrumentation.

7.2 Theoretical Approach to the LSPR in the Case of a Nanosphere

In this section, we aim at deriving how an electromagnetic plane wave $\boldsymbol{E}_0\exp[i(\boldsymbol{k}\cdot\boldsymbol{r}-\omega t)]$ interacts with a metallic nanosphere of diameter D (see Fig. 7.2). We want to know how a sphere scatters this incident wave. Scattering is an essential phenomenon in everyday life as soon as light interacts with tiny particles of sizes close to the optical wavelength: This is what makes fumes visible, clouds appear white, and milk a white liquid. Scattering by a metallic nanosphere shows a different behavior, as we will see. We also want to know what proportion of the energy of the incoming wave is absorbed and therefore that of the transmitted light. This is called the extinction efficiency of the nanosphere. We will show that the electric field inside the nanoparticle undergoes a pronounced resonance at a given wavelength, which implies a strong optical absorption at this same wavelength and therefore a pronounced color change.

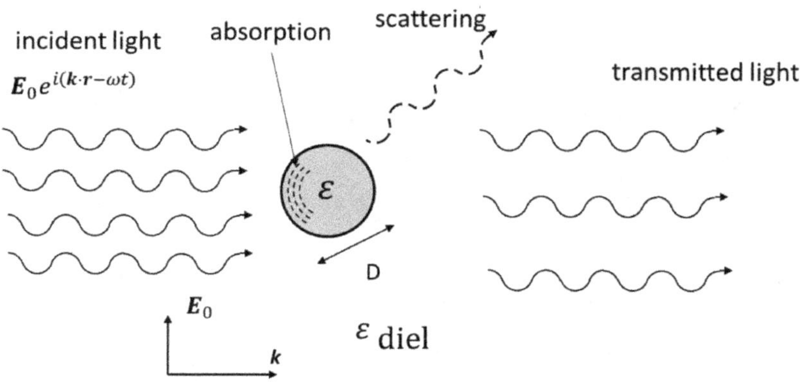

Fig. 7.2. Representation of the problem: A sphere of dielectric permittivity ε is excited by a plane wave of wavelength λ and wave vector $\boldsymbol{k}$. The sphere is placed in a dielectric medium of dielectric permittivity $\varepsilon_{\text{diel}}$. A part of the light is absorbed by the particle, and another part is scattered. The remaining part of light is transmitted.

7.2.1 *Rayleigh scattering, Mie theory, and the electrostatic approach*

A first approach to the problem of scatterers was proposed by Lord Rayleigh in 1871 for the case of $D \ll \lambda$. Such a condition implies that the electromagnetic field is homogeneous across the entire nanosphere. For this reason, it is also called the *electrostatic approximation*. The nanosphere behaves like a dipole, with the positive surface charges and the negative surface charges being symmetric relative to the center of the sphere. As an example, Rayleigh scattering is a suitable model for explaining how light is scattered by ionized molecules in the atmosphere and why the color of the sky is blue. This approach is valid as long as $D < \lambda/100$.

However, when D grows closer to λ, a more accurate approach is provided by the Mie theory developed by Gustav Mie in 1908 [8]. It gives the exact solution of a plane wave interacting with a nanosphere. The driving idea of Mie theory is to express the incoming plane wave with the spherical coordinates centered on the nanosphere. It leads to an expansion over a series of multipolar contributions: dipolar, quadrupolar, octupolar, etc. The expansion coefficients are found by applying the correct boundary conditions for electromagnetic fields at the interface between the metallic sphere

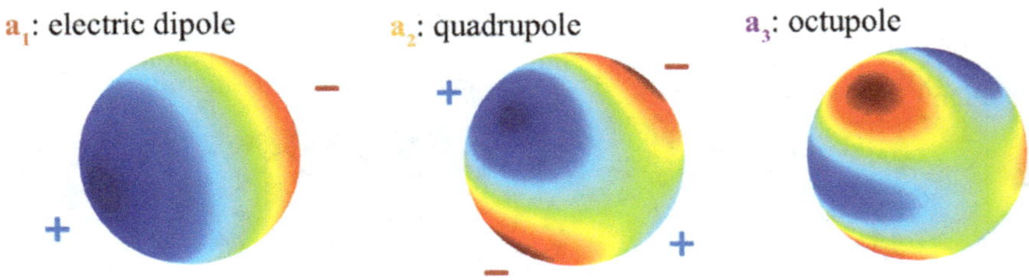

Fig. 7.3. Charge oscillations in a nanosphere submitted to a quasi-uniform electric field. The dipolar contribution corresponds to one positive pole and one negative pole. The quadrupolar contribution exhibits two positive poles and two negative ones. And the octupolar contribution has four positive and four negative poles.

Source: Reproduced from Ref. [13].

and its surroundings. Figure 7.3 shows the result of such a calculation, with the repartition of the surface charges on the nanosphere. Only the first three contributions are given: the dipolar, quadrupolar, and octupolar contributions [13]. The overall charge repartition is a combination of these various contributions, and they are oscillating over time. It is clear that this calculation requires a delicate treatment that will not be detailed here. In the current chapter, we limit ourselves to the dipolar effects. More details will be discussed in the following chapter regarding the other multipolar contributions. The full discussion can be found in reference textbooks, such as those by Born and Wolf [14] and Bohren and Huffmann [15].

For small metallic particles ($D < 20$ nm), it is often sufficient to limit the multipole expansion to its first term, which is the dipolar contribution, and this is the *Rayleigh limit*, or the *electrostatic approximation* (sometimes also termed *quasistatic approximation*). In this chapter, we develop this approximation, which is sufficient to grasp the working principles of localized plasmon resonance and account for a number of experimental cases.

7.2.2 *Electrostatic approach to the LSPR*

7.2.2.1 *Electrostatic potential generated by a metallic sphere*

In the electrostatic limit, the electronic polarization is exactly in phase with the excitation field (no retardation effect), and the electrons are displaced as

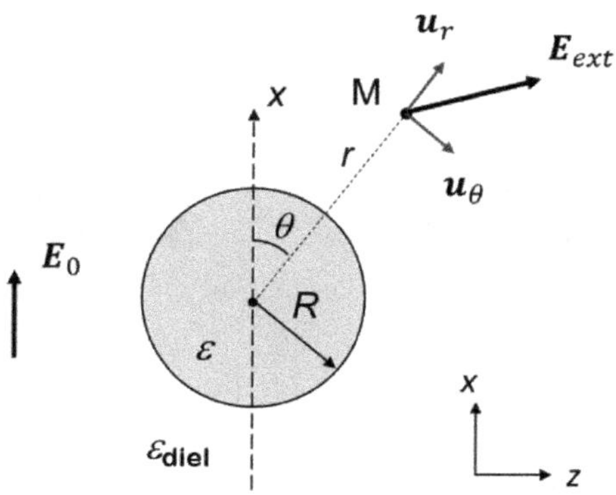

Fig. 7.4. The spherical coordinates used for calculating the electric field generated by a metallic nanosphere excited by a plane wave polarized along the x-axis.

a whole. This is the major approximation of the electrostatic approach (see Fig. 7.4). Therefore, the charge distribution in the particle can be treated as if it were a static distribution [15, 16]. The electric field should obey the Laplace equation:

$$\Delta V = 0, \tag{7.1}$$

where V is the electric potential linked to the electric field by the gradient operator $\boldsymbol{E} = -\boldsymbol{\nabla} V$.[1]

Due to the symmetry of the problem, spherical coordinates are used. Moreover, the x-axis is an axis of symmetry so that the third coordinate, usually required for spherical coordinates, becomes useless. Therefore, Equation (7.1) written in spherical coordinates becomes

$$\frac{1}{r}\frac{\partial}{\partial r^2}(rV) + \frac{1}{r^2 \sin\theta}\frac{\partial}{\partial\theta}\left(\sin\theta\frac{\partial V}{\partial\theta}\right) = 0. \tag{7.2}$$

The solutions to this differential equation are the spherical harmonics with the following form:

$$V(r,\theta) = \sum_{n=0}^{\infty}\left(A_n r^n + \frac{B_n}{r^{n+1}}\right) P_n(\cos\theta), \tag{7.3}$$

[1]In Cartesian coordinates, the Laplacian operator is written as $\Delta V = \partial^2 V/\partial x^2 + \partial^2 V/\partial y^2 + \partial^2 V/\partial z^2$.

where A_n and B_n are coefficients to be determined. P_n are the Legendre polynomials which often appear in physics problems expressed in spherical coordinates.[2]

This electric potential has two different forms: $V_{int}(r, \theta)$ and $V_{ext}(r, \theta)$, inside and outside the metallic particle, respectively. Moreover, it should obey the following boundary conditions:

- The electric field should be properly defined at $r = 0$, which implies that $\frac{\partial V_{int}}{\partial r}\Big|_{r=0}$ exists.
- The electric field far away from the particle should match the excitation field, which is uniform in the electrostatic approximation, so that $\lim_{r\to\infty} V_{ext} = 0$.
- At the interface, the electric fields obey the continuity equation applied to the particle surface: $\varepsilon(\omega)E_{int}(r = R) = \varepsilon_{diel}(\omega)E_{ext}(r = R)$.
- Finally, the potential should be continuous at the particle surface: $V_{int}(r = R) = V_{ext}(r = R)$.

Applying these conditions allows us to determine the coefficients for the electric potentials inside and outside the particle. For the external potential, one can determine that all the A_n coefficients are null except one: $A_{1\,ext}$. Similarly, for the B_n coefficients, the only non-zero coefficient is $B_{1\,ext}$ so that

$$A_{1ext} = -E_0 \text{ and } B_{1ext} = R^3 \frac{\varepsilon - \varepsilon_{diel}}{\varepsilon + 2\varepsilon_{diel}} E_0. \tag{7.4}$$

Regarding the internal potentials, A_{1int} is the only non-zero coefficient since $B_{n\,ext}$ must be zero for preventing V_{int} from diverging at $r = 0$. Therefore,

$$A_{1int} = -\frac{3\varepsilon_{diel}}{\varepsilon + 2\varepsilon_{diel}} E_0. \tag{7.5}$$

---

[2]The Legendre polynomials are obtained as solutions to the Legendre equations or the nth element of the recurrence relation of Bonnet. The first four terms are $P_0(X) = 1$, $P_1(X) = X$, $P_2(X) = 3/2X^2 - 1/2$, and $P_3(X) = 5/2X^3 - 3/2X$.

The resulting electric potentials are written as

$$V_{ext} = -E_0 r \cos\theta + R^3 \frac{\varepsilon - \varepsilon_{diel}}{\varepsilon + 2\varepsilon_{diel}} E_0 \frac{\cos\theta}{r^2}, \tag{7.6}$$

$$V_{int} = -\frac{3\varepsilon_{diel}}{\varepsilon + 2\varepsilon_{diel}} E_0 r \cos\theta. \tag{7.7}$$

7.2.2.2 *Dipolar moment and polarizability of the metallic sphere*

Equation (7.6) appears as a sum of two terms: The first one is nothing but the external applied field since $-\boldsymbol{E}_0 = -\boldsymbol{\nabla}(-E_0 r \cos\theta)$ in spherical coordinates, and the second one is the potential generated by an electric dipole $\boldsymbol{p}$ given that [17]

$$V_{dipole} = \frac{1}{4\pi\varepsilon_0} \cdot \frac{1}{r^2} \boldsymbol{p} \cdot \boldsymbol{u_r}, \tag{7.8}$$

and provided that $\boldsymbol{p}$ can be written as

$$\boldsymbol{p} = 4\pi\varepsilon_0 R^3 \frac{\varepsilon - \varepsilon_{diel}}{\varepsilon + 2\varepsilon_{diel}} E_0 \boldsymbol{u_x}. \tag{7.9}$$

It is easy to check that the combination of relations (7.8) and (7.9) results in the second term of relation (7.6).

The quantity $\boldsymbol{p}$ is the dipolar moment induced by the electric field $E_0 \boldsymbol{u_x}$ that tends to polarize the sphere, as depicted in Fig. 7.5. The proportionality coefficient between $\boldsymbol{E}_0$ and $\boldsymbol{p}$ is, by definition, the polarizability α [17]:[3]

$$\boldsymbol{p} = \alpha \boldsymbol{E}_0 = \alpha \boldsymbol{E}_0 \boldsymbol{u_x}. \tag{7.10}$$

In our case, the identification of Equations (7.9) and (7.10) leads to

$$\alpha = 4\pi\varepsilon_0 R^3 \frac{\varepsilon - \varepsilon_{diel}}{\varepsilon + 2\varepsilon_{diel}}. \tag{7.11}$$

The polarizability α plays a central role in understanding the response of a nanosphere to an electromagnetic field, and we will show that most of the properties of the plasmon resonance are extracted from α.

---

[3] Some authors, such as Maier and Bohren, define the polarizability with the relationship $\boldsymbol{p} = \alpha \varepsilon_r \varepsilon_0 \boldsymbol{E_0}$, with $\varepsilon_r = 1$ in air. In Equation (7.9), this definition would lead to the removal of the ε_0 factor.

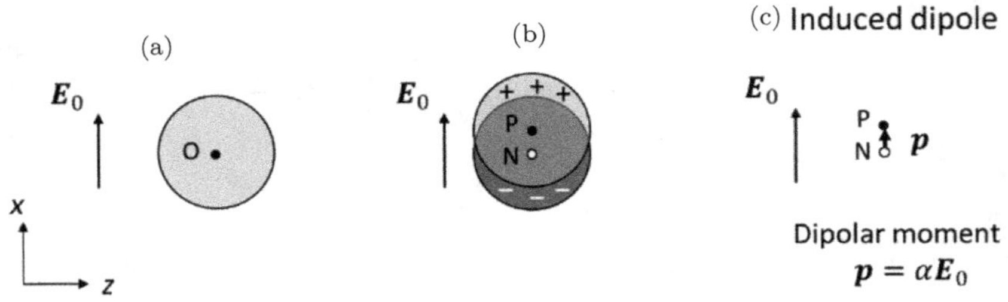

Fig. 7.5. A metallic sphere is subjected to an electrostatic field in (a). The sphere immediately acquires a polarization, with positive charges (P) slightly pushed away from the negative charges (N), as shown in (b). This is equivalent to a simple electric dipole $\boldsymbol{p}$, which is linked to the electrostatic field E_0 by the polarizability α (c).

Meanwhile, we can rewrite Equation (7.6) as a function of α:

$$V_{ext} = -E_0 r \cos\theta + \frac{1}{4\pi\varepsilon_0} \cdot \alpha E_0 \frac{\cos\theta}{r^2}. \tag{7.12}$$

Once the electric potential is obtained, the electric field can be deduced by using the gradient operator: $\boldsymbol{E} = -\boldsymbol{\nabla} V$.

After some calculations, one obtains the following expression for the electric field outside of the nanoparticle, which is a sum of the incident field and the radiated field produced by the particle:

$$\boldsymbol{E}_{ext} = \boldsymbol{E}_0 - \frac{\alpha E_0}{4\pi\varepsilon_0}\left[-2\frac{\cos\theta}{r^3}\boldsymbol{u}_r - \frac{\sin\theta}{r^3}\boldsymbol{u}_\theta\right].$$

Given that $\boldsymbol{E}_0 = E_0 \cos\theta \boldsymbol{u}_r - E_0 \sin\theta \boldsymbol{u}_\theta$, the relation can also be written as

$$\boldsymbol{E}_{ext} = \boldsymbol{E}_0 + \frac{1}{4\pi\varepsilon_0}\frac{1}{r^3}\left[3\left(\boldsymbol{p}\cdot\boldsymbol{u}_r\right)\boldsymbol{u}_r - \boldsymbol{p}\right]. \tag{7.13}$$

The electric field inside the particle is derived in a similar way from relation (7.7):

$$\boldsymbol{E}_{int} = \frac{3\varepsilon_{diel}}{\varepsilon + 2\varepsilon_{diel}}\boldsymbol{E}_0. \tag{7.14}$$

The expression of $\boldsymbol{E}_{ext}$ given by Equation (7.13) corresponds to the scattered field (within the electrostatic approach). This field decreases quickly as $1/r^3$ when moving away from the particle and tends to merge with $\boldsymbol{E}_0$.

More interesting is the fact that this field is proportional to the polarizability α. If α goes through a resonance, the scattered field also reaches a maximum. This is known as the Fröhlich condition.

7.2.2.3 *The Fröhlich condition of resonance*

It is clear that the external field will go through a maximum when the polarizability is maximized. In Equation (7.11), ε_{diel} can be considered a constant in the visible range (ε_{diel} is linked to the optical index: $\varepsilon_{diel} = n^2$), and ε takes mostly negative values for a metal and depends on the frequency ω (see Chapter 2). Therefore, $|\alpha|$ is maximized when the following relationship is fulfilled:

$$|\varepsilon + 2\varepsilon_{diel}| \text{ is minimum.} \tag{7.15}$$

This condition is the Fröhlich condition.

If we consider a free-electron metal, described by the Drude permittivity (as in Equation (2.7)), ε is real and is expressed as $\varepsilon = 1 - \omega_p^2/\omega^2$. Then, condition (7.15) becomes simply $\varepsilon + 2\varepsilon_{diel} = 0$ and leads to the following LSPR condition:

$$\omega_{LSPR}^{Drude} = \frac{\omega_p}{\sqrt{1 + 2\varepsilon_{diel}}} \text{ or } \lambda_{LSPR}^{Drude} = \lambda_p\sqrt{1 + 2\varepsilon_{diel}}. \tag{7.16}$$

This formula applied to the case of a metallic sphere in air shows that $\omega_{LSPR}^{Drude} = \omega_p/\sqrt{3}$ or $\lambda_{LSPR}^{Drude} = \lambda_p\sqrt{3}$. If this same sphere is placed in water, $\lambda_{LSPR}^{Drude} = 2.13 \cdot \lambda_p$. This shows that the higher the optical index of the surrounding, the higher the wavelength of the LSPR. This redshift (wavelength increase) of the LSPR is confirmed by experiments. However, the values of the LSPR are not correctly reproduced since the Drude model is often too crude, as discussed in Chapter 2. Therefore, it is highly preferable to use more accurate values for ε.

Figure 7.6 shows a plot of the absolute value of the polarizability for a gold sphere in air. The maximum is observed at 520 nm, which does not match the experimental value of 505 nm. Note that the calculation of λ_{LSPR}^{Drude}, according to $\lambda_{LSPR}^{Drude} = \lambda_p\sqrt{3}$, yields $\lambda_{LSPR}^{Drude} = 239$ nm, which is even further away from the expected value of 505 nm. Again, the contribution

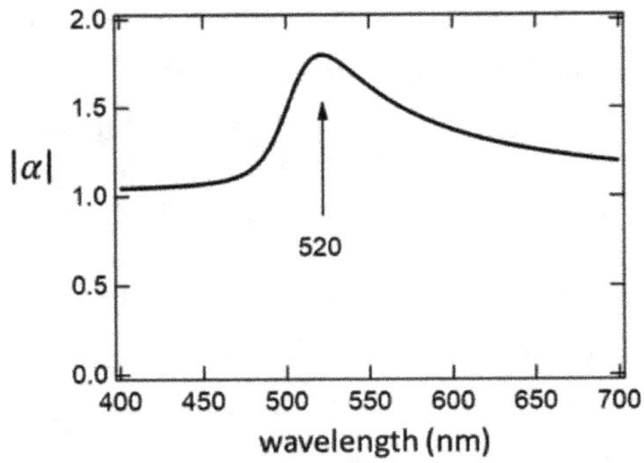

Fig. 7.6. Plot of the absolute value of the polarizability of a gold sphere in air as a function of the wavelength of the excitation wave. Gold is described by the values of the dielectric permittivity given in Chapter 2 (measured by ellipsometry).

from the interband transition in gold prevents us from using the Drude model for gold, and this formula is not suitable in this case.

Therefore, the Fröhlich condition is a rough first approach for predicting the LSPR and cannot be used for drawing numerical conclusions. Actually, this condition applies to the polarizability $\alpha(\lambda)$, which is never directly measured. The quantity that is measured is the extinction cross section, where α appears multiplied by $1/\lambda$, which tends to shift the peak position (see Section 1.2.4).

7.2.3 *Luminous power, flux, and intensity: Important remarks on photometry*

Let's start by defining some key notions from photometry to understand the interaction of nanoparticles with light. Photometry is important for understanding the efficiency of a photovoltaic cell that converts the light flux into electrical power or for understanding how the lighting industry calculates the power of light sources.

Let's first recall that the **power flow**, denoted by the Greek letter $\mathbf{\Pi}$, is the power per unit area ($\mathrm{W/m^2}$) transported by a wave and is calculated with the Poynting vector, as explained in Chapter 1.

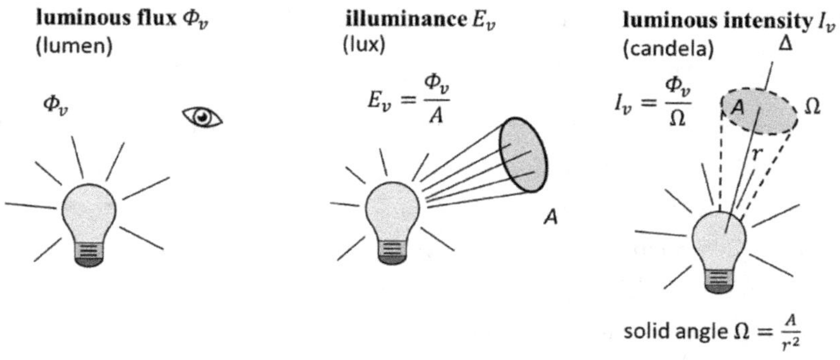

Fig. 7.7. Main quantities used in photometry: The luminous flux is the fraction of the power of the light source which is visible to the human eye. The illuminance corresponds to the luminous flux that falls on the surface of area A. The luminous intensity is the luminous flux radiated within the solid angle Ω. The solid angle is a two-dimensional angle defined by $\Omega = A/r^2$.

The **radiant power** Φ_e is the total amount of light produced by a luminous point source and is expressed in *Watt* (W). However, not all this power is visible to the human eye, but just a fraction of it, within the visible spectrum. The **luminous flux** or **luminous power** Φ_v is this visible part of the luminous power. It has the same dimension as Φ_e and is measured in *lumen* (lm): 1 W = 683 lm (at 555 nm for day vision).

The **illuminance** E_v is the luminous flux that falls on a given area A, expressed in *lux*, a unit equivalent to lm/m^2.

The **luminous exitance** M_v is useful when the light source is no longer punctual. The luminous flux emitted by a portion dS of this source is $\Phi_v = dS\, M_v$. M_v is in lm/m^2.

The **luminous intensity** I_v is the luminous flux per unit solid angle expressed in *lumen per steradian* (lm/sr). The special unit for the intensity is the *candela* (cd). See Fig. 7.7 for graphical presentations of these definitions.

For example, a candle produces a luminous intensity of 1 candela (historical definition of the candela). Therefore, this candle emits a total luminous flux over all the solid angles of $4\pi \times 1$, which is 12.6 lm and corresponds approximately to a luminous power of 18 mW.

7.2.4 Optical cross sections: Absorption, scattering, and extinction

7.2.4.1 Definition of the absorption and scattering cross sections

Up to now, we have restrained ourselves to electrostatics and calculated the polarizability of the sphere induced by a permanent field $\boldsymbol{E}_0$ and the resulting static electric fields $\boldsymbol{E}_{int}$ and $\boldsymbol{E}_{ext}$. If we now consider an oscillating field $\boldsymbol{E}_0 e^{i\omega t}$, we can simply generalize the results, and the induced dipole becomes $\boldsymbol{p} = \alpha e^{i\omega t}\boldsymbol{E}_0$ as long as the radius of the particle remains much smaller than the wavelength ($R \ll \lambda$) so that the retardation effect on the nanoparticle volume can be neglected. The electromagnetic fields $\boldsymbol{E}$ and $\boldsymbol{H}$ radiated by an oscillating dipole are calculated in the classical textbooks of electromagnetism authored by Griffith [17], Jackson [18], and Born and Wolf [14] and briefly recalled by Maier [19].

However, the most relevant feature of the interaction of a nanoparticle with light is its capacity to absorb and radiate the incoming energy. This radiated power flow (power per unit area) is calculated with the Poynting vector given as $\boldsymbol{\Pi} = \langle \boldsymbol{E} \times \boldsymbol{B}/\mu_0 \rangle$ (see Chapter 1). $\Pi_0 = \langle \boldsymbol{\Pi} \cdot \boldsymbol{S}_0 \rangle$ is the time-averaged power flow (W/m^2) crossing the unit of surface $\boldsymbol{S}_0$ (see Fig. 7.8). The nanosphere scatters a power P_{sca}, which is proportional to the incoming power flow. P_{sca} is the power integrated over all the directions. The coefficient of proportionality is called the *scattering cross section* σ_{sca} and defined by

$$P_{sca} = \sigma_{sca}\Pi_0. \tag{7.17}$$

Part of the incoming flux is also absorbed by the nanosphere and converted mostly into heat. The absorbed power P_{abs} is also proportional to the incoming flux so that we can define the *absorption cross section* σ_{abs} as

$$P_{abs} = \sigma_{abs}\Pi_0. \tag{7.18}$$

Note that σ_{sca} and σ_{abs} have the dimension of a surface since P_{sca} and P_{abs} are powers (W) and Π_0 is a power per unit area (W/m^2). Here is an easy comparison to understand the meaning of the cross section. If the light rays were like rifle bullets, only those touching the nanosphere would bounce against it and be diffused. The area of this *target* seen by the bullet is πR^2, which is the *geometrical cross section* of a sphere. Therefore, it is

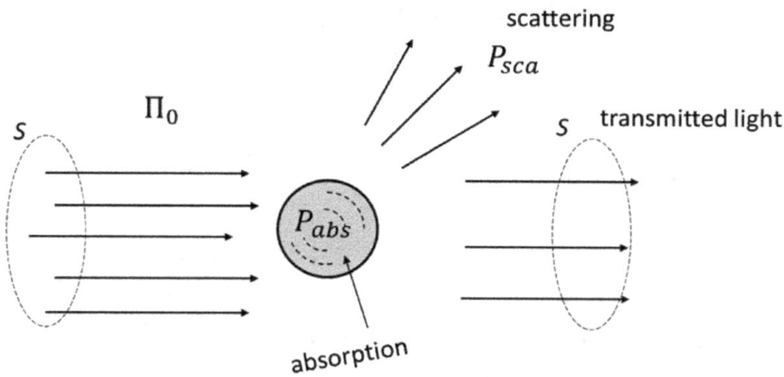

Fig. 7.8. Representation of the interaction of the incoming light with a nanoparticle. The incoming power flow Π_0 (power per surface area) is either scattered or absorbed.

meaningful to compare the cross sections σ_{sca} and σ_{abs} with πR^2. The corresponding quantities are the efficiencies Q_{sca} and Q_{abs} explained in the following. We calculate later that a nanoparticle may have a cross section larger than its geometrical cross section πR^2.

7.2.4.2 *Extinction cross section*

It is useful to also define the *extinction cross section* σ_{ext}, which represents the total power attenuated after the incoming beam had interacted with the nanoparticle:

$$\sigma_{ext} = \sigma_{sca} + \sigma_{abs}. \tag{7.19}$$

In practice, σ_{ext} is the quantity measured in the absorption spectra with an optical spectrometer.

The analytical expressions for σ_{sca} and σ_{abs} have been established by Bohren and Huffman using the scattering matrix approach within the electrostatic approach (see Chapter 5 of their book [15]). This is a complex derivation but with easy and practical results, so we skip it in this book:

$$\sigma_{sca} = \frac{k^4}{6\pi} |\alpha|^2, \tag{7.20}$$

$$\sigma_{abs} = k \operatorname{Im}(\alpha), \tag{7.21}$$

where $\mathrm{Im}(\alpha)$ is the imaginary part of the polarizability and k the wave vector of the wave in the surrounding medium of dielectric permittivity ε_{diel}. Do not forget that k is linked to the wavelength in free space λ_0 by $k = 2\pi n_{diel}/\lambda_0 = 2\pi\sqrt{\varepsilon_{diel}}/\lambda_0$. Using Equation (7.11), the cross sections can be detailed as [15]

$$\sigma_{sca} = \frac{8}{3}\pi R^6 k^4 \left|\frac{\varepsilon - \varepsilon_{diel}}{\varepsilon + 2\varepsilon_{diel}}\right|^2 = \frac{128}{3}\pi^5 R^6 \frac{\varepsilon_{diel}{}^2}{\lambda_0^4}\left|\frac{\varepsilon - \varepsilon_{diel}}{\varepsilon + 2\varepsilon_{diel}}\right|^2, \tag{7.22}$$

$$\sigma_{abs} = 4\pi R^3 k\, \mathrm{Im}\left(\frac{\varepsilon - \varepsilon_{diel}}{\varepsilon + 2\varepsilon_{diel}}\right) = 8\pi^2 R^3 \frac{\sqrt{\varepsilon_{diel}}}{\lambda_0}\, \mathrm{Im}\left(\frac{\varepsilon - \varepsilon_{diel}}{\varepsilon + 2\varepsilon_{diel}}\right). \tag{7.23}$$

These last two relations are fundamental for understanding the optical behavior of nanoparticles.

7.2.4.3 *Important remarks*

First, the scattering exhibits a dependence of $1/\lambda_0^4$, which is the signature of Rayleigh scattering. This behavior shows a rapid decrease in the scattering efficiency with the wavelength: The blue region of the visible spectrum is more strongly scattered than the red one. If we consider scattering particles with negligible dependence on the visible wavelengths, they would have a scattering efficiency of $1/\lambda_0^4$. This is the case for the oxygen and nitrogen molecules in the atmosphere, and this explains why the sky appears blue. However, Rayleigh scattering (as it was described by Lord Rayleigh in 1871) does not account for absorption.

Second, both relations (7.22) and (7.23) obey the Fröhlich condition of resonance discussed above, and one might expect a maximum for a given wavelength.

The third comment on the expressions (7.22) and (7.23) is related to their dependence on the nanoparticle radius R. Absorption grows with R^3, whereas scattering grows with R^6. Therefore, we can anticipate that absorption will dominate over scattering for small values of R and that for larger nanoparticles, scattering will be dominant. This property is essential for understanding the color of nanoparticles. Colors of small nanoparticles and molecules are controlled by the resonant lines in absorption. It means that the color is given by the missing wavelength which is the resonance for extinction. On the other hand, since R^6 grows faster than R^3, larger nanoparticles will mostly scatter light, and their color will correspond to the scattering resonance.

Finally, it is regrettable that (7.22) and (7.23) do not provide any simple expression for the extinction cross section. However if one computes the values of σ_{abs} and σ_{sca} for small metallic spheres, one will notice that the scattering is negligible, which is in agreement with the previous remark. For example, for a 15 nm spherical gold nanoparticle in air at 505 nm (position of the plasmon resonance), $\sigma_{abs} = 5.5 \times 10^{-17}\text{m}^2$ and $\sigma_{sca} = 5.5 \times 10^{-20}\text{m}^2$. Therefore, as long as nanoparticles remain much smaller than the wavelength ($D < \lambda/100$), we can consider that

$$\sigma_{ext} \approx \sigma_{abs},$$

which yields

$$\sigma_{ext} \approx 8\pi^2 R^3 \frac{\sqrt{\varepsilon_{diel}}}{\lambda_0} \text{Im}\left(\frac{\varepsilon - \varepsilon_{diel}}{\varepsilon + 2\varepsilon_{diel}}\right). \tag{7.24}$$

Note that the condition $R \ll \lambda$ is also the validity condition for the electrostatic approximation used for deriving the cross sections.

7.2.5 *Scattering, absorption, and extinction efficiencies*

Instead of the cross sections, some authors prefer using the scattering, absorption, and extinction efficiencies, Q_{sca}, Q_{abs}, and Q_{ext}, respectively:

$$Q_{sca} = \frac{\sigma_{sca}}{\pi R^2}, \quad Q_{abs} = \frac{\sigma_{abs}}{\pi R^2}, \text{ and } Q_{ext} = \frac{\sigma_{ext}}{\pi R^2}. \tag{7.25}$$

Q_{sca}, Q_{abs}, and Q_{ext} compare the cross sections with the geometrical cross section. For example, if $Q_{sca} > 1$, it means that the nanoparticle perturbs the electromagnetic field further than its physical volume in space. On the contrary, if $Q_{sca} < 1$, the nanosphere has an optical signature smaller than its volume and might be difficult to detect. Examples are given in the following section.

7.3 Experimental Verifications in Simple Cases of Spherical Nanoparticles

7.3.1 *Beer–Lambert law: An empirical law*

If we consider a single nanoparticle in an incident light power flow Π_0, the amount of power removed by the presence of the nanoparticle is linked to the extinction cross section by $P_{ext} = \sigma_{ext}\Pi_0$. However, in most experimental cases, nanoparticles are not single but are found in ensembles of several billions of nanoparticles in suspension or dispersed in a bulk material. For example, the well-known Turkevich method used to prepare 15 nm spherical gold nanoparticles yields an approximate number of 2,400 billion nanoparticles (2.4×10^{12}) in a 20 mL beaker. Therefore, a light beam interacts with many nanoparticles. How to evaluate the remaining power after this light beam has traversed the beaker?

The attenuation of the transmitted flux is measured in practice with a spectrometer and expressed in terms of *absorbance* A as

$$A = -\log\left(\Pi_{NP}/\Pi_0\right), \tag{7.26}$$

where Π_{NP} is the power flow transmitted through the medium with nanoparticles and Π_0 the power flow transmitted by the same medium without nanoparticles (the supporting medium). The supporting medium is mostly transparent and can be a liquid (solvent or suspension) or a solid, such as glass or polymer (see Fig. 7.9).

The Beer–Lambert law is an empirical law that states that A is proportional to the length l (in cm) of the medium traversed by the optical beam and the concentration of the absorbing species C (in mol/L), with a proportionality coefficient ε_{mol} called the molar extinction coefficient (in $\text{mol}^{-1}.\text{cm}^{-1}$):

$$A = \varepsilon_{\text{mol}} \cdot l \cdot C \text{ (Beer–Lambert law).} \tag{7.27}$$

By grouping relations (7.25) and (7.26), we can express the Beer–Lambert law as

$$\Phi_{NP} = \Phi_0 \times 10^{-\varepsilon_{\text{mol}} \cdot l \cdot C} = \Phi_0 \times e^{-\ln 10 \cdot \varepsilon_{\text{mol}} \cdot l \cdot C}. \tag{7.28}$$

It shows that the light flux decreases exponentially with the length of the medium.

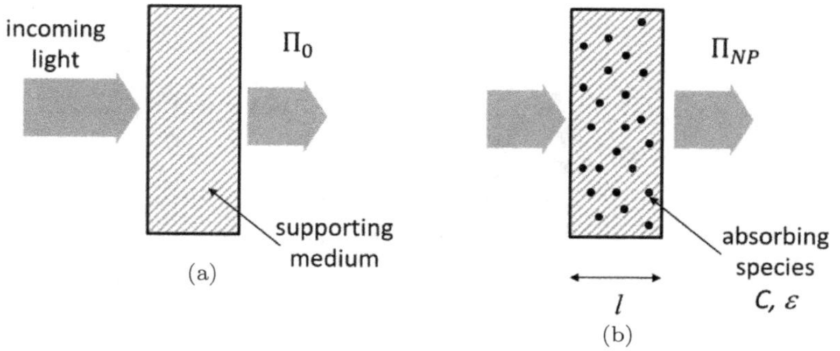

Fig. 7.9. Principle of measurement of the absorbance of absorbing species. (a) A first measurement is recorded through the supporting medium, and (b) a second measurement is carried out with the absorbing species embedded (solid) or dissolved (liquid) in the supporting medium. The absorbing species are characterized by their concentration C and their molar extinction coefficient $\varepsilon_{\mathrm{mol}}$.

7.3.2 *Demonstration of the Beer–Lambert law from the extinction cross section*

Although Beer–Lambert law is empirical, it can be derived in the case where we consider a homogeneous assembly of absorbers of volumic density n, with a known cross-section extinction σ_{ext}. n is the number of particles per unit volume expressed in m^{-3}. The light travels in the x direction and in the medium, with the absorbing species extending from $x = 0$ to $x = l$. We consider a power flow Π_0 at the entrance of the medium, and this power is $\Pi(x)$ at the position x. We consider an area S, which is illuminated by this beam, as shown in Fig. 7.10.

Let's write the balance of luminous power in the small section between x and $x + dx$. The power exiting is $S\Pi(x + dx)$, which is equal to the amount of incident power $S\Pi(x)$ minus the amount of power absorbed by the embedded species. Since there are $n \cdot S \cdot dx$ absorbers, this balance equation can be written as

$$S\Pi(x + dx) = S\Pi(x) - n \cdot S \cdot dx \cdot \sigma_{ext} \cdot \Pi(x). \tag{7.29}$$

Since $\Pi(x + dx) - \Pi(x) = (d\Pi/dx) \cdot dx$, (7.28) simplifies into

$$\frac{d\Pi}{dx} = -n\sigma_{ext}\Pi(x). \tag{7.30}$$

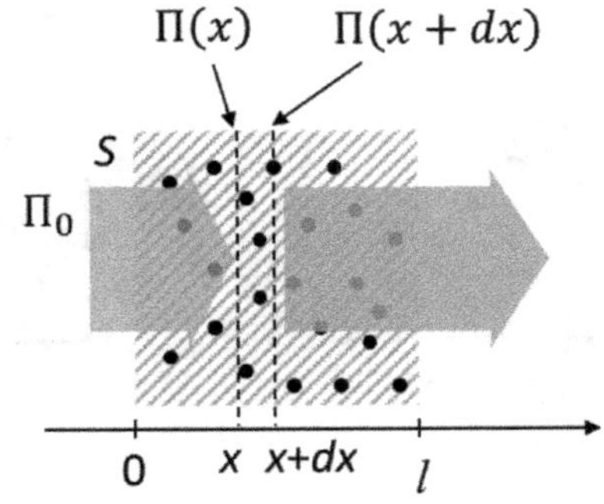

Fig. 7.10. Sketch for deriving the Beer–Lambert law. Each particle at position x absorbs an amount $\sigma_{ext}\Pi(x)$ from the power flow.

The solutions to these first-order differential equations can be written as $\Pi(x) = B\exp(-n\sigma_{ext}x)$, where B is an integration constant. Knowing that $\Pi(0) = \Pi_0$, B is determined and the solution is: $\Pi(x) = \Pi_0\exp(-n\sigma_{ext}x)$. After a length l, the light flux is

$$\Pi(l) = \Pi_0\exp(-n\sigma_{ext}l), \tag{7.31}$$

which is identical to the empirical Beer–Lambert law (7.27). The identification leads to

$$n \cdot \sigma_{ext} = \ln 10 \cdot \varepsilon_{\mathrm{mol}} \cdot C. \tag{7.32}$$

In Equation (7.32), we have to be cautious with the units since n is the number of absorbing species in m^{-3} and C is the concentration of the absorbing species in mol/L.

7.3.3 *Comparison of the calculated and experimental absorbance spectra for spherical gold nanoparticles*

From Section 7.2.4, we know that a spherical nanoparticle has an extinction cross section given by $\sigma_{ext} = \sigma_{abs} + \sigma_{scat} \approx \sigma_{abs}$, and from relation (7.24),

$$\sigma_{ext} \approx 8\pi^2R^3\frac{n_{diel}}{\lambda_0}\mathrm{Im}\left(\frac{\varepsilon - \varepsilon_{diel}}{\varepsilon + 2\varepsilon_{diel}}\right). \tag{7.33}$$

If we consider spherical gold nanoparticles of diameter 15 nm dispersed in water with a concentration $C = 2 \times 10^{-9}$ mol/L, it is easy to compute the expected absorbance spectrum and compare it to experimental data.

According to (7.32), the absorbance is given by

$$A = n\sigma_{ext}l/\ln 10 = n \cdot l \cdot \frac{8\pi^2}{\ln 10} R^3 \frac{n_{diel}}{\lambda_0} \operatorname{Im}\left(\frac{\varepsilon - \varepsilon_{diel}}{\varepsilon + 2\varepsilon_{diel}}\right). \tag{7.34}$$

If we consider that ε_{diel} is real, (7.34) slightly simplifies into a formula with no more complex values and easier to plot:

$$A = n \cdot l \cdot \frac{8\pi^2}{\ln 10} R^3 \frac{\varepsilon_{diel}^{3/2}}{\lambda_0} \frac{\operatorname{Im}(\varepsilon)}{\left|\varepsilon + 2\varepsilon_{diel}\right|^2}. \tag{7.35}$$

A plot of (7.35) along with experimental values is shown in Fig. 7.11(a).

The agreement between the experimental data and the calculation carried out with the electrostatic model is excellent since there is no adjustable parameter. There is a slight shift in the position of the LSPR since the calculation predicts 519 nm and the experiment yields 523 nm. This slight discrepancy has two main causes: the presence of surfactant molecules around the nanoparticles (citrate in this case) and the insufficiently accurate values used for the dielectric permittivity of gold. Figure 7.11(b) is a photo of the corresponding solution. Since this solution exhibits a strong absorbance

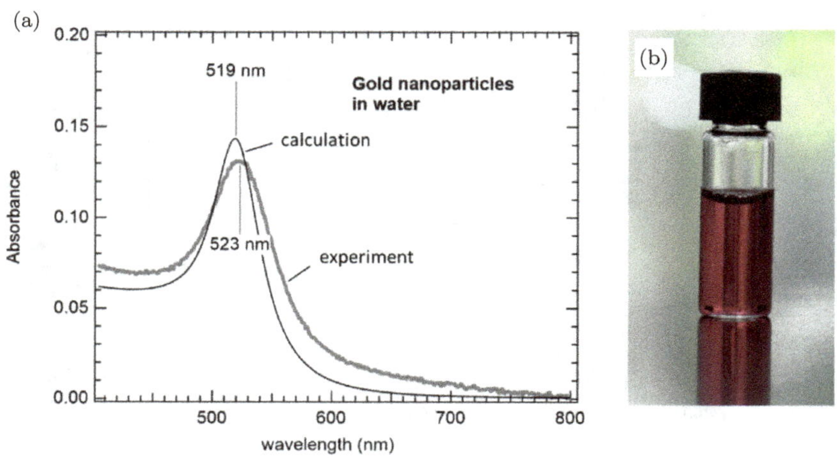

Fig. 7.11. (a) Comparison of measured and calculated extinction spectra of 15 nm gold nanoparticles prepared by the Turkevich method. The experimental spectrum is taken in a l = 1 mm thick cuvette and the volumic density of gold nanoparticles is $n = 1.76 \times 10^{18}$ m^{-3}. (b) Photo of the corresponding solution. Copyright © O. Pluchery.

at 523 nm, which is green, we expect the complementary color, which is red, traditionally called *ruby red* in the case of gold nanoparticles.

7.3.4 *Calculations of the cross sections of gold, silver, and copper nanoparticles*

Formula (7.24) allows calculating the extinction cross sections for nanoparticles of different metals. In Fig. 7.12, we plot the extinction efficiencies $Q_{ext} = \sigma_{ext}/\pi R^2$ for gold, silver, and copper nanoparticles of 15 nm diameter ($R = 7.5$ nm) in water.

Figure 7.12 shows that the extinction efficiency of silver exhibits a sharp resonance at 387 nm. This absorption of the blue wavelengths explains why silver colloidal solutions are yellow, as shown in Fig. 7.12(b) (yellow is the complementary color of blue). The extinction efficiency of silver nanoparticles is much greater than 1 and reaches 19. It means that at 387 nm, silver nanoparticles are able to "capture" photons that hit a surface 19 times larger than their individual geometrical cross section πR^2. For the gold nanoparticle, the efficiency reaches its maximum at 523 nm, with a value of

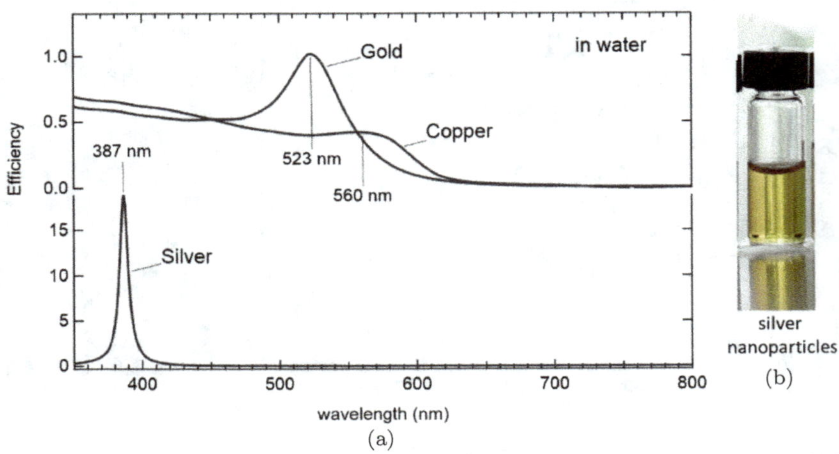

Fig. 7.12. (a) Calculated extinction efficiencies of spherical gold, copper, and silver nanoparticles of 15 nm diameter dispersed in water. An example of a suspension of silver nanoparticles is shown in (b). Copyright © O. Pluchery.

Q_{ext}= 1.02. Copper exhibits a weaker resonance at 560 nm and an efficiency of only 0.42.

We mentioned earlier that the scattering efficiencies were much smaller than the absorption efficiencies for small particles. Here, we can verify this fact. Indeed, $Q_{sca} = 1.6$ for silver, $Q_{sca} = 4 \times 10^{-3}$ for gold, and $Q_{sca} = 1 \times 10^{-3}$ for copper at their respective resonances.

7.3.5 *LSPR shift induced by the modification of the surrounding medium*

In experimental situations, the easiest measurement of the LSPR is carried out by recording the absorbance, which is directly linked to the extinction cross section, as demonstrated by relation (7.25). The medium in which the nanoparticles are embedded directly influences the resonance position. The Fröhlich condition (7.16) expresses this dependence, but as we already noted, it is very approximative and provides only a trend, which is expressed as $\lambda_{LSPR}^{Drude} = \lambda_p\sqrt{1 + 2\varepsilon_{diel}}$. It is more accurate to use the expression (7.25) to obtain a precise evaluation of the LSPR shift. Here, however, we push a bit further and anticipate what will be discussed in the next chapter. We use the software MiePlot based on the full Mie theory. It is developed and shared by Philip Laven [20]. Values are slightly different from those obtained with (7.24), and both results are given in Table 7.2.[4] It is remarkable that this simple electrostatic approximation properly predicts the fundamental trends of the LSPR shift. The results are given in Fig. 7.13 and Tables 7.1 and 7.2.

7.3.5.1 *LSPR shift and molecular sensors*

In the case of gold nanospheres in media whose optical index increases from 1 (air) to 1.66 (alumina), the plasmon resonance is redshifted from 505.6 to 541.7 nm, respectively. A similar trend is observed with silver

[4]The differences in the two calculations are due to two factors. The software MiePlot takes into account the full Mie theory and not only the dipolar contribution of the electrostatic approximation. For nanoparticles of 15 nm of diameter, the dipolar calculation is off by 1 nm. The second factor comes from a slight difference in the values of the gold dielectric permittivity.

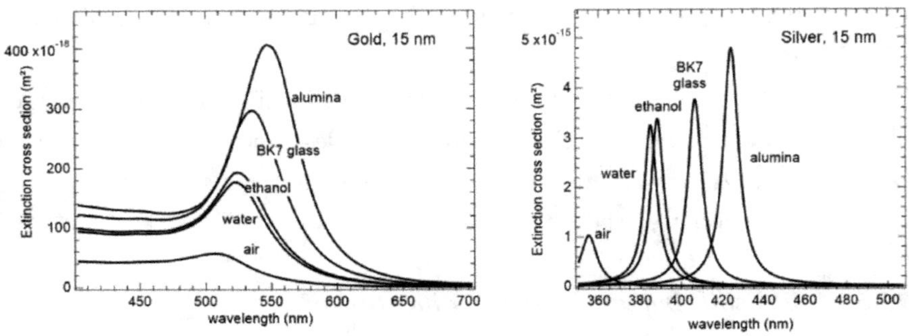

Fig. 7.13. Calculated extinction cross sections of gold and silver spherical nanoparticles in different media obtained using the MiePlot software.

Table 7.1. Values of the plasmon resonance along with the extinction cross section for 15 nm **gold nanoparticles** in various surrounding media. These values are obtained with the MiePlot software, except for λ'_{LSPR}, which is obtained with the dipolar approximation of relation (7.24).

Surrounding material	Air	Water	Ethanol	Glass (BK7)	Alumina
n	1	1.33	1.36	1.52	1.66
λ'_{LSPR} (nm)	502.9	518.3	520.1	531.6	541.7
λ_{LSPR} (nm)	505.6	522.5	524.2	543.2	555.2
$\sigma_{ext}(\lambda_{LSPR})$ (nm^2)	57	177	193	297	405

Table 7.2. Values of the plasmon resonance along with the extinction cross section for 15 nm **silver nanoparticles** in various surrounding media. These values are obtained with the MiePlot software.

Surrounding material	Air	Water	Ethanol	Glass (BK7)	Alumina
n	1	1.33	1.36	1.52	1.66
λ_{LSPR} (nm)	355.1	385.0	388.4	406.6	424.2
$\sigma_{ext}(\lambda_{LSPR})$ (nm^2)	1032	3262	3380	3775	4796

(see values in Table 7.2). This shift is a fundamental property of the LSPR: the higher the surrounding index, the higher the wavelength of the plasmon resonance. The dependence of λ_{LSPR} on the index of the surrounding material is not linear in the range $n = 1 - 1.66$, but a linear approximation can be used for small variations of n:

$$\Delta\lambda_{LSPR} = m \cdot \Delta n, \tag{7.36}$$

where m is refractive index sensitivity, which is the slope of a plot of λ_{LSPR} as a function of the refractive index (plot of the quantities given in Tables 7.1 and 7.2). For example, with 22 nm gold nanoparticles in water, $m = 69\,\text{nm/RIU}$, where RIU stands for refractive index unit [21].

In the case a local change in optical index caused by the adsorption of molecules, Van Duyne proposed the following model [22, 23]:

$$\Delta\lambda_{LSPR} = m \cdot \Delta n \left[1 - \exp\left(-\frac{2d}{l_d}\right)\right], \tag{7.37}$$

where m is the refractive index sensitivity with the same value as in Equation (7.36), Δn is the change in the refractive index induced by the presence of the adsorbate, d is the thickness of the adsorbate, and l_d is the electromagnetic decay length (see next section and also Ref. [21]).

Relation (7.37) is especially important for monitoring chemical reactions occurring at the surface of nanoparticles [24]. For example, dodecanethiol molecules with formula $(CH_2)_{12}$-SH) have a refractive index of 1.46. When they are dispersed at a low concentration (100 μM) in ethanol, they do not modify the refractive index of ethanol, which retains the value of 1.36. However, when dodecanethiol eventually binds to the nanoparticle, the local index increases to $\Delta n = 0.10$. A full coverage of the nanoparticle will typically yield a shift of λ_{LSPR} from 2 to 4 nm.

7.3.5.2 *Biosensing by monitoring the variation of absorbance*

Note also that the extinction cross section also depends on the index change, as shown in Fig. 7.13: the higher the index, the more intense the resonance peak. This is another possibility for tracking chemical functionalization of nanoparticles since the absorbance is directly linked to the extinction cross section, as given by relation (7.34). What is more delicate with this approach is to make sure that the number of probed nanoparticles remains identical during the experiment, since it is very common for the functionalization to be accompanied by destabilization and aggregation of the colloidal solution.

These sensing properties linked to the LSPR will be further explored in Chapter 9.

7.4 Optical Near Field Radiated by a Spherical Nanoparticle

7.4.1 *Introduction to near-field optics, with the example of the dipole*

In this section, we introduce the optical near field based on the electric dipole. The electric dipole can be considered as an elementary source of radiation, and the concepts that are derived can be applied far beyond the dipole case. Moreover, we have also stressed at the beginning of this chapter that the oscillating dipole was an excellent approximation of a metallic nanosphere under illumination.

We consider an electric dipole made of two charges $\pm q$ positioned along the y-axis at a distance d from each other. The dipole is oscillating at the frequency ω, and $k = 2\pi/\lambda$ is the wave vector. We suppose that the dipole is much smaller than the wavelength. In the following, the $e^{-i\omega t}$ factor is implicit. We want to know the electric field at a position M situated at a distance r from the center of the dipole and in a direction $\boldsymbol{u}$ (see Fig. 7.14). This calculation can be found in classical textbooks: see, for example, Chapter 9 in the textbook by Jackson [18], where $\boldsymbol{E}(\boldsymbol{r})$ is derived from the vector potential $\boldsymbol{A}(\boldsymbol{r}, t)$ or Chapter 2 in the book by Born and Wolf [14] based on the Hertz vectors. See also Chapter 8 in the book by Novotny [25] who starts from the Green dyadic functions.

The radiated electric field is written as

$$\boldsymbol{E}(\boldsymbol{r}) = \frac{k^2}{4\pi\varepsilon_0}\frac{e^{ikr}}{r}\left\{[\boldsymbol{p} - (\boldsymbol{p}\cdot\boldsymbol{u})\,\boldsymbol{u}] - \frac{1}{ikr}[\boldsymbol{p} - 3(\boldsymbol{p}\cdot\boldsymbol{u})\,\boldsymbol{u}] - \frac{1}{k^2r^2}[\boldsymbol{p} - 3(\boldsymbol{p}\cdot\boldsymbol{u})\,\boldsymbol{u}]\right\}. \tag{7.38}$$

$\boldsymbol{E}(\boldsymbol{r})$ is the sum of three contributions, as shown in relation in square brackets in (7.38). The first term decays as $1/r$ and is the **far-field dipolar radiation**. Usually, this is the only term that is retained after

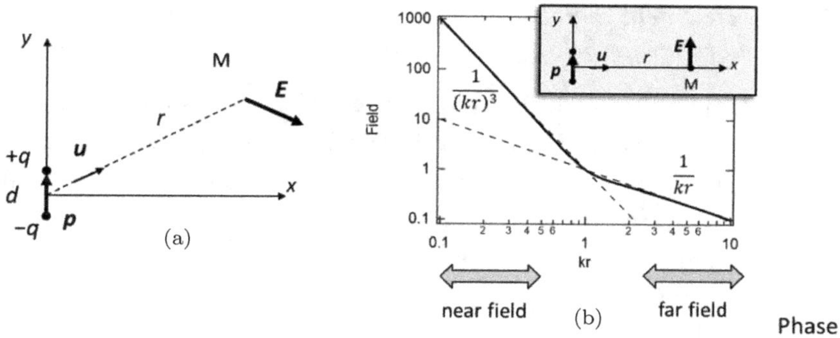

Fig. 7.14. (a) An oscillating dipole is placed at the origin, and the electric field is evaluated at a distance r from the center of the dipole. (b) Logarithmic plot of the amplitude of the electric field as a function of the distance from the center when moving on the x-axis. The plot clearly shows the two regimes of the radiation of the dipole: the near field with a decay of $1/kr^3$ and the far field with a decay of $1/kr$.

deriving at the first order and is the unique contribution to the radiated energy flux. The third term in relation (7.38) decays as $1/r^3$ and is the **near-field** contribution, which is dominant at short distances. This term is exactly the one obtained previously with the electrostatic approximation in Equation (7.13). This near-field term is essential for understanding nano-optics. There is also an intermediate term containing $1/r^2$, which is sometimes called the induction field. At what distance does the transition between the near and far fields occur? This can be answered by plotting the evolution of $\boldsymbol{E}(\boldsymbol{r})$ along the x-axis, as shown in Fig. 7.14(b). Along this axis, $\boldsymbol{E}(\boldsymbol{r})$ is written as

$$\boldsymbol{E}(r) = \frac{k^3}{4\pi\varepsilon_0} e^{ikr} \boldsymbol{p} \left\{ \frac{1}{kr} + \frac{i}{(kr)^2} - \frac{1}{(kr)^3} \right\} \quad \text{(along } x\text{)}. \tag{7.39}$$

Figure 7.14(b) clearly shows that the transition between the near and far fields occurs for $kr = 1$, which is $r = \lambda/2\pi$. As long as we are interested in phenomena at distances closer than $\lambda/2\pi$ from the dipole, we deal with the near field, and the contribution of $1/r^3$ is dominant.

In practice, for a 20 nm diameter nanosphere, in the visible range, the near-field regime is active as long as we do not go beyond 50 nm.

7.4.2 Electric field radiated by a dipole: Transition from the near field to the far field

We can also plot the intensity of the electric field on successive planes above the dipole, as shown in Fig. 7.15. This calculation was done by Carminati for a glass dipole, and it also shows nicely the transition from the near-field to the far-field structure [26]. At $z = 0.05\lambda$ above the dipole, the field is structured and presents sub-wavelength lateral variations. At $z = \lambda$, the far field is already established and exhibits spatial oscillations with the expected period of λ.

When the observation distance r is much smaller than the wavelength $\lambda (kr \to 0)$, the electric field can simply be written as

$$\boldsymbol{E}(r) = \frac{1}{4\pi\varepsilon_0 r^3}\left[3\left(\boldsymbol{p}\cdot\boldsymbol{u}\right)\boldsymbol{u} - \boldsymbol{p}\right]. \tag{7.40}$$

This expression is the same as the one obtained in the electrostatic limit in Section (7.2). This is coherent since from a mathematical point of view,

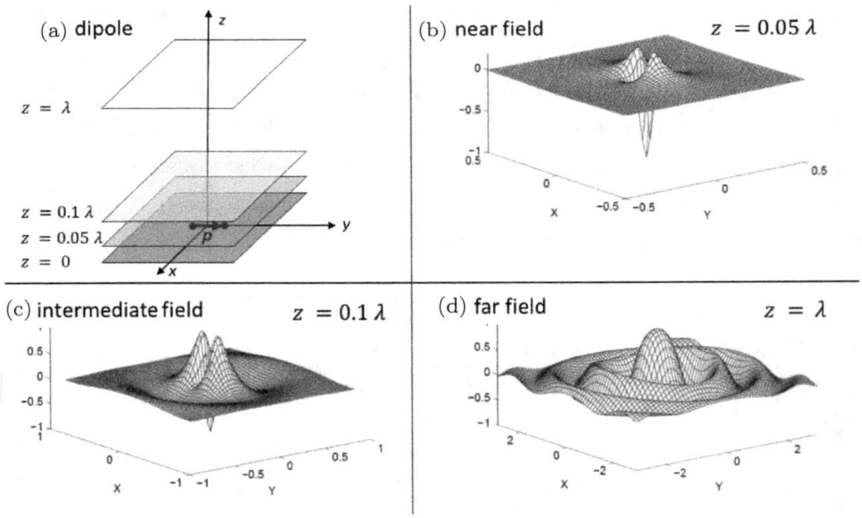

Fig. 7.15. Electric field radiated by a point dipole oriented along the y-axis at $z = 0$, as shown in (a). The intensity of the field is plotted on successive planes above the dipole at distances (b) $z = 0.05\lambda$, (c) $z = 0.1\lambda$, and (d) $z = \lambda$. The x and y axes are normalized by the wavelength $\lambda = 633$ nm.

Source: Reproduced with permission from Ref. [26]. Copyright © R. Carminati.

considering $kr \to 0$ is equivalent to considering an infinite speed of light, $c \to \infty$, with both r and ω fixed. In other words, in the near-field limit, retardation effects are neglected. This gives a more general framework for the electrostatic approximation considered in this chapter, which corresponds to cases where retardation effects play no role.

7.4.3 *Near field around a nanosphere*

Within the dipole approximation, the field immediately around a plasmonic nanoparticle can be calculated. This is relation (7.13) or (7.38), with the proper expression of the polarizability α given by (7.11). This calculation has been carried out for a 12 nm gold nanoparticle in water at the plasmon resonance (520 nm) [27]. The nanoparticle is illuminated by a plane wave with field amplitude $\boldsymbol{E_0}$ and wave vector $\boldsymbol{k}$, as shown in Fig. 7.16. The electric field undergoes a strong amplification in the vicinity of the nanoparticle along the direction of $\boldsymbol{E_0}$. The near-field enhancement factor is defined as $|E/E_0|^2$ and reaches values above 20, as shown in Fig. 7.16. This amplification region is very limited and expands no further than 4–5 nm away from the nanoparticle.

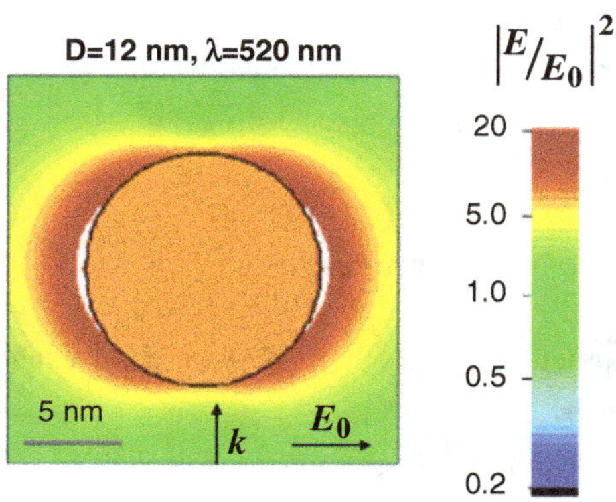

Fig. 7.16. Near-field enhancement by a gold nanoparticle in water when illuminated at its plasmon resonance ($\lambda = 520$ nm). The intensity of the electric field is enhanced by a factor of 20 in the vicinity of the nanoparticle surface.

Source: Reproduced with permission from Ref. [27]. Copyright © 2006 American Chemical Society.

7.4.4 *Concept of nano-antenna from the case of a nanosphere*

In the radio frequency domain, an antenna is a mediator or a transducer between the local currents in conductive elements and the far-field radiation. An antenna can work in both directions: as a receiver or as an emitter. Analogically, a plasmonic particle is a transducer between far-field waves and near-field waves. As discussed above, a nanoparticle is a great receiver and can efficiently collect the far field of a plane wave. It transforms into the near field within a confined region of space and special directions. The directionality of the plasmonic nano-antenna is very important, and this property is not fully used with spherical particles that are too symmetric.

A nano-antenna acts as an emitter when it is coupled to a fluorophore: It can strongly enhance the decay rate of the fluorophore, which increases its brightness. A relatively bad fluorophore can be converted into a super emitter when the plasmon resonance and the fluorophore–nanoparticle distance are well adjusted. This adjustment is crucial. If not, the coupling of the fluorophore with a nanoparticle can also *quench* the fluorescence because the metallic particle offers a non-radiative path for the excited electron of the fluorescent molecule. In this case, the decay of the electron will not go through the emission of a photon. Plasmonic nano-antenna are also used in surface-enhanced Raman scattering (SERS), and the electric field enhancement is often linked to the so called *hot spot* occurring in tailored nanostructure geometries. These aspects are discussed in Chapter 9.

7.5 Introduction to Nanothermics

Plasmonic nanoparticles exhibit a strong absorption at the LSPR frequency, which is expressed by the absorption cross section discussed in Section 7.3 above. The absorbed energy is transformed into heat. This photothermal process is complex and occurs at ultra-short timescales [5] The heat generation after absorption of light pulse develops in three stages. During the first step, photons will excite the free electrons of the nanoparticle, driving the electron cloud out of equilibrium. The electron cloud will then thermalize within 10–100 fs (1 fs $= 10^{-15}$ s). The electron temperature reaches several thousands of Kelvin. The second step starts simultaneously and corresponds to electron–phonon collisions. This transfers heat to the crystal lattice over a time duration of 1 ps (10^{-12} s), which is 100 times larger. The third step is thermal energy exchange through the nanoparticle surface toward the embedding medium, which can be air, a surface, a solution, etc. This last step is strongly dependent on the heat capacity and thermal conductance of the surrounding medium. It ranges from picoseconds to nanoseconds.

The stationary regime of such plasmonic heat nano-source is usually reached after a timescale of nanoseconds.

In the simple case, the temperature increase can be computed [28]. For example, at a distance r from a spherical nanoparticle, the temperature increase at the thermal equilibrium is simply

$$\Delta T\left(r\right)=\frac{\phi\sigma_{abs}}{4\pi\kappa r},\quad r>R, \tag{7.41}$$

where ϕ is the illuminance (W/m^2) illuminating the nanoparticle and κ is the thermal conductivity of the surrounding medium. The time needed to reach half the maximum final temperature can be estimated at $\tau = R^2/\alpha$, with α being the thermal diffusivity of the surroundings.

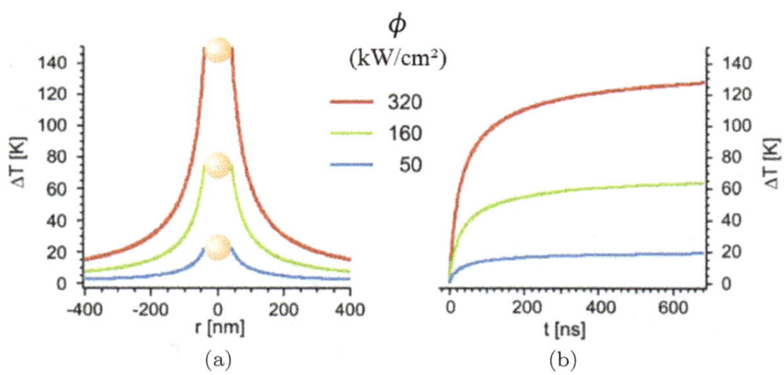

Fig. 7.17. Photothermal heating of 80 nm gold nanoparticles in water: (a) temperature increase is given as a function of the distance from the nanoparticle; (b) temperature increase as a function of time for a CW laser.
Source: Reproduced with permission from Ref. [28]. Copyright © 2011 The Royal Society of Chemistry.

For example, if we consider a 80 nm gold nanoparticle in water, Fig. 7.17 shows that with a power flux of 320 kW/cm^2 at $\lambda = 532$ nm, the temperature within 10 nm of the nanoparticles rises by more than 100 K [28]. Such a power flux is easily reached with a 10 mW laser focused on a 1 mm spot.

7.6 Conclusion

The key properties of LSPR have been discussed in the simplest case of the metallic sphere whose diameter is much smaller than the wavelength of the exciting wave. The plasmonic oscillation in the metallic nanosphere within the electrostatic approximation allows us to draw important conclusions about plasmonics despite its relative simplicity: The position of the LSPR peak can be fairly well predicted for particles smaller than 20 nm. Its dependence on the surrounding optical index is also captured. The three optical cross sections (absorption, scattering, and extinction) were also expressed as functions of the polarizability α. However, this approximation does not predict any peak position dependence on the nanoparticle size. Moreover, the absorption spectra of nanoparticles with shapes other than a sphere cannot be predicted. This will be discussed in the following chapter.

References

[1] Brill R. H. 1965. *The Chemistry of the Lycurgus Cup* (Brussels, Belgium: National Institute of Glass), pp. 223.1–223.12.
[2] Barber D. J. and Freestone I. C. 1990. An investigation of the origin of the colour of the Lycurgus cup by analytical transmission electron microscopy. *Archaeometry* **32**, 33–45.
[3] Freestone I., Meeks N., Sax M. and Higgitt C. 2007. The Lycurgus Cup — A Roman nanotechnology. *Gold Bulletin* **40**, 270–277.
[4] Hornyak G. L., Patrissi C. J., Oberhauser E. B., Martin C. R., Valmalette J. C., Lemaire L., Dutta J. and Hofmann H. 1997. Effective medium theory characterization of Au/Ag nanoalloy-porous alumina composites. *Nanostructured Materials* **9**, 571–574.
[5] Louis C. and Pluchery O. (eds.) 2017. *Gold Nanoparticles for Physics, Chemistry and Biology (2nd Ed.)* (London: World Scientific).
[6] Faraday M. 1857. Lecture — Experimental relations of gold (and other metals) to light. *Philosophical Transactions* **147**, 145–181.
[7] Carbert J. 1980. Gold-based enamel colours. *Gold Bulletin* **13**, 144–150.

[8] Mie G. 1908. Beiträge zur Optik trüber Medien, speziell kolloidaler Metallösungen. *Annalen der Physik* **25**, 377–445.

[9] Pines D. 1956. Collective energy losses in solids. *Reviews of Modern Physics* **28**, 184–198.

[10] Haynes C. L., Schatz G. C. and Weiss P. S. 2020. Virtual issue in honor of Prof. Richard Van Duyne (1945–2019). *Analytical Chemistry* **92**, 4165–4166.

[11] Kreibig U. and Vollmer M. 1995. *Optical Properties of Metal Clusters* (Berlin: Springer Verlag).

[12] Kreibig U. and Genzel L. 1985. Optical-absorption of small metallic particles. *Surface Science* **156**, 678–700.

[13] Tzarouchis D. and Sihvola A. 2018. Light scattering by a dielectric sphere: Perspectives on the Mie resonances. *Applied Sciences* **8**, 184.

[14] Born M. and Wolf E. 1999. *Principles of Optics: Electromagnetic Theory of Propagation, Interference and Diffraction of Light* (Cambridge: Cambridge University Press).

[15] Bohren C. F. and Huffman D. R. 1998. *Absorption and Scattering of Light by Small Particles* (New York: Wiley-Interscience).

[16] Novotny L. and Hecht B. 2006. *Principles of Nano-Optics* (Cambridge: Cambridge University Press).

[17] Griffiths D. J. 2017. *Introduction to Electrodynamics* (New York: Cambridge University Press).

[18] Jackson J. D. 1998. *Classical Electrodynamics. 3rd Ed* (New York: Wiley).

[19] Maier S. A. 2007. *Plasmonics: Fundamentals and Applications* (New York: Springer New York).

[20] Laven P. 2019. MiePlot 4.3.05. Calculation of Mie scattering from a sphere. http://www.philiplaven.com/mieplot.htm.

[21] Dileseigres A. S., Prado Y. and Pluchery O. 2022. How to use localized surface plasmon for monitoring the adsorption of thiol molecules on gold nanoparticles? *Nanomaterials* **12**, 292.

[22] Willets K. A. and Van Duyne R. P. 2007. Localized surface plasmon resonance spectroscopy and sensing. *Annual Review of Physical Chemistry* **58**, 267–297.

[23] Unser S., Bruzas I., He J. and Sagle L. 2015. Localized surface plasmon resonance biosensing: Current challenges and approaches. *Sensors* **15**, 15684–15716.

[24] Messersmith R. E., Nusz G. J. and Reed S. M. 2013. Using the localized surface plasmon resonance of gold nanoparticles to monitor lipid membrane assembly and protein binding. *The Journal of Physical Chemistry. C, Nanomaterials and Interfaces* **117**, 26725–26733.

[25] Novotny L. and Hecht B. 2012. *Principles of Nano-Optics (2nd Ed.)* (Cambridge: Cambridge University Press).

[26] Carminati R. 2005. *Optics Beyond the Diffraction Limit.* https://www.institut-langevin.espci.fr/download.

[27] Rodríguez-Fernández J., Pérez-Juste J., García de Abajo F. J. and Liz-Marzán L. M. 2006. Seeded growth of submicron Au colloids with quadrupole plasmon resonance modes. *Langmuir* **22**, 7007–7010.

[28] Coronado E. A., Encina E. R. and Stefani F. D. 2011. Optical properties of metallic nanoparticles: Manipulating light, heat and forces at the nanoscale. *Nanoscale* **3**, 4042.
[29] Liu X., Atwater M., Wang J. and Huo Q. 2007. Extinction coefficient of gold nanoparticles with different sizes and different capping ligands. *Colloids and Surfaces B: Biointerfaces* **58**, 3–7.

Exercises

(1) Consider a metallic sphere under the excitation of an electric field E_0, as in Section 7.2.2. The expression of the electric potential is given by Equation (7.12). Calculate the electric field generated by the sphere and derive Equation (7.13).

(2) Use the Fröhlich condition to determine λ_{LSPR} of metallic nanoparticles in air and in water in the case where the metals are gold and silver with their dielectric permittivities given in the following figure.

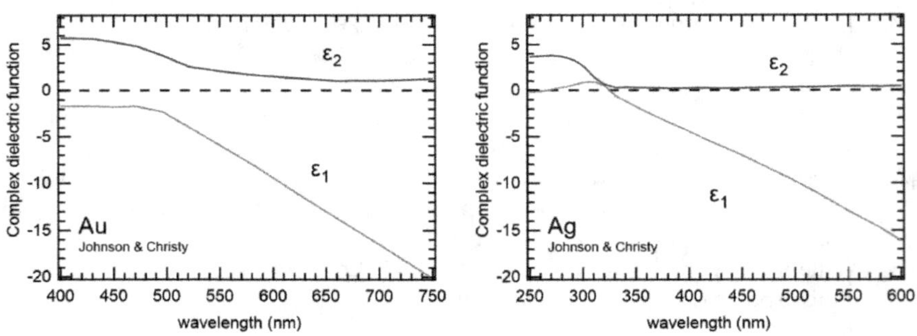

Complex dielectric functions of gold and silver measured by Johnson and Christy (1972). *Source*: https://refractiveindex.info/.

(3) We now consider aluminum that is characterized by its Drude permittivity with good accuracy. Predict at what wavelength the LSPR of spherical aluminum particles is expected when they are embedded in glass ($n = 1.52$).

(4) Consider a punctual light source S. Demonstrate that the power measured by a detector of area A positioned at a distance r from S decreases as $1/r^2$.

(5) **Beer–Lambert law:** In their article published in 2007, Liu and colleagues measured the molar extinction ε_{mol} of gold nanoparticles stabilized by citrate of various diameters [29]. They are all suspended in an aqueous solution. The values of ε_{mol} for three diameters are given in the following table.

Diameter (nm)	8.6	20.6	34.5
ε $(\text{M}^{-1}\text{cm}^{-1})$	5.1×10^7	8.8×10^8	6.1×10^9

(a) Deduce the values of the extinction cross sections for these three diameters.

(b) Use the software MiePlot and calculate the three extinction cross sections [20]. Compare the values with those found in (a).

(6) For a gold nanosphere of radius R in water ($n = 1.33$), compare the scattering cross sections σ_{sca} and the absorption cross section σ_{abs} obtained within the electrostatic approximation. Plot them as a function of R. Is there a value for R where σ_{sca} grows larger than σ_{abs}? If yes, give the value and comment. Use $\varepsilon_{Au} = -3.9 + 2.6i$ (value at 520 nm).

(7) Same question as (6) but for silver. Use $\varepsilon_{Ag} = -3.16 + 0.19i$ (value at 385 nm).

(8) **LSPR sensitivity factor:** Consider 15 nm gold nanoparticles in a water–ethanol mixture.

(a) Use the values given in Table 7.1 and Figure 7.13 and calculate the sensitivity factor m that links the LSPR shift with the optical index variation.

(b) Use the software MiePlot and calculate λ_{LSPR} when the optical index of the surrounding medium takes the values from 1.30 to 1.60 (increment of 0.05). Plot λ_{LSPR} as a function of n. Is it a linear relationship?

(c) Deduce the value of the sensitivity factor m if the nanoparticles are in water or in an organic solvent of index 1.5.

(9) **Electromagnetic decay length:**

(a) Express the formula of the evolution of the electric field generated by a plasmonic nanoparticle of radius R within the dipolar approximation. What is the dependence of the field on r?

(b) Plot this field decay as a function of the position x away from the dipole (see Fig. 7.14). Fit this function with an exponential.

(c) Show that the field decays almost as $\exp(-(r-R)/l_D)$, where l_D is the electromagnetic decay length. Show that $l_D = 0.45R$. *Here, we ask for a "numerical demonstration" and not an analytical derivation.*

(10) **Near field radiated by a metallic sphere:** Let's consider a light wave of wavelength λ_0 in vacuum and a field amplitude of $E_0\boldsymbol{u}_x$. A spherical gold nanoparticle of radius R, much smaller than λ_0, is placed in this incoming field. Give the expression of the electric field radiated by this nanoparticle within the electrostatic approximation. Use the cylindrical coordinates oriented along the x direction.

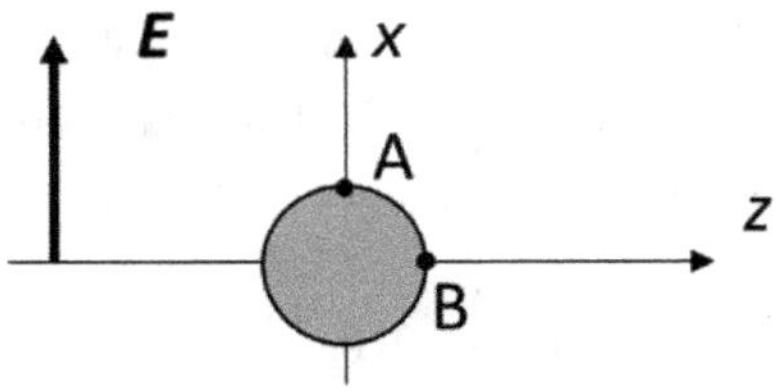

(a) Calculate the expression of the electric field E_{ext}(A) at point A. Calculate the numerical values by expressing E_{ext}(A) as a function of E_0.

(b) Same question as (a) but for point B.

(c) Conclude on the nano-antenna effect around a nanoparticle.

8

Localized Plasmon Resonance of Complex Nanostructures

If plasmonic nanoparticles are larger than $\lambda/2\pi$ or if they are not spherical, the electrostatic approximation does not hold when calculating the optical spectra. In these cases, the **Mie theory** provides significant improvements. The Mie theory will be explained but not in full detail, and its results will be discussed for **large spheres**, **ellipsoids**, and **spherical core–shell** nanoparticles. We also present some of the popular computational approaches, such as **discrete dipole approximation (DDA)** used for predicting the scattering, absorption, and extinction cross sections of complex nanostructures. Finally, we explain what ***hot spots*** are in plasmonics and **dark plasmon modes**.

8.1 Introduction

In the previous chapter, we have focused on the localized plasmon resonance generated by spherical nanoparticles with a diameter much smaller than the optical wavelength. This approach allows using the electrostatic approximation and deriving the principle properties of the LSPR. It helps in establishing the key concepts of the LSPR, by which everything can be understood and calculated analytically. In this chapter, the complexity is raised by one step: We consider non-spherical nanoparticles, and we investigate how the near field of two nanoparticles interferes and affects the LSPR response. We also discuss what happens in assemblies of nanoparticles or nanoparticles supported by various surfaces. Even if this leads to non-analytical models and more sophisticated approaches, the qualitative trends can be easily understood and are discussed.

8.2 Ellipsoidal, Core–Shell Nanoparticles: Analytical Approaches to the LSPR Beyond the Electrostatic Model

In many cases, the goal is to assess the optical response of nanoparticles in the far field. This is the case when chemists want to relate the optical extinction spectra of a suspension of nanorods with the shape they observe by transmission electronic microscopy. It is important in this case to be able to evaluate the extinction cross section of a nanorod. In this section, we calculate the extinction, absorption, and scattering cross sections of nanoparticles of various shapes. As mentioned in the previous chapter and detailed by Bohren and Huffman in their book [1], these cross sections are directly obtained from the polarizability α by

$$\sigma_{abs} = k\,\mathrm{Im}(\alpha), \quad \sigma_{sca} = \frac{k^4}{6\pi}\,|\alpha|^2, \text{ and } \sigma_{ext} = \sigma_{abs} + \sigma_{abs}. \qquad (8.1)$$

Therefore, we mostly focus here on discussing the evolution of α.

8.2.1 *The case of large spherical nanoparticles: Multipolar resonances*

8.2.1.1 *Experimental evidence of the dependence of LSPR on particle diameter*

The electrostatic model yields an expression for the polarizability α where its dependence on the radius of the particle is only a multiplication by the factor R^3 (see Equation (7.11) in the previous chapter). Therefore, this model does not predict any dependence of the plasmon resonance with the wavelength. However, experiments demonstrate the opposite. In Fig. 8.1 taken from Ref. [2], the LSPR maximum is plotted as a function of the nanoparticle diameter in the case of spherical gold nanoparticles in water. This maximum is measured at 520 nm for 20 nm nanoparticles and undergoes a redshift of up to 570 nm for a diameter of 100 nm. This demonstrates that even if the electrostatic model is accurate for gold nanoparticles of

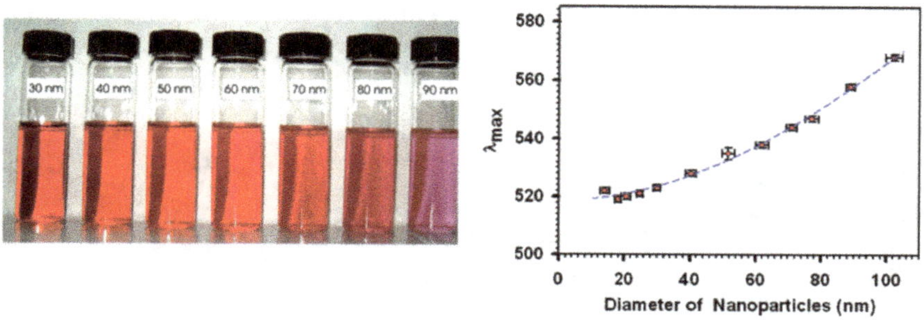

Fig. 8.1. Series of suspensions of monodispersed spherical AuNPs with diameters between 20 and 100 nm. The LSPR shifts from 520 to 570 nm, especially for diameters higher than 40 nm. This shift is a clear demonstration of the quadrupolar effects arising for diameters above ~30 nm. The color of this suspension is systematically red.

Source: Reproduced with permission from Ref. [2]. Copyright © 2007 American Chemical Society.

15 nm diameter, as discussed in the previous chapter (see Fig. 7.11), it fails for larger diameters.

This dependence of the LSPR on the wavelength is the consequence of the fact that R is no longer negligible against the wavelength. The electric field on the two opposite ends of the nanoparticle is not strictly uniform anymore, and the induced dipole turns into an induced quadrupole or even higher orders poles. A graphical illustration was given in Fig. 7.3. The resulting polarizability will not be given explicitly.

8.2.1.2 *Rayleigh vs. Mie scattering*

Consider a plane wave traveling in a transparent medium (dielectric medium) of optical index n_{diel} at frequency ω irradiating a spherical particle characterized by its radius R, complex index $\tilde{n}$, and dielectric permittivity ε. The notations are given in Fig. 8.2. $\beta = 2\pi R/\lambda$ is the size parameter and is dimensionless. λ is the wavelength in the transparent medium. If $\beta \ll 1$, the radiation is properly described by the oscillating dipole model, and Rayleigh scattering can be used. The consequences for the LSPR have been discussed in the previous chapter. For example, for metallic nanoparticles of 20 nm in water, at $\lambda_0 = 500\,\text{nm}$, $\beta = 0.17$. It is normal for Rayleigh scattering to become inaccurate for larger size particles, as shown in Fig. 8.1.

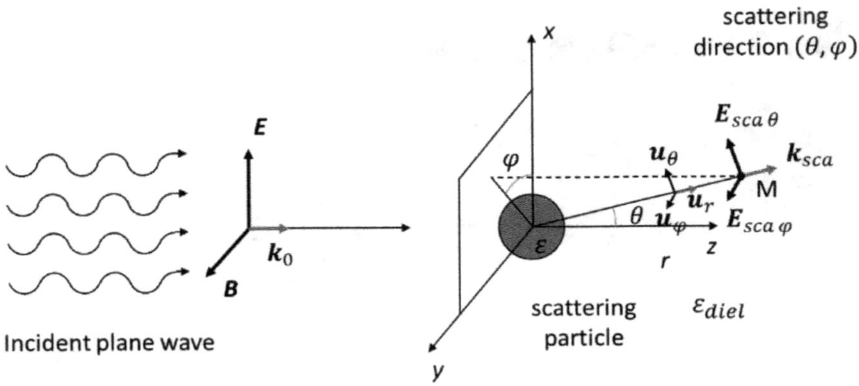

Fig. 8.2. Geometry for the Mie scattering calculation: A spherical particle (dielectric permittivity ε) is placed in an incident plane wave. The scattered wave is examined in the direction $\boldsymbol{k}_{sca}$, and the resulting field is decomposed in the spherical coordinates represented by (r, θ, φ). The electric field is decomposed into $\boldsymbol{E}_{\mathrm{sca}\theta}$ and $\boldsymbol{E}_{\mathrm{sca}\varphi}$.

When $\beta \ll 1$ is no longer true, Mie theory comes into play. Mie theory is the exact solution to the problem of a plane wave scattered by a spherical particle of any diameter and of any material. This solution is derived using the spherical coordinates and is based on vectorial calculations and, above all, on Bessel functions. It was developed by Gustav Mie in 1908 [3]. A fairly detailed summary can be found in the textbook by Bohren and Huffman explained over 48 pages in Chapter 4 [1]. We only reproduce here the principle of this derivation (see also Ref. [4] for a complete set of equations).

The incident plane wave shown in Fig. 8.2 is simply $\boldsymbol{E} = E_0 \boldsymbol{u}_x \exp i(k_0 z - \omega t)$ in the cartesian coordinates. The first task consists in expressing this field in the spherical coordinates centered on the scattering particle. At a given point of space $\mathrm{M}(x, y, z)$, the incident field is expressed with the three unit vectors $\boldsymbol{u}_r$, $\boldsymbol{u}_\theta$, and $\boldsymbol{u}_\varphi$ and the three coordinates (r, θ, φ). This change in the system of coordinates is achieved by using the spherical Bessel function, and it results in the expansion (infinite sum) over the spherical harmonics. The decomposition of one given harmonic is called a *mode*. Quoting Bohren and Huffmann: "*The desired expansion of a plane wave in spherical harmonics was not achieved without difficulty. This is undoubtedly the result of the unwillingness of a plane wave to wear a guise in which it feels uncomfortable; this is somewhat like trying to force square peg into a round hole*" (Ref. [1], p. 92).

The next task is much more simple. The electromagnetic field scattered in a given direction $\boldsymbol{k}_{\mathrm{sca}}$ will have the spherical symmetry of the scattering sphere. Since the spherical harmonics are orthogonal and form a basis, it is sufficient to treat each harmonic individually. Applying the boundary conditions at the sphere surface allows to calculate the scattered field for each mode. The scattered electric field will be expressed with the two unit vectors $\boldsymbol{u}_\theta$ and $\boldsymbol{u}_\varphi$, and the components $\boldsymbol{E}_{sca\theta}$ and $\boldsymbol{E}_{\mathrm{sca}\varphi}$ are an infinite sum over the successive modes. This allows obtaining the decomposition of the scattering cross section for the intensity as a function of the scattering coefficients a_n and b_n:

$$\sigma_{sca} = \frac{\lambda^2}{2\pi} \sum_{n=1}^{\infty} (2n+1)(|a_n|^2 + |b_n|^2). \tag{8.2}$$

The extinction cross section is

$$\sigma_{ext} = \frac{\lambda^2}{2\pi} \sum_{n=1}^{\infty} (2n+1)\, Re(a_n + b_n). \tag{8.3}$$

The scattering coefficients a_n and b_n are functions of $2\pi a/\lambda$ and expressed as combinations of the Ricatti–Bessel functions of the nth order, which we will not develop herein. For each n, there are two modes: one corresponding to coefficient a_n, where there is no radial magnetic field component. It is called the *transverse magnetic mode,* or the electric-type wave (E-wave). The second mode corresponds to the b_n factors and does not have any radial electric field component. It is called the *transverse electric mode*, or the magnetic-type wave (H-wave). The complete mathematical formula can be found in Chapter 4 of the book by Bohren and Huffman or in the short manuscript by Pr. David H. Hahn [4]. In his original 1908 publication, Mie proposed a representation of these modes, where the electric field lines of the transverse components were plotted on a sphere centered on the particles and fairly away from it. Figure 8.3 shows the first two modes (with coefficients a_1, a_2, b_1, and b_2).

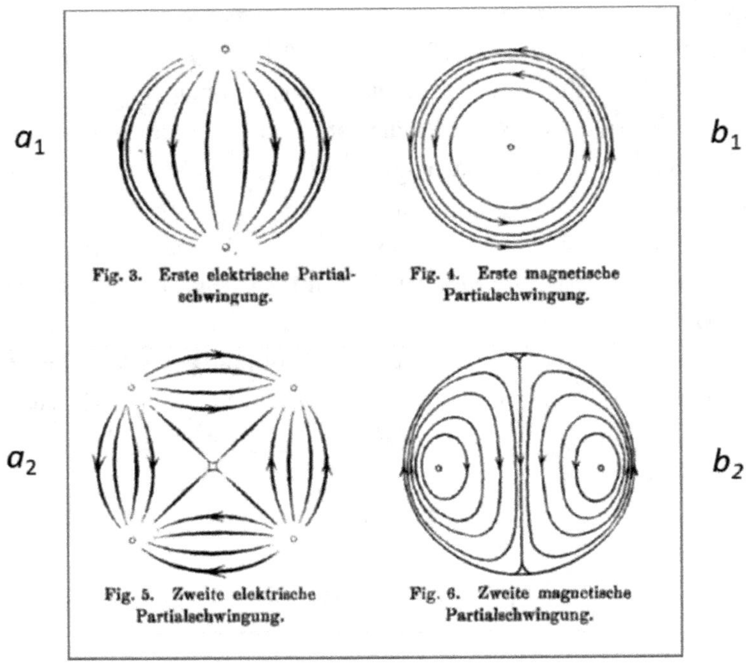

Fig. 8.3. Representation of the dipolar and quadrupolar modes of the field scattered by a sphere based on the Mie theory and taken from the original article of 1908. Only the first two modes are shown, where a_1, a_2, b_1, and b_2 are the weighting coefficients of the corresponding spherical harmonics. The lines represent only the transverse components of the electric field (radial components are difficult to plot simultaneously in a 2D sketch).

The a_1 component corresponds to the dipolar contribution of the Rayleigh approximation. In Fig. 8.3, the a_1 mode clearly shows the dipolar structure of the electric field. The charges are also polarized in a dipolar way with the negative charge on the north pole of the sphere and the positive charge on the south pole.

8.2.1.3 *How to implement the Mie calculation?*

The Mie calculation involves a series of an infinite number of terms.

An example of how to implement this calculation is given in Appendix I of the book by Bohren and Huffmann [1]. The program is written in FORTRAN.

MiePlot, a ready-to-use Windows program, has been released by Philip Laven and can be downloaded from his website [5]. This software is rather complete, allowing the calculations of the optical cross sections of spheres of silver, gold, or copper and various surrounding media, such as water, air, or a user-defined optical index. This software will be repeatedly referred to in this book, including the exercises.

Another software is MieLab developed by O. Peña-Rodríguez in 2011 (http://scattering.sourceforge.net/downloads.php). But this is not maintained.

A python module has also been developed by Benjamin Sumlin and can be downloaded from https://pypi.org/project/PyMieScatt/.

There are also several websites with an online Mie scattering calculator: www.bichromatics.com or https://nanocomposix.com/pages/mie-theory-calculator [6, 7].

8.2.1.4 *Retardation effects and quadrupolar modes*

Mie theory, which was briefly presented above, is an exact model used for calculating the scattering and extinction by spherical particles of any material and of any diameter. However, the physics is somewhat hidden behind the heavy mathematical approach. In this section, we decompose the three main complexity levels in describing the scattering and extinction cross sections of a metallic sphere: radiation of a point dipole, retardation effects, and inclusion of the quadrupolar and higher order modes.

For a given particle of radius R, when the size parameter $\beta = 2\pi R/\lambda$ is small, the particle behaves like a point dipole that oscillates at a frequency ω (case of $\beta < 0.2$). This case corresponds to the electrostatic approximation that was largely discussed in the previous chapter and is depicted in Fig. 8.4(a). Within this approximation and for an observation point M far from the dipole, we had the expression (7.37), to which the oscillating factor $e^{i\omega t}$ has been added. This factor is always dropped in the electrostatic approach:

$$\boldsymbol{E}(\boldsymbol{r}) = \frac{k^2}{4\pi\varepsilon_0} \left\{\boldsymbol{p} - (\boldsymbol{p}.\boldsymbol{u})\boldsymbol{u}\right\} \frac{e^{i(kr-\omega t)}}{r}. \tag{8.4}$$

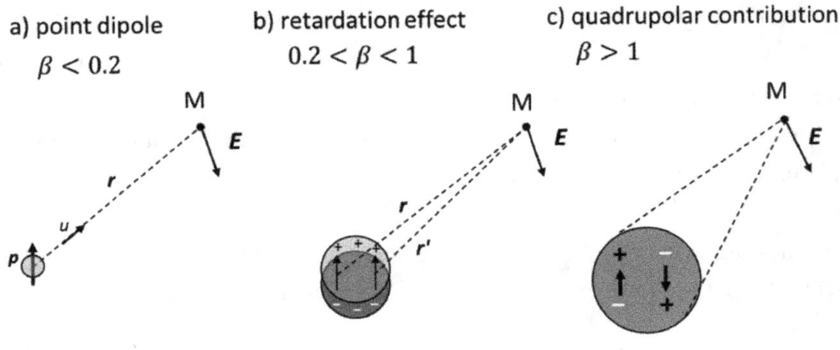

Fig. 8.4. Sketch of the various contributions at play when the particle size (radius R) increases. The size parameter is $\beta = 2\pi R/\lambda$. For small β (case a), the radiation is purely dipolar. For intermediate values of β ($\beta \in [0.2; 1]$), the retardation effects must be included in the calculations (case b). Finally, for β larger than 1, the multipolar effects become important (case c).

When β grows above 0.2, retardation effects start becoming significant. The particle can be seen as an ensemble of dipoles, oscillating in phase at frequency ω. At the observation point M, all these dipolar contributions have to be added up. Figure 8.4(b) shows two of these elemental dipoles at distances r and r' from M. The propagation term is written as $e^{i(kr-\omega t)} = e^{ik(r-t/c)}$. Since $r \neq r'$, the two contributions might be dephased at point M, and the integral over all the dipoles produces destructive additions. The propagation factor needs to be taken into account when integrating over all the dipolar contributions. This is termed the *retardation effect*, and after the integration has been carried out, it results in a redshift of the plasmon resonance.

When β grows further ($\beta > 1$), the particle no longer behaves like a dipole since the excitation electric field cannot be considered homogeneous over the particle. This is shown in Fig. 8.4(c), where a sketch of a quadrupolar charge distribution is shown. These effects are summarized in Table 8.1.

8.2.1.5 *Experimental evidence of the retardation and quadrupolar contributions*

This behavior is well illustrated in a combined experimental–theoretical short study by Liz-Marzan *et al.* [8]. They prepared several sets of gold

Table 8.1. Schematic presentation of the effects that need to be taken into account in the calculation of the scattered and absorbed light when the size parameter increases. The corresponding nanoparticle diameter is given for a nanoparticle in water ($\lambda = \lambda_0/n_{water}$ and for a 500 nm wavelength.

Size parameter $\beta = 2\pi R/\lambda$	-------- 0.2	-------- 1	--------
Nanoparticle diameter D	20 nm	120 nm	
LSPR calculation method	Point dipole radiation	Include retardation effects	Include multipolar effects

nanospheres in an aqueous solution using the seeded growth method. The seeds are CTAB-stabilized nanoparticles of 12 nm diameter. The seeds were purified to eliminate the non-spherical particles (rods, hexagonal platelets, and nano-triangles) obtained as by-products. Starting from the 12 nm seeds, further growth steps were carried out by reducing $HAuCl_4$ with ascorbic acid in the presence of CTAB at 35°C. TEM images of the resulting nanoparticles are shown in Fig. 8.5. The extinction spectra of these six sets of nanoparticles were recorded with a conventional optical spectrophotometer and exhibit a remarkable agreement with the calculated spectra that were obtained using the Mie theory (read Liz-Marzan's work [8] for details). Figure 8.5(f) shows the extinction cross sections of these sets of nanoparticles. In the calculation, the authors could choose the number of modes they want to include in the expansion: keeping only the first term ($l = 1$) corresponds to the dipolar radiation. Including the second one is the quadrupolar ($l = 2$), then comes the octupolar ($l = 3$), and so on. Two observations need to be made: First, when the size increases, the LSPR shifts to a higher wavelength (see Table 8.2 for the values). This shift is ascribed to the retardation effects and is accompanied by a clear resonance broadening. Second, when the diameter becomes larger than 138 nm, a shoulder is visible on the spectrum that grows into a clear peak at 550 nm for the 181 nm diameter. This peak is due to the quadrupolar mode, as demonstrated by the calculations: Including the quadrupolar mode in the Mie calculations allows reproducing this peak at 545 nm.

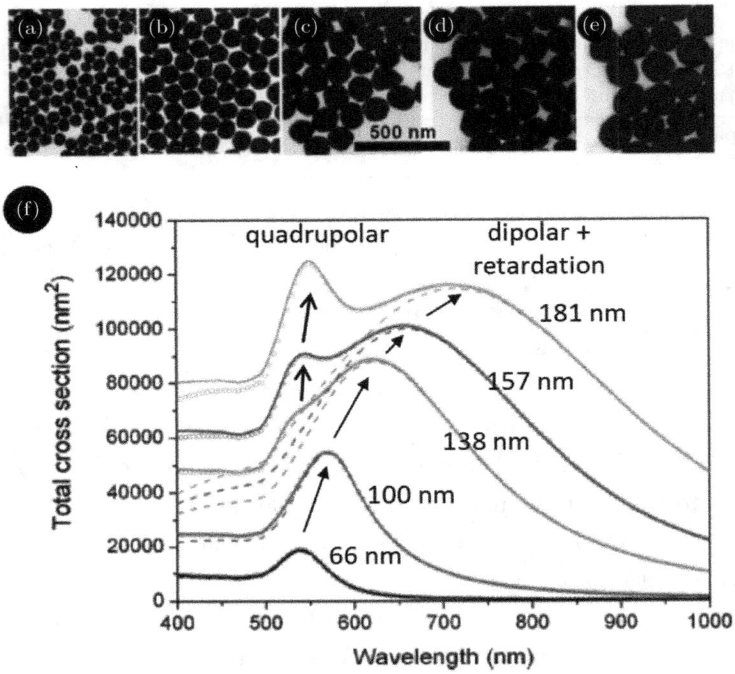

Fig. 8.5. Representative TEM images of the gold nanospheres prepared by Liz-Marzan *et al.*, with diameters ranging from 12 nm (not shown) to 181 nm. From (a) to (e), the diameters are respectively: 66, 100, 139, 157, and 181 nm. In (f), the calculated absorption cross sections of these gold nanoparticles in water are shown. Full line plots show the Mie calculation including the multipolar effects until $l = 12$. Dashed lines correspond to the sole dipolar contribution (including the retardation effects), and symbols correspond to the dipolar + quadrupolar contributions (shown only for 157 and 181 nm diameters).

Source: Reproduced with permission from Ref. [8].

Table 8.2. Evolution of the LSPR peaks for gold spherical nanoparticles in water when their diameter is increased from 12 to 181 nm. The size parameter $\beta = 2\pi R/\lambda$ serves as a control for indicating which effects are contributing. The quadrupolar effects become visible when $\beta > 1$, first as a shoulder, then as a real peak.

Diameter (nm)	12	66	100	138	157	181
β	0.10	0.55	0.83	1.16	1.31	1.51
LSPR dipolar (nm)	523	540	570	618	655	707
LSPR quadrupolar (nm)				sh.	545	551

8.2.1.6 *Scattering cross sections of large spherical nanoparticles*

Let's now compare the scattering and extinction cross sections of spherical nanoparticles. Extinction cross section is responsible for the color of the nanoparticle suspension when observed in transparency, whereas scattering cross sections controls the spectrum of the diffused light.

For subwavelength nanoparticles, the scattering efficiency is largely smaller than that of the extinction. Therefore, nanoparticles of 10–30 nm are majorly absorbers at the plasmon frequency (see Fig. 8.6(a)). The appearance of the suspension will be dominated by this absorption, and the color will be the complementary color of the plasmon maximum. When the size increases, the scattering is no longer negligible, as shown for 50 nm silver nanoparticles in vacuum in Fig. 8.6(b).

This trend was already pointed out in Chapter 7 and is already taken into account by the electrostatic approach summarized in relations (7.22) and (7.23). See also Exercises 6 and 7 in Chapter 7.

8.2.2 *The case of small spherical nanoparticles*

When the diameter of the nanoparticle decreases below ~15 nm, two effects will modify the LSPR, compared to the simple dipolar approach discussed previously. The first one is due to the confinement of the electrons in the nanoparticle that leads to changes in the dielectric permittivity. The second effect is the modification of the band structure of the metal at small sizes

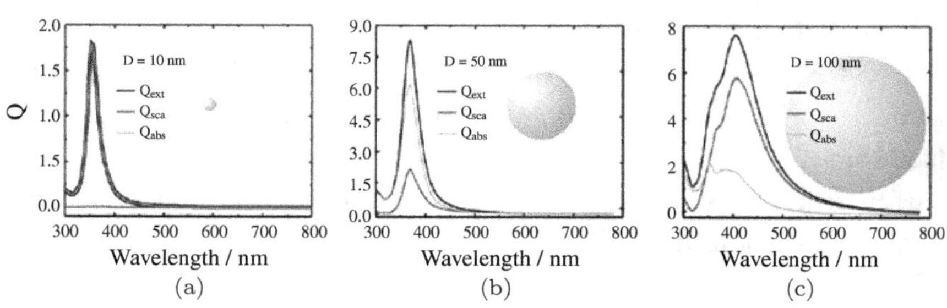

Fig. 8.6. Extinction, absorption, and scattering efficiencies (Q) of spherical silver nanoparticles (in vacuum) with increasing diameter (D).

Source: Reproduced with permission from Ref. [9]. Copyright © 2011 The Royal Society of Chemistry.

that will couple the plasmonic modes with the interband absorption and eventually annihilate the resonance.

8.2.2.1 *Modified dielectric function in small nanoparticles*

Electrons in a bulk metal have a mean free path l_{bulk} of the order of magnitude of 30 nm. The mean free path is inversely proportional to the relaxation frequency γ, which is a parameter of the Drude model for calculating the dielectric permittivity ε (see Equation (2.6) in Chapter 2 and Equation (8.5)). Within the classical theory of electricity, γ results from various contributions related to the interactions of the free electrons with phonons, impurities, and lattice imperfections. When the nanoparticle size decreases, a supplemental contribution is added to γ that takes into account the bouncing of the electrons on the edges of the nanoparticle. This contribution depends on the particle radius R. Therefore, to solve the discrepancy between measured and calculated extinction cross sections, the values of ε should be revisited. The complex values of ε used for describing the bulk metal are not suitable anymore. This is clear in Fig. 8.6, where the absorbance spectrum of a suspension of 13 nm gold nanoparticles in water is presented. The intensity of the experimental spectrum is less than half of the calculated one, and it exhibits a stronger broadening of the LSPR [10].

Since plasmonics deals most of all with the free electrons of metals, we use the Drude model for their description. It is characterized by a plasma frequency ω_p and a relaxation frequency γ, as explained in Chapter 2. If we assume that only the electrons near the Fermi surface (with Fermi velocity v_F) contribute to the conductivity, the mean free path l_{bulk} is related to γ through $\gamma = v_F/l_{bulk}$. For gold, l_{bulk} amounts to 37.2 nm. Since the sizes of nanoparticles are markedly smaller than this value, the effective mean free path is reduced considerably (see Table 8.3). Subsequently, the relaxation frequency is modified into

$$\gamma_{NP}(R) = \gamma + A\frac{v_F}{R}, \tag{8.5}$$

where $A = 4/3$, as given by theory. This diameter dependent value of γ leads to a modification of the dielectric permittivity: $\varepsilon_{NP}(R) = \varepsilon + \Delta\varepsilon(R)$.

Table 8.3. Values of the Fermi velocity v_F, relaxation frequency γ, and mean free path of electrons in bulk metallic crystal l_{bulk} obtained at room temperature.

Element		Crystal structure	v_F (10^5 m/s)	γ ($10^{13}\,s^{-1}$) @RT	l_{bulk} (nm) RT
Silver	Ag	fcc	14.48	2.72	53
Copper	Cu	fcc	11.09	2.77	40
Gold	Au	fcc	13.82	3.66	38
Aluminum	Al	fcc	15.99	8.47	19
Tungsten	W	bcc	9.71	6.25	16

Source: Reproduced from Ref. [11].

In Chapter 2, we have shown that the bulk dielectric permittivity can be written within the Drude model as

$$\tilde{\varepsilon} = 1 - \frac{\omega_p^2}{\omega^2 + i\gamma\omega}. \tag{8.6}$$

We have already discussed that this model is very inaccurate for gold, copper, and silver, and we cannot simply replace γ with γ_{NP}. Instead, we add to the experimental values of ε a correction factor $\Delta\varepsilon(R)$ that can be expressed as a difference between two Drude factors:

$$\Delta\varepsilon(R) = \frac{\omega_p^2}{\omega}\left[\frac{1}{\omega + i\gamma} - \frac{1}{\omega + i\gamma_{NP}(R)}\right]. \tag{8.7}$$

The correction ε introduced by Equations (8.5) and (8.7) results in an excellent agreement between the calculations and experiments (see Fig. 8.7). Note that Kooij and his colleagues used only the electrostatic model for these calculations.

8.2.2.2 *Coupling with interband transitions in smaller nanoparticles*

For small nanoparticles of noble metals, we cannot ignore the quantum structure of the crystal. The LSPR is ascribed to the oscillation of the *s*-electrons of the conduction band. In silver, gold, and copper, the *d*-electrons form a close shell situated not far below the conduction band so that visible light can trigger *interband transitions*. For gold and copper, the interband transition edge occurs at 2 eV for the bulk metal, which is

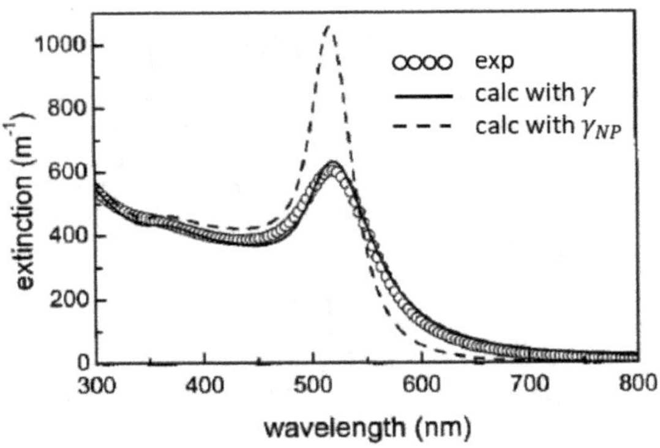

Fig. 8.7. Extinction coefficients of an aqueous suspension of gold nanoparticles of 13 nm diameter. The calculated LSPR spectrum with the bulk value of γ (relaxation frequency) is twice as intense as the one with the experimental values. The use of the corrected value γ_{NP} yields an almost perfect agreement with the experiments.
Source: Reproduced with permission from Ref. [10]. Copyright © 2002 American Chemical Society.

very close to the LSPR energy (507 nm or 2.44 eV for gold and 555 nm or 2.23 eV for copper in air). For gold nanoparticles, the decrease in size results in a blue shift so that the LSPR couples more and more with the interband transitions and eventually vanishes, as shown in Fig. 8.8 [12]. There is no more detectable plasmon for gold nanoparticles smaller than 2.5 nm. Silver is a simpler case since the interband transition is higher at 4 eV, and the LSPR is at 3.48 eV (355 nm) so that the coupling is minimized. For silver, there is only a slight blueshift when the size decreases, and the plasmon is detected for clusters as small as 20 atoms (read the discussion in Ref. [13], for example).

For small nanoparticles, a typical quantum phenomenon that influences the optical response is the spill out of the electrons. It occurs at the surface of a metal and corresponds to the fact that the density of the electron clouds does not undergo an abrupt transition. For spherical nanoparticles, for example, the radius of the electron sphere is slightly larger than the radius of the positively charged sphere constituted with the cationic cores. The amount with which the electrons spill out is of the order of magnitude of a few atomic units (0.529 Å), which is not negligible anymore for small nanoparticles and leads to a slight blueshift in the case of gold [14].

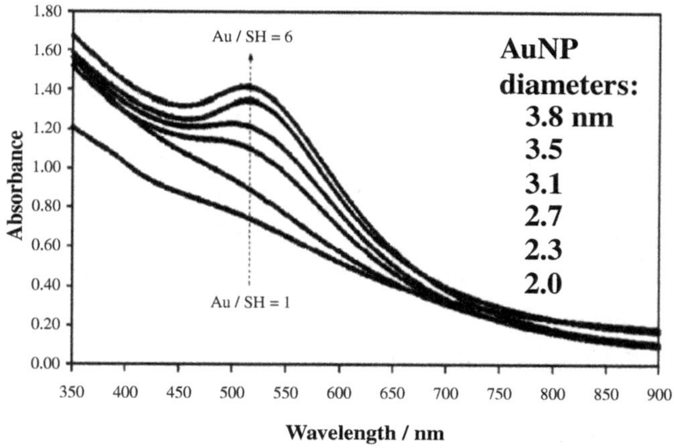

Fig. 8.8. Experimental absorbance spectra of small AuNPs showing the uptake of the LSPR when the diameter of nanoparticles become greater than 2.5 nm.

Source: Reproduced with permission from Ref. [12]. Copyright © 2008 Elsevier.

8.2.3 *Ellipsoids*

We now leave the case of the sphere, for which we have discussed in length the calculation of the LSPR. The issue that we face now is that there are very limited cases where analytical approaches are possible. For instance, an exact analytical model to calculate the polarizability α is only available in the case of spheres, ellipsoids. [15], and infinite cylinders [16].

We focus on the case of ellipsoids [17, 18]. They possess three plasmon resonances corresponding to the oscillation of electrons along the three principal semi-axes, denoted by a, b, and c. By changing the axis lengths, the LSPR can be tuned from $\lambda = {\sim}500\,\text{nm}$ in the case of gold up to the infrared range. It is obvious that the excitation of one of these plasmon resonances depends on the direction of the impinging electric field. Therefore, it depends on the direction of the light beam and its polarization. Since they cannot be probed as individual objects except with dedicated experimental setups [19–21], their optical response is obtained as an average over all the sizes and orientations present in the assembly. This may result in a smoothing or vanishing of the plasmon resonances if the nanoparticles are not sufficiently monodispersed.

For an individual metallic ellipsoid, the polarizability along the axis i ($i = x, y$, or z) is given by

$$\alpha_i = \frac{V_e}{4\pi} \times \frac{\varepsilon - \varepsilon_m}{\varepsilon_m + L_i(\varepsilon - \varepsilon_m)}. \tag{8.8}$$

ε is the dielectric function of the metal, which is a complex number and a function of the wavelength, and ε_m is the dielectric function of the surrounding medium, which can be considered as a real number, independent of the wavelength in the visible range of interest as long as this medium is transparent with negligible absorption.

V_e is the volume of the ellipsoids given by $V_e = 4\pi/3abc$, with a, b, and c being the semi-axes of the ellipsoids.

L_i is the depolarization factor, which is a purely geometric factor.

For simplicity, we consider revolution ellipsoids with two equal axes. These ellipsoids correspond to the majority of the experimental cases and belong to two families: they can be prolate (*rugby ball*–like with $a > b = c$) or oblate (*pumpkin*-like with $a = b > c$). Among the three proper modes, two are degenerated, and the two depolarization factors are equal. They appear as functions of the eccentricity e of the corresponding ellipse along axis i. In case of prolate spheroids, the depolarization factors are given by [18, 22]

$$L_x = \frac{1-e^2}{2e^3}\left(\ln\frac{1+e}{1-e} - 2e\right),$$

$$L_y = L_z = 1/2(1 - L_x) \text{ and } e = \sqrt{1 - b^2/a^2}. \tag{8.9}$$

And for oblate spheroids,

$$L_z = \tfrac{1+e^2}{e^3}(e - \tan^{-1} e),$$

$$L_x = L_y = 1/2(1 - L_z) \text{ and } e = \sqrt{a^2/c^2 - 1}. \tag{8.10}$$

We label the two modes as longitudinal modes (LMs) when the electric field is along the symmetry axis and transverse modes (TMs) when it is perpendicular to this axis. Note that this model assumes that the dipolar approximation is acceptable (nanoparticles of moderate size) and that no

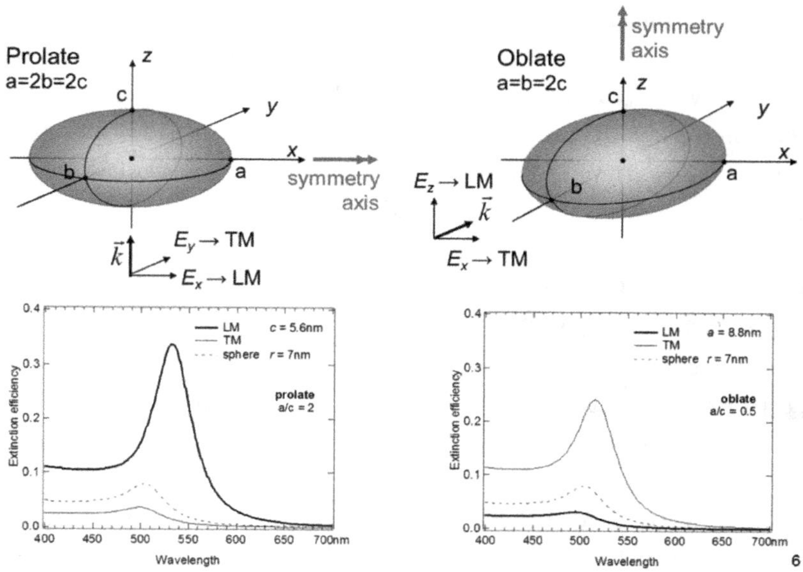

Fig. 8.9. Calculated extinction efficiency of prolate and oblate ellipsoids within the dipolar approximation. Ellipsoids with an aspect ratio of 2 are compared to a sphere of equal volume. The longitudinal mode corresponds to an excitation field parallel to the symmetry axis, and the transverse mode is perpendicular to this axis. In case of prolate ellipsoids, the LM is the most sensitive to the change in shape, whereas for oblate particles, the TM is the most sensitive.

quadrupolar modes enter into play. For a sphere ($a = b = c$), the three resonances are degenerated and the depolarization factors are all equal to 1/3. In this case, it is easy to check that Equation (8.8) gives the polarizability of the electrostatic model of relation (7.11) for spheres.

This model allows for a fully analytical calculation of the extinction cross section, assuming a dipolar model for the ellipsoid. Some results are presented in Fig. 8.9 in the case of particles in air and compared to spherical gold particles with a radius of 7 nm whose plasmon wavelength is at 504 nm. These two graphs plot the extinction efficiency, which is the ratio of the extinction cross section to the geometrical cross section. The ellipsoid is chosen so that its volume is the same as the volume of a sphere with $r = 7\,\text{nm}$. For example, in the case of the prolate ellipsoid of aspect ratio

$a/c = 2$, the dimensions are the following: $a = 11.2\,\text{nm}$, $b = 5.56\,\text{nm}$, and $c = 5.56\,\text{nm}$. In this case, the LM is shifted to 532 nm and its intensity is enhanced by a factor of almost 5. The TM is slightly blueshifted down to 500 nm and partially damped. By increasing the aspect ratio, this trend is amplified. For example, elongated ellipsoids of aspect ratio of 10 exhibit a longitudinal mode in the near infrared at 1037 nm, whereas the transversal mode stays at 495 nm. Some other results, as well as the values of the depolarization factors, are given in Table 8.4. The case of oblate ellipsoids is similar, with the slight difference that the TM is now the most sensitive to the change in particle shape and shifts to higher wavelengths when the aspect ratio is increased. For example, an oblate ellipsoid of aspect ratio of 2 exhibits a transverse plasmon at 515 nm. Other values are summarized in Table 8.5.

For an electric field of random orientation relative to the symmetry axis, the extinction spectrum is a combination of the two modes and will exhibit two resonances [18].

Table 8.4. Calculated plasmon resonance wavelengths for **prolate** gold spheroids ($a > b = c$) of different aspect ratios in air.

Prolate spheroids							
Aspect ratio (a:b)	1:1	2:1	4:1	6:1	8:1	10:1	20:1
L_x	0.3333	0.1735	0.0754	0.0432	0.0284	0.0203	0.0067
$L_y = L_z$	0.3333	0.4132	0.4623	0.4784	0.4858	0.4898	0.4966
λ_{LM} (nm)	**504**	**532**	**634**	**760**	**897**	**1037**	**>1700**
λ_{TM} (nm)	**504**	**500**	**498**	**496**	**495**	**495**	**495**

Table 8.5. Calculated plasmon resonance wavelengths for **oblate** gold spheroids ($a = b > c$) of different aspect ratios in air.

Oblate spheroids							
Aspect ratio (a:c)	1:1	2:1	4:1	6:1	8:1	10:1	20:1
$L_x = L_y$	0.3333	0.2363	0.1482	0.1077	0.0845	0.0695	0.0369
L_z	0.3333	0.5272	0.7036	0.7846	0.8308	0.8608	0.9262
λ_{TM} (nm)	**504**	**515**	**545**	**579**	**615**	**650**	**807**
λ_{LM} (nm)	**504**	**495**	**492**	**490**	**490**	**490**	**488**

8.2.4 *Core–shell nanoparticles*

Core–shell nanoparticles are structures made of two concentric materials, e.g. a spherical gold core covered by a silica shell (Fig. 8.10(a)). The classical nomenclature is core@shell. There are many cases where the core–shell structures are relevant. For example, in the synthesis of metallic nanoparticles, there is often a shell of molecular surfactants that stabilizes the nanoparticle. Sometimes, a silica shell can be synthesized around nanoparticles in order to prevent the plasmonic nanoparticles from approaching too closely and modifying the plasmon response. Moreover, core–shell nanoparticles are interesting because they allow a wide tuning of the LSPR wavelength, which can be critical for applications such as cancer therapy (see Chapter 10). This is the case of the SiO2@Au nanoparticles prepared by the group of Halas [23, 24]. In this section, we review the main plasmonic properties of core–shell nanoparticles.

The calculation of the plasmon resonance of spherical core–shell nanoparticles can be carried out analytically with the same approach as

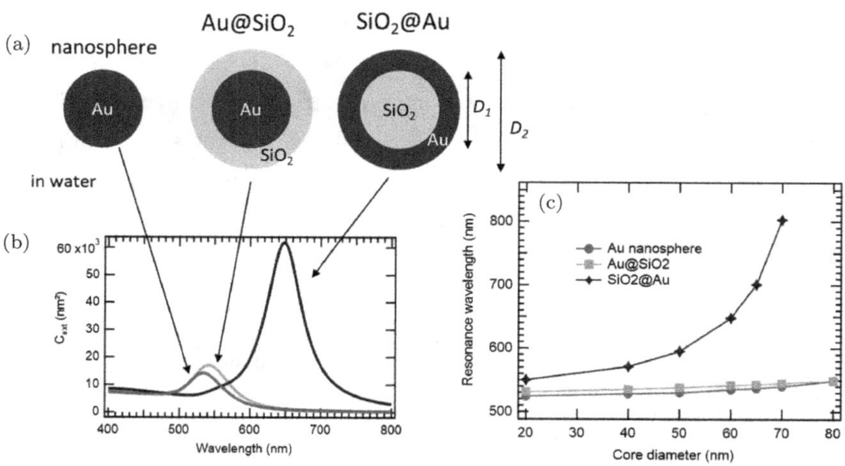

Fig. 8.10. (a) Geometry of three types of spherical nanoparticles: a gold sphere, a gold-silica core–shell nanoparticle, and a silica–gold core–shell nanoparticle. D_1 is the core diameter and D_2 the total diameter, which is fixed at 80 nm in this example. (b) Plot of the extinction cross section when the core has $D_1 = 60\,\text{nm}$ for the three types of nanoparticles in water. (c) Plot of the evolution of the maximum of the plasmon peak as a function of the core diameter D_1 for the three types of nanoparticles in water.

the Mie theory and is developed in the textbook by Bohren and Huffman in Chapter 8 [1]. The authors called them *coated spheres* and proposed a FOTRAN code for programming the calculation. Many authors have also implemented this method for simulating the extinction, absorption, and scattering cross sections of core–shell nanoparticles. Herein, we used the values obtained with the Python module that was implemented on the Bichromatics website [7]. We will now discuss the main trend of this plasmon resonance.

8.2.4.1 *Examples of gold–silica and silica–gold nanospheres*

Let's consider a nanoparticle in water with a total diameter $D_2 = 80\,\text{nm}$ (shell diameter) and a core with diameter varying between 20 and 70 nm. We consider the two cases: either a gold core and a silica shell or the opposite, as shown in Fig. 8.10(a). We focus on the extinction cross section σ_{ext} since it is directly accessible for measurements with a UV–visible spectrometer when nanoparticles are dispersed in water and placed into a cuvette. As given by Equation (7.34), the absorbance measured by the spectrometer is proportional to σ_{ext}. The values of the wavelength for the maximum of σ_{ext} are given in Table 8.6. The case of $Au@SiO_2$ in water exhibits little variation in λ_{max} when the inner diameter grows from 20 to 70 nm. When $D_1 = 20$ nm, the capping silica layer has a thickness of 30 nm. This case is very similar to the case of AuNPs embedded in a silica matrix and $\lambda_{max} = 531$ nm. When $D_1 = 70$ nm, with a capping layer of just 5 nm, the

Table 8.6. Plasmon peak of the extinction cross section for spherical core–shell nanoparticles in water. The total diameter D_2 of the particles is fixed at 80 nm. D_1 is the core diameter. Calculations are presented for a gold nanosphere, gold–silica core–shell nanoparticles, and a silica–gold nanoparticle.

D1 (nm)	20	40	50	60	70	80
λ_{max} for Au (nm)	524	528	530	535	540	549
λ_{max} for $Au@SiO_2$	531	535	538	541	545	549
λ_{max} for $SiO_2@Au$	550	571	595	648	803	no reson.

plasmon resonance shifts to 545 nm (see Table 10.4). The values are compared to the case of a pure gold nanosphere in water, and the two evolutions are very similar. Figure 8.10(b) shows that the extinction cross sections are very close indeed. Note that the optical indices of water and silica are fairly close: $n_{\text{water}} = 1.33$ and $n_{\text{silica}} = 1.44$. The presence of the silica shell helps in controlling the distance between nanoparticles when they aggregate, which is critical for using them in colorimetric immuno-assays, for example [25]. This is also used for positioning fluorescent molecules at the right distance from the metallic core to avoid the quenching of luminescence and simultaneously maximize the antenna effect [26].

The case of SiO_2Au is very different. Figure 8.10(b) shows the extinction cross section when the core diameter is 60 nm and the shell is 10 nm thick ($D_2 = 80\,\text{nm}$), and the resonance is three times more intense and strongly redshifted. The resonance shows up at 648 nm compared to 541 nm for Au@SiO_2. Moreover, the evolution of λ_{max} with the inner diameter is much more pronounced (see Fig. 8.10(c)). The thinner the gold shell, the higher the plasmon wavelength. This core–shell geometry offers a more adjustable LSPR in the infrared in the case where the metal is gold.

8.2.4.2 *Simple analytical approach for calculating the LSPR of core–shell nanoparticles*

The calculation of the polarizability of a spherical core–shell nanostructure can be carried out rigorously using the Mie approach [1]. However, it is also possible to restrict the Mie expansion to the first term and develop the electrostatic approximation in a similar way as explained in Chapter 7. Some details can be found in the work of Tzarouchis [27,28]. We assume that the particle is defined by its total radius R_1 and its dielectric permittivity ε_1. R_2 and ε_2 are the equivalent parameters for the inner core ($R_2 < R_1$). The electric potential outside the nanoparticle is expressed as (see notations in Fig. 7.4 of previous chapter)

$$V_{sca} = \frac{B_1}{r^2} \cos\theta, \tag{8.11}$$

where B_1 is the first non-zero coefficient of the Legendre polynomials given by

$$B_1 = \frac{C\varepsilon_1 - \varepsilon_{diel}}{C\varepsilon_1 - 2\varepsilon_{diel}} R_1^3 E_0, \tag{8.12}$$

$$\text{with } C = -2 + 3/\left[1 - \frac{\varepsilon_2 - \varepsilon_1}{\varepsilon_2 - 2\varepsilon_1}\eta^3\right] \text{ and } \eta = \frac{R_2}{R_1}. \tag{8.13}$$

The polarizability is written as

$$\alpha = 4\pi \frac{C\varepsilon_1 - \varepsilon_{diel}}{C\varepsilon_1 - 2\varepsilon_{diel}} R_1^3. \tag{8.14}$$

The resonances will be given by the poles in Equation (8.12) or (8.14). The poles are the values of λ for which the denominator becomes zero: $C\varepsilon_1 - 2\varepsilon_{diel} = 0$. This is the analog to the Fröhlich condition discussed in Chapter 7. This condition is far less obvious here, and we briefly discuss it in Exercises. See Refs. [27–29] for examples and experimental verifications. This core–shell model includes both cases: the case where the core is the plasmonic metal and the shell is a dielectric and the reverse case where the plasmonic metal is the outer shell. It also includes the case of hollow-shell nanostructure with a metallic shell and a spherical vacuum hole in its center. In this case $\varepsilon_2 = 1$ and $\varepsilon_{diel} = 1$.

8.3 LSPR of Nanorods, Nanocubes, Nanostars, and Other Complex Nanostructures

8.3.1 *Numerical approaches to calculating the LSPR of complex nanostructures*

In this section, we focus on the calculation of the LSPR of nanostructures where there is no analytical approach. We focus mainly on the prediction of the extinction cross section since this is the easiest one to be experimentally obtained.

Thanks to the strong computational power of computers nowadays, it has become possible to implement very effective numerical methods on regular computers. There are dozens of numerical methods, each one being more suited to a given situation depending on the geometry of the structure, its size compared to the wavelength (nano, macro), its periodicity, the

Table 8.7. Quick presentation of the 10 numerical methods used for calculating the electromagnetic response of plasmonic nanostructures. The meanings of the column headers are the following: *Space grid* distinguishes the type of mesh used in the method. *Domain* stands for either the temporal (T) or the frequency (F) domain in which the method operates. *Order* stands for the expansion order of the method: dipole, multipole, or full numerical.

Method		Space grid	Domain	Order
FDTD	finite-difference time-domain	V	T	Full
FEM	finite-element method	V	F	Full
DG	discontinuous Galerkin	V	T/F	Full
DDA	discrete-dipole approximation	Vs	F	Dip
MMP	multiple multipole	Vs	F	M
G.Dyadic	Green Dyadic	Vs	F	Full
BEM	boundary element method	S	F	Full
SIE	surface integral equations	S	F	Full
SIEMoM	SIE with method of moments	S	F	Full
T-Matrix	transfer matrix	S	F	Full

Source: Reproduced from Ref. [30].

quantities desired (near field vs. far field), or the time evolution (steady-state, transient field, or ultrashort optical response, such as femtosecond). Table 8.7, taken from Ref. [30], gives the 10 principal numerical methods, and we describe with some more details the discrete dipole approximation (DDA) and the finite-difference time-domain (FDTD) methods.

From a numerical point of view and depending on how the discretization is achieved, three groups of methods can be proposed:

(1) **Volume methods** (marked ***V*** in Table 8.7), in which the whole space where the electromagnetic field is needed has to be discretized.
(2) **Scatterer volume methods** (marked ***Vs***), in which only the volume containing the plasmonic nanostructure (the scatterers) is discretized.
(3) **Scatterer surface methods** (marked ***S***), in which the discretization can be limited to the surface of the plasmonic nanostructure (surface of the scatterers).

8.3.2 *Discrete dipole approximation*

This method serves to compute absorption and scattering by a structure of arbitrary shape and at any distance from it (far field or near field can be calculated). It is based on the decomposition of the nanostructure into a square lattice of dipoles that mimics as closely as possible the original object, which is called *the target*. The programmer must introduce two key elements to ensure that the DDA calculation is accurate: the 3D grid of dipoles and the polarizability of these dipoles. The first one is a model of the shape and the second one a model of the nature of the material.

The method was first introduced in 1973 by Purcell and Pennypacker for studying the optical properties of interstellar dust [31] and was improved in 1993 by Draine *et al.* [32].

8.3.2.1 *Defining a computational grid*

The object should be divided into a set of N identical cubes, whose size has to be much smaller than the wavelength. Each box contains one dipole. The accuracy of the computation depends on how close the grid represents the object. In Fig. 8.11, an example is given in two dimensions for simplicity: a disk can be discretized into $N = 16$ dipoles, and the DDA model will look like a cross rather than like a disk. A better representation is obtained

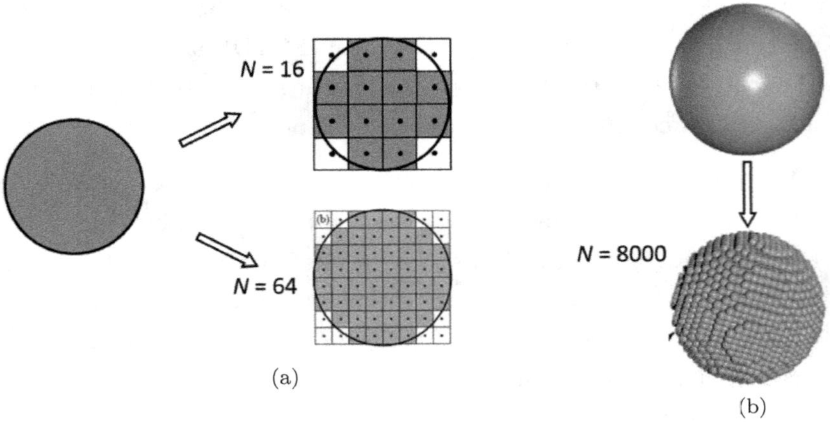

Fig. 8.11. Design of the computation grid for the DDA calculations. A disk (a) can be represented by a small number of dipoles ($N = 16$) with a lack of realism or a higher number of dipoles ($N = 64$). The grey boxes correspond to the inside of the object where the refraction index is set to be that of the material and the white boxes are the *void dipoles*. In the case of a sphere (b), a discretization with $N = 8000$ dipoles starts providing an accurate model.

with $N = 64$ dipoles. A dipole is placed at each point of this grid, as represented by the black point in Fig. 8.11(a). A refractive index is assigned to each dipole: the material refractive index n_{mat} for the grey boxes and the void refractive index for the white boxes. For a 3D object, the number of dipoles needed for a good result grows very quickly. For example, modeling a sphere requires 8,000 dipoles, as shown in Fig. 8.11(b). If the *target* is not in vacuum but in a surrounding medium of n_{med}, the material index is simply replaced by the relative index (n_{mat}/n_{med}), and the wavelength is the one in the medium $(\lambda_{vacuum}/n_{med})$.

8.3.2.2 *The refractive index and dielectric permittivity*

This is the second important ingredient for the DDA method. The dipolar moment at each box positioned at $\boldsymbol{r}_i$ depends on the polarizability $\boldsymbol{\alpha}_i$ and the local field $\boldsymbol{E}_{loc}(\boldsymbol{r}_i)$ at this position:

$$\boldsymbol{P}_i = \alpha_i \cdot \boldsymbol{E}_{loc}(\boldsymbol{r}_i). \tag{8.15}$$

The local electric field is generated by the incident field and all the other dipoles. In the DDA approach, this interaction of dipole i with all the other dipoles j is described by a $3N \times 3N$ matrix whose elements result from the physics of the dipole–dipole interaction [33]. This matrix allows for the calculation of the local field at any position $\boldsymbol{r}_i$. The polarizability is obtained from the Clausius–Mossotti formula, which is an exact result in the case of a square lattice of parameter d and is written as

$$\alpha_i = \frac{3d^3}{4\pi}\left(\frac{\varepsilon_i - 1}{\varepsilon_i + 2}\right). \tag{8.16}$$

All the variables are linked together, and the problem needs to be solved self-consistently by iteration. The time needed for solving the case of a 40 nm nanoparticle is in the order of magnitude of minutes to hours.

8.3.2.3 *Examples of DDA calculations*

The DDA calculation is an exact method in the sense that there is no approximation for calculating the radiated field of the interacting dipoles. The approximations lie in the accuracy of representing the object with

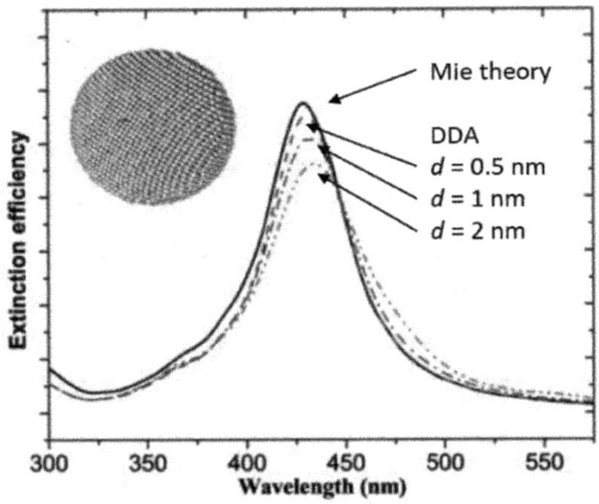

Fig. 8.12. Extinction spectra of 40 nm silver spherical nanoparticles in an organic solution ($n_{medium} = 1.447$) calculated using the Mie theory (solid line, resonance at 429.2 nm) or with DDA and an increasing grid parameter d. For this 40 nm diameter nanoparticle, the grid parameter is set at $d = 0.5$ nm ($N = 512{,}000$), $d = 1$ nm ($N = 64{,}000$), and $d = 2$ nm ($N = 8{,}000$). The inset shows the representation of the grid for $d = 1$ nm.

Source: Reproduced with permission from Ref. [33]. Copyright © 2013 Springer Nature.

a square lattice and in the accuracy of the model for the refractive index. A typical accuracy of 5% for calculating extinction spectra is easily accessible with DDA.

In the case of a sphere, it is possible to check the accuracy of the DDA against the Mie theory. Figure 8.12 shows an example with a 40 nm silver nanoparticle in an organic solvent of refractive index 1.447 [33]. The Mie theory predicts a plasmon resonance at 429.2 nm. The dipole grid for the DDA is made denser and denser. A grid parameter of $d = 2$ nm for a volume of $(40 \text{ nm})^3$ corresponds to a number of dipoles of $N = 8{,}000$. In this case, the DDA predicts a LSPR with 6 nm deviation at 434.9 nm. When the grid is tightened to $d = 1$ nm ($N = 64{,}000$), the LSPR shows up at 432.4 nm and at 430.7 nm for $d = 0.5$ nm ($N = 512{,}000$). With the densest grid, DDA is exact at 1.5 nm.

The major interest of DDA is to model more complex shapes, such as a silver cube, as shown in Fig. 8.13. A cube exhibits six different resonances indicated in the figure.

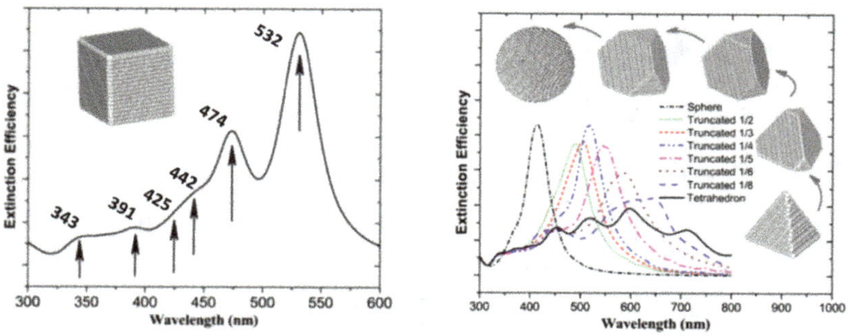

(a) (b)

Fig. 8.13. (a) Extinction efficiency of a silver cubic nanoparticle in organic solution calculated with the DDA method. Six LSPRs are predicted and marked by arrows. (b) Extinction efficiency of a silver tetrahedron that is progressively transformed into a sphere of 40 nm diameter. These particles are dispersed in water.

Source: Reproduced with permission from Ref. [33]. Copyright © 2013 Springer Nature.

8.3.2.4 *Open-source DDA codes*

Various source codes and program are freely available and can be run on regular computers. We list six of them in Table 8.8, referenced from [34].

8.3.3 *Finite-difference time-domain method*

The FDTD method is another method based on discretization, where the nanostructure is represented by a spatial grid of small mesh size. It was introduced in 1966 by Kane S. Yee. In contrast to DDA, not only space is discretized but also time. It is used to solve the time-dependent Maxwell equations (see Chapter 1 of this book). It is an explicit time-marching algorithm based on the special structure of the Maxwell equations: The spatial derivatives of the electric field $\boldsymbol{E}$ are linked to the time derivatives of the magnetic induction $\boldsymbol{B}$ and vice versa. Therefore, the components of the electric field in a given volume of space are solved at a given instant t, then the components of $\boldsymbol{B}$ are solved at the next instant of time $t+\Delta t$ in the same volume. This method is repeated until the evolution of the system has been described sufficiently. It can be used for transient evolutions as well as steady-state phenomena. It also applies to studying both the near-and

Table 8.8. Examples of programs that can be downloaded and freely used for carrying out calculations of the scattering and of the absorption of light by particles. These programs are all based the discrete dipole approximation (DDA).

Name	Authors	Website	Language	Updated	Features
DDSCAT	B.T. Draine P.J. Flatau	http://ddscat.wikidot.com/	Fortran	2016	First free distribution of DDA code
DDscat.C++	V. Choliy	http://space.univ.kiev.ua/Choliy/DDscatcpp/	C++	2017	DDSCAT translated into C++, with additional features
ADDA	M.A. Yurkin A.G. Hoekstra *et al.*	https://github.com/adda-team/adda/	C	2018	Includes the case of object near a planar substrate
OpenDDA	J. MacDonald		C	2009	Focused on computational efficiency
DDA-GPU	S. Kiess	https://github.com/steffen-kiess/dda	C++	2016	Code optimized for running on Graphics Processing Units
VIE-FFT	W.E.I Sha	www.zjuisee.zju.edu.cn/weisha/SourceForge/sourceforge.html		2019	DDA that also calculates near field

far-field electromagnetic responses of heterogeneous materials of arbitrary geometry [9, 35].

However, as input parameters, FDTD requires the values of the permittivity for a wide range of frequencies, and this is usually done by using the Drude model, which is accurate for a limited frequency range, as we have seen (see Chapter 2). Moreover, the far field is not directly calculated and requires some postprocessing. Nevertheless, FTDT methods have been attracting more and more interest because it is an explicit method without matrix inversion, which usually is a limitation in the number of fields to be solved. Nowadays (since 2020), FDTD simulations can solve several billions of unknowns: They can be field components at different points of the space grid and at different instants of time. They are also very suited for parallel calculations and benefit from the highly efficient, graphical-processor-unit-based computers of today. Its structure allows visualizing the transient evolution of very short pulses throughout a plasmonic nanostructure.

As an example, we show the results published by Berini *et al.*, where they investigate a classical plasmonic nano-antenna called the *bowtie nanoantenna*, made of two triangles facing each other [36]. The tips of the two triangles are placed at 20 nm from each other, and the total size of the structure is 160 nm (see Fig. 8.14). In their study, they are interested in mapping the electric field, and they discuss the strengths and weaknesses of the FDTD simulations as a function of some key parameters. The study was conducted on a supercomputer developed by IBM, namely the third generation of the *Blue Gene* family (the *Blue Gene/Q*) that was run on 2,048 nodes for this study. Each node consists of 16 cores (i.e. physical processors) and 16 GB of RAM.

They have studied the effect of an ultrashort light pulse of 1 fs duration of this nanostructure. They monitor with the FDTD calculations the evolution of the electric field radiated by the nanostructure over a timescale of 25 fs. Figure 8.14 shows that at 1.3 fs, a strong field is localized at the gap. At 3.1 fs, the field goes to zero and a second oscillation ($N = 2$) can be detected at 7.3 fs. The third oscillation occurs at 12.9 fs (not shown) and a fourth one at 17.7 fs. We can also see that the external tips of the triangles also concentrate some field. These positions where the field is strongly

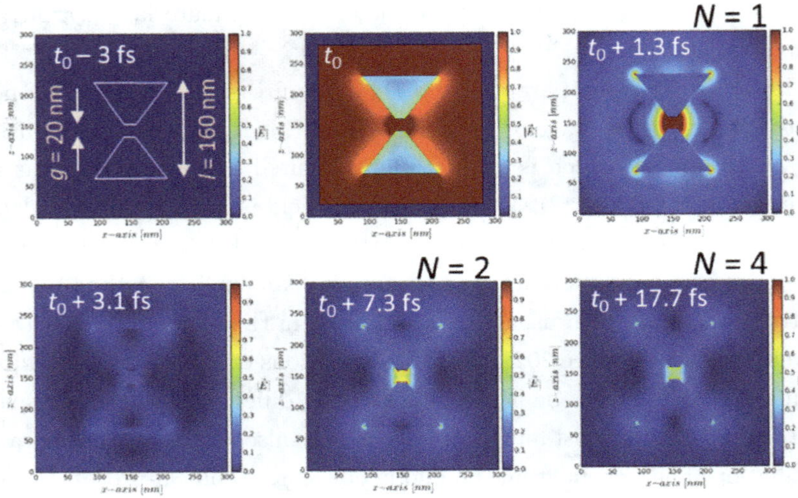

Fig. 8.14. Snapshots extracted from a video sequence of 25 fs, where the electric field is calculated with FDTD. The nanostructure is a gold bowtie in air with a total length of 160 nm, width of 120 nm, and thickness of 40 nm. The gap between the two triangles is 20 nm. A light pulse of 1 fs duration hits the nanostructure at t_0 (polarized along z-axis and propagating along the y-axis i.e. from below). The evolution of the scattered field is plotted in false color at 1.3, 3.1, 7.3, and 17.7 fs after the excitation.

Source: Reproduced with permission from Ref. [36]. Copyright © 2015 Optica Publishing Group.

concentrated are called *hot spots*, and in the bowtie nanostructure, the main one is located in the gap.

FDTD is the method of choice for calculating the amplitude of the local fields $\boldsymbol{E}$ and $\boldsymbol{H}$.

8.4 Plasmonic Modes in Nanoparticles

8.4.1 *Definition of the normal modes*

The notion of modes is central to many situations in physics where a system is subjected to a periodic excitation of circular frequency ω. The system is described by a variable, which can be the deviation angle for a pendulum, the displacement of a mass attached to a spring, the displacement of a vibrating membrane on a drum skin, or the intensity of light inside an optical fiber. In plasmonics, these variables are the components of the electric field, the components of the magnetic fields or the local charge density. When the system is excited at a variable frequency, the amplitude of the variables of interest reaches a maximum at a frequency called *eigenfrequency*. An *eigenmode* is the full description of the corresponding oscillation: It includes the frequency and the values of the parameters of interest throughout the whole system. For example, the *eigenmode* of a spherical nanoparticle includes the resonance frequency and the mapping of the electric field throughout the entire nanoparticle. The intensity of the field is generally not uniform over the entire nanoparticle, but it is in phase. A system will typically exhibit several discrete eigenfrequencies, ω_0, $\omega_1, \ldots, \omega_N$. Eigenfrequencies are equivalent to resonance frequencies even though the concept of resonance implies that there is an external harmonic excitation for triggering the system (harmonic regime), whereas the normal modes can be probed with a percussion (transient regime).

The denomination "normal modes" indicates that each mode can be excited independently from the other ones: The oscillation at one frequency does not generate an oscillation of any other normal mode. In mathematical terms, normal modes are orthogonal to each other. Therefore, an arbitrary motion can be expressed as a linear combination of different normal modes (as long as the system is linear). Each mode can be characterized by the amount of energy it can store. The calculation of the mode of a system is the resolution of the Maxwell equations, taking into account the boundary conditions of the nanostructure and the excitation conditions. A critical discussion of electromagnetic modes in plasmonic resonators can be found in the article by Kristensen and Hughes [37].

Modes can be viewed as optical discrete states. However, they always include a finite width, which is linked to the energy dissipation of this mode. A more complete view can be obtained by considering the optical local density of states (LDOSs) of the photons in a given nanostructure: This is the optical LDOS or photonic LDOS. It is the equivalent conception to the local density of states of electrons in condensed matter physics [38]. This gives rise to the analog formalism of hybridization taken from the linear combination of atomic orbitals (LCAOs) theory in chemistry, which is explained in the following.

8.4.2 *Probing plasmonic modes with electron energy loss spectroscopy and cathodoluminescence*

This optical LDOS can be probed by one of the "two faces" of the plasmon oscillations: either by photons or by electrons. Photons are used to probe these optical LDOSs when the optical spectra (absorbance, scattering, etc.) are recorded. Electrons can also be used since a plasmon is a charged oscillation, and this is achieved with electron microscopes with the well-known technique of electron energy loss spectroscopy (EELS). This technique has recently regained strong focus due to the progress in electron spectrometers and electron focusing lenses. EELS is usually performed using a transmission electron microscope (TEM) or a scanning transmission electron microscope (STEM), where a high energy electron beam (typically 200 keV) is focused on a sample with a spot size below 1 nm. The beam is raster-scanned across the sample and probes different structures of the sample with a resolution as low as 0.1 nm (see Fig. 8.15). In the case of EELS, the sample must be thin enough so that electrons can traverse the sample. But in doing so, they lose a part of their energy when coupling with plasmon-polariton modes of the sample. The analysis of this energy loss reveals these plasmonic modes, as shown in Fig. 8.15(b) [39].

Another way to access the plasmon-polariton modes is to place an optical spectrometer above the sample and collect the signals of cathodoluminescence. The signals are generated by the electrons forced to oscillate inside a confined nanostructure so that eventually an electromagnetic wave is radiated. A demonstration was given by Garcia de Abajo and Kociak, where the electron beams of EELS and cathodoluminescence probe the same modes as those probed by the photons of optical spectroscopies [40].

Exciting plasmonic nanostructures with electrons offers a series of advantages. This is a way of overcoming the diffraction limits imposed by photons. Lateral resolution is better than 1 nm with EELS, which is incomparably better than the classical 400 nm limit of photonics. It also allows recording hyperspectral maps: At each position (x, y), a full spectrum is recorded, revealing the geometry of each mode (see example in the next section). Finally, since the excitation of the plasmonic modes is triggered

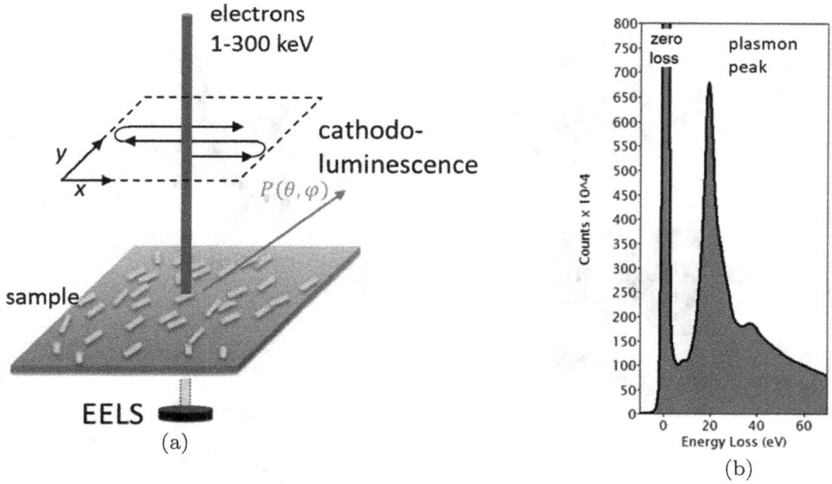

Fig. 8.15. Principle of EELS and cathodoluminescence for probing the plasmonic modes of a sample. (a) The electron beam scans the sample and eventually excites plasmonic modes. (b) An electron spectrometer measures the energy lost by the electrons and produces the EEL spectrum. The zero loss peak corresponds to the elastic diffusion of electrons. The plasmon peak reflects the energy lost by the incoming electrons due to their interaction with the plasmons.

by the passing of an impulsion of electrons through the sample, it is possible to use ultrashort electron pulses and probe the temporal structures of the plasmonic modes [39]. On the negative side, the cost of a TEM equipped with an EELS spectrometer is 50 times higher than the equivalent in optics (optical microscope and spectrometer).

8.4.3 *Example of plasmonic modes in a nanorod*

We now give an illustration of the use of EELS for probing the charge oscillations of a silver nanorod shown in Fig. 8.16(a) [41]. The 270×42 nm nanorod is imaged with TEM in the angular darkfield mode. Figure 8.16(b) shows the experimental mapping at 0.9 eV energy loss when the electron beam is scanned over the nanorod. It clearly shows two areas of maximal intensity at the apices of the nanorod. This is interpreted as a maximum coupling of the incoming electron beam with the plasmon-polariton modes of the nanorod. It is confirmed by DDA simulations shown in Fig. 8.16(d),

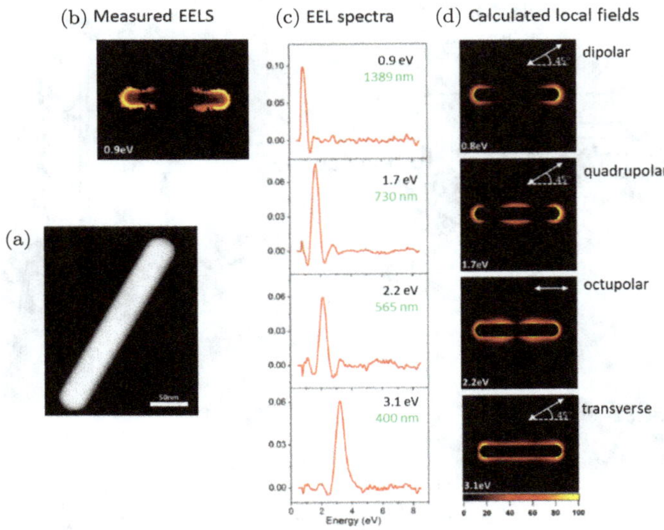

Fig. 8.16. Plasmonic modes of a silver nanorod of 270 nm length and 42 nm diameter obtained with EELS: (a) The nanorod is imaged with TEM in the angular dark field (ADF) mode. (b) False-color mapping of electron loss intensity at 0.9 eV, showing the maximal excitation at the two ends. (c) Spectral intensity integrated over the whole nanorods and are decomposed by multivariate statistical analysis into four contributions at 0.9, 1.7, 2.2, and 3.1 eV. (d) DDA calculations allow plotting the field intensity maps of these four modes when the nanorod is excited by a polarized plane wave, as indicated on the images.

Source: Reproduced with permission from Ref. [41].

where the local electric field is plotted in false colors for different energies of the exciting plane wave. It shows that the mode at 0.9 eV (1390 nm) is the dipolar mode of the nanorod. The nanorod also accommodates other higher order modes (quadrupolar, octupolar, etc.). The authors have carried out a multivariate statistical analysis (MVA) of the EELS data and were able to identify 12 components, some of them overlapping. In Fig. 8.16(c), only the principal modes are shown.

The systematic comparison of the EELS data and simulations allows assigning the different modes of a nanorod. In Fig. 8.16(b), only one EELS image is shown, but the authors have recorded four images [41]. The four modes are found at the following energies: 0.9 eV (1390 nm), 1.7 eV (730 nm), 2.2 eV (565 nm), and 3.1 eV (400 nm). The 1390 nm mode is the usual dipolar mode. This mode has the strongest intensity (see EEL spectra). The 730 and 565 nm modes are the quadrupolar and octupolar modes, respectively. And the 400 nm mode corresponds to the transverse mode where electrons oscillate perpendicular to the rod axis.

Nanorods offer a good example of the wealth of plasmonic modes in nanostructures. The number of modes is strongly dependent on the symmetry of the structure: A sphere exhibits one single mode, whereas an ellipsoid with axial symmetry has two (longitudinal and transverse). Nanorods display these two modes and all the higher order modes.

8.4.4 *Coupled nanoparticles and dimers*

When two nanoparticles come close to each other, their LSPRs can interact via their optical near fields so that their modes couple. New resonances arise, which strongly depend on the interparticle distance. The new plasmonic modes are generally redshifted. The prediction of the position of the resonance demands heavy calculations. As an example, Fig. 8.17 shows the case of a dimer of spherical silver nanoparticles [9]. An isolated nanoparticle of 25 nm diameter has an LSPR at 365 nm. A dimer of these same nanoparticles separated by 5 nm exhibit two plasmon modes: at 355 and 414 nm. The mode at 355 nm is the transverse mode, arising from the

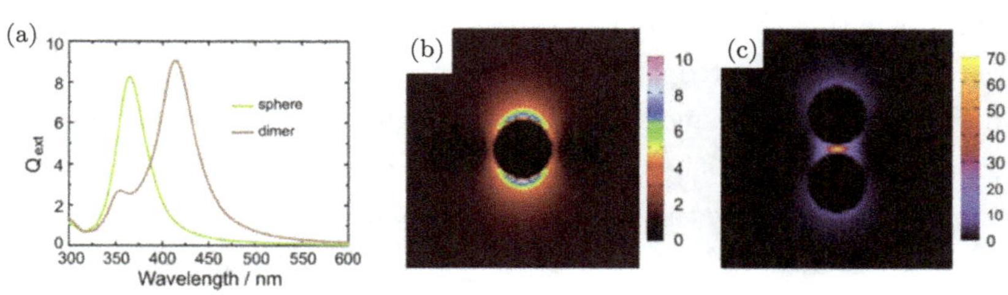

Fig. 8.17. Plasmonic coupling of silver nanoparticles of 25 nm diameter: (a) Calculated extinction spectra of an isolated sphere (λ_{max} = 365 nm) and of a nanoparticle dimer (λ'_{max} = 355 nm and λ''_{max} = 414 nm). (b) and (c) represent the local field enhancement. Calculations performed with the generalized multiparticle Mie theory.

Source: Reproduced with permission from Ref. [9]. Copyright © 2011 The Royal Society of Chemistry.

lift of degeneracy of the spherical symmetry. The mode at 414 nm is the longitudinal mode, which is highly sensitive to the interparticle distance.

Figure 8.17 also displays the intensity of the near-field enhancement $|E| / |E_0|$ around the nanoparticles. For an isolated nanoparticle, the enhancement is of the order of magnitude of ~10, as already noted in the previous chapter. In the case of the dimer, however, the enhancement is maximized in the gap and reaches a value as high as ~60. This phenomenon is known as *hot spots* and is crucial for using nanoparticle assemblies as biosensors.

8.4.5 *The plasmon hybridization model*

The description of LSPR in terms of modes and optical LDOS has strong analogy with the molecular orbital theory in chemistry, where the coupling of two molecular orbitals generate a bonding orbital and an antibonding orbital. For plasmon, the model was introduced by Halas and Nordlander in 2003 [42] and rigorously demonstrated in 2004 [43]. We briefly explain the plasmon hybridization model in the case of two core–shell nanoparticles brought close to each other so that their plasmon modes hybridize, following a study by Halas' group published in 2008 [20]. They study core–shell nanoparticles SiO_2@Au with a silica core of 80 nm diameter and a total diameter of 110 nm (gold shell thickness is 15 nm). An isolated nanoshell has a strong dipolar plasmonic mode at 592 nm ($l = 1$) and much weaker quadrupolar ($l = 2$) and octupolar ($l = 3$) modes. In Fig. 8.18(a), these three modes are pictured with the outer horizontal levels labeled $l = 1, 2$, and 3. When the two nanoparticles are brought closer but still at a distant of $D = 25$ nm, modes slightly hybridize into bonding and antibonding levels. In this regime, only the lowest order dipolar, $l = 1$, dimer mode (mode **1**) is excitable by light because only this mode has a strong dipole moment. The resonance shows up at 623 nm, a lower energy than the initial dipolar mode (see Fig. 8.18(c)).

When the nanoshells are pushed closer to each other, the coupling parameter grows and the splitting becomes greater. It leads to an energy decrease in mode **1** (redshift for the LSPR). The value of the coupling parameters was calculated by Nordlander and Stockman in the case of two

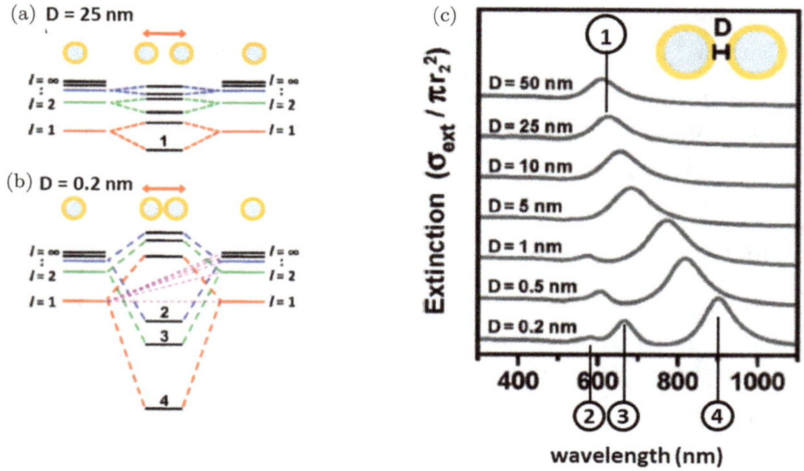

Fig. 8.18. The plasmon hybridization model: the isolated nanoshell exhibits a dipolar mode at 592 nm. (a) Upon weak coupling of two nanoshells ($D = 25\,\text{nm}$), the bonding level is redshifted to 623 nm (mode 1). (b) In the case of strong coupling, the hybridized modes appeared at 579 (mode 2), 662 (mode 3), and 902 nm (mode 4). The red double arrows in (a) and (b) stands for the polarization of the exciting electric field. The resulting extinction spectra are shown in (c) for decreasing interparticle distances from 50 to 0.2 nm. The modes 1, 2, 3 and 4 are indicated.

Source: Reproduced with permission from Ref. [20]. Copyright © 2008 American Chemical Society.

interacting gold nanospheres and is a strenuous integral with a dependence of $1/d^3$ on the separation distance for the dipolar modes ($l = 1$). More generally, the splitting grows as $1/d^{(l+l'+1)}$ [43]. For example, the coupling parameters of the quadrupolar modes grows as $1/d^5$ when $d \to 0$. In the case of very closely packed nanoparticles, this leads to a mixture of modes, as shown in Fig. 8.18(b), where two nanoshells have been pushed as close as 0.2 nm to each other. The dipolar mode is redshifted to 902 nm (mode **4**), the quadrupolar to 662 nm (mode **3**), and the octupolar to 579 nm (mode **2**). Interestingly, the coupling of the two nanoshells makes visible these higher order modes, which were not active for the isolated particle. The evolution of the coupling strength occurs at $1/d^5$ and $1/d^7$ for the quadrupolar and octupolar modes, respectively. This means that they grow more quickly than the dipolar mode, which is the reason for their emergence.

The plasmon hybridization model applies to any kind of interacting nanoparticles and is not limited to only two objects. It was applied to nanorods of different geometries interacting with each other [44] and to assemblies of coupled nanoparticles with an approach similar to that of solid state physics (Bloch's theorem) [45].

8.4.6 *Dark plasmon modes*

In most of the examples explained in this book, the plasmon modes are excited by a plane wave, which is the ideal configuration for exciting the dipolar modes, which in turn, can be coupled to higher order modes (quadrupolar, octupolar). However, for some morphologies of nanoparticles, the net dipole moment is zero, and there is no dipolar plasmon mode. Other modes are allowed, corresponding to resonant oscillations of the electron density. These modes are not radiative and are called *dark plasmon modes*. An example of such modes is given by close-by nanoparticles

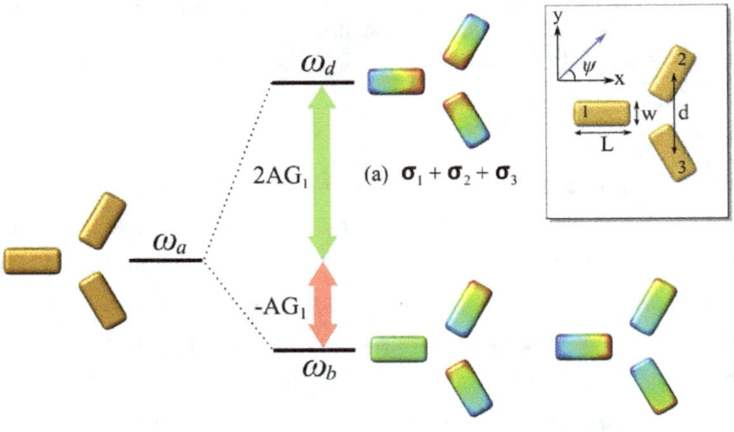

Fig. 8.19. The assembly of three nanorods into a triangular shape lifts the degeneracy of the dipolar mode. Three modes are generated: a doubly degenerated mode ω_b of lower energy that can be excited by two linearly polarized waves. The third mode ω_d corresponds to electron oscillations in radial symmetric modes. This mode can only be excited by radially polarized waves. This is a dark plasmonic mode.
Source: Reproduced with permission from Ref. [46]. Copyright © 2013 American Chemical Society.

whose dipolar modes couple. This is achieved by placing three identical nanorods in a triangular geometry, as shown in Fig. 8.19(a) and using a radially polarized light. The interaction of the dipolar modes of the three nanorods can be carried out with the plasmon hybridization model. The dipolar mode of the nanorod ω_a splits into three modes [46]. Two of these modes are degenerate $\omega_b = \omega_a - AG$, and the other one is the dark mode: $\omega_d = \omega_a + 2A \cdot G$, where G is the coupling constant between the three nanorods (decrease in $1/d^3$ with the interparticle distance) and A is a constant. The charge density of these modes is shown on Fig. 8.19. The two degenerate modes ω_b correspond to dipolar modes, with linearly polarized light emission in the orthogonal directions. The last mode ω_d is not dipolar and can only be excited by a radial polarization.

The excitation of the dark plasmon mode can be monitored with the scattering spectra of the nanostructure upon two different optical excitation conditions. With radially polarized light, the maximum of extinction is detected at 535 nm and at 755 nm with a linearly polarized light, as shown in Fig. 8.20. If a "usual" linearly polarized light is used, the dark mode at 535 nm is not seen. A radially polarized light can be created with a radial polarization converter. In this case, the dark plasmon is excited at 535 nm.

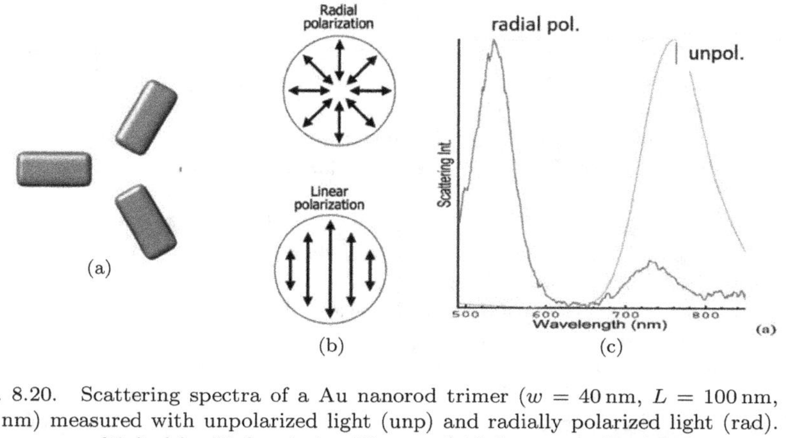

Fig. 8.20. Scattering spectra of a Au nanorod trimer (w = 40 nm, L = 100 nm, d = 110 nm) measured with unpolarized light (unp) and radially polarized light (rad). The latter was multiplied by 20 for clarity. The two-fold degenerate bright plasmon modes are detected at 755 nm, and the dark plasmon mode is at 535 nm.

Source: Reproduced with permission from Ref. [46]. Copyright © 2013 American Chemical Society.

8.5 Conclusion

This chapter developed important properties of plasmonic nanoparticles beyond the spherical model and pushed as far as possible the Mie theory. The Mie theory provides the exact calculation for large nanospheres, ellipsoids, and core–shell spherical nanostructures. As soon as more elaborate nanoparticles are considered, it is almost necessary to use a computational approach for computing the exact optical cross sections and calculating the optical spectra. We also used the dipolar model and showed how it was an efficient approach to calculate the electric field radiated by a nanostructure: It describes the near-field as well as the far-field contributions. It shows as well that the frontier between these two domains lies at distances of $\lambda/2\pi$. We also pointed out the interesting concepts of *hot spots*, where the electric field is greatly enhanced. This ability of complex nanostructures to manipulate and control this near-field contribution is of great interest for sensors and for developing new scientific approaches based, for example, on surface-enhanced Raman scattering. This will be further developed in the following chapter.

References

[1] Bohren C. F. and Huffman D. R. 1998. *Absorption and Scattering of Light by Small Particles* (New York: Wiley-Interscience).

[2] Njoki P. N., Lim I. I. S., Mott D., Park H. Y., Khan B., Mishra S., Sujakumar R., Luo J. and Zhong C. J. 2007. Size correlation of optical and spectroscopic properties for gold nanoparticles. *Journal of Physical Chemistry C* **111**, 14664-9.

[3] Mie G. 1908. Beiträge zur Optik trüber Medien, speziell kolloidaler Metallösungen. *Annalen der Physik* **25**, 377–445.

[4] Hahn D. W. 2009. *Light Scattering Theory.* ed U o Arizona.

[5] Laven P. 2019. MiePlot 4.3.05. Calculation of Mie scattering from a sphere. http://www.philiplaven.com/mieplot.htm.

[6] Oldenburg S. 2000. nanoComposix, Mie Theory Calculator.

[7] Watkins W. L. and Pluchery O. 2019. Bichromatics, spectrum calculator.

[8] Rodríguez-Fernández J., Pérez-Juste J., García de Abajo F. J. and Liz-Marzán L. M. 2006. Seeded growth of submicron au colloids with quadrupole plasmon resonance modes. *Langmuir* **22**, 7007–7010.

[9] Coronado E. A., Encina E. R. and Stefani F. D. 2011. Optical properties of metallic nanoparticles: manipulating light, heat and forces at the nanoscale. *Nanoscale* **3**, 4042.

[10] Kooij E. S., Wormeester H., Brouwer E. A. M., van Vroonhoven E., van Silfhout A. and Poelsema B. 2002. Optical characterization of thin colloidal gold films by spectroscopic ellipsometry. *Langmuir* **18**, 4401–4413.

[11] Gall D. 2016. Electron mean free path in elemental metals. *Journal of Applied Physics* **119**, 085101.

[12] Rance G. A., Marsh D. H. and Khlobystov A. N. 2008. Extinction coefficient analysis of small alkanethiolate-stabilised gold nanoparticles. *Chemical Physics Letters* **460**, 230–236.

[13] Cottancin E., Celep G., Lerme J., Pellarin M., Huntzinger J. R., Vialle J. L. and Broyer M. 2006. Optical properties of noble metal clusters as a function of the size: Comparison between experiments and a semi-quantal theory. *Theoretical Chemistry Accounts* **116**, 514–523.

[14] Dhara S. 2016. *Reviews in Plasmonics 2015.* ed C Geddes.

[15] Asano S. and Yamamoto G. 1975. Light scattering by a spheroidal particle. *Applied Optics* **14**, 29–49.

[16] Lind A. C. and Greenberg J. M. 1966. Electromagnetic scattering by obliquely oriented cylinders. *Journal of Appplied Physics* **37**, 3195–3203.

[17] Lifshitz E., Landau M. and Pitaevskii L. D. 1984. *Electrodynamics of Continuous Media* (Burlington, MA: Butterworth-Heinemann).

[18] Noguez C. 2007. Surface plasmons on metal nanoparticles: The influence of shape and physical environment. *Journal of Physical Chemistry C* **111**, 3806–3819.

[19] Berciaud S., Cognet L., Tamarat P. and Lounis B. 2005. Observation of intrinsic size effects in the optical response of individual gold nanoparticles. *Nano Letters* **5**, 515–518.

[20] Lassiter J. B., Aizpurua J., Hernandez L. I., Brandl D. W., Romero I., Lal S., Hafner J. H., Nordlander P. and Halas N. J. 2008. Close encounters between two nanoshells. *Nano Letters* **8**, 1212–1218.

[21] Novo C., Funston A. M. and Mulvaney P. 2008. Direct observation of chemical reactions on single gold nanocrystals using surface plasmon spectroscopy. *Nature Nanotechnology* **3**, 598–602.

[22] Landau L. D., Pitaevskii L. P. and Lifshitz E. M. 1984. *Electrodynamics of Continuous Media, Second Edition: Volume 8 (Course of Theoretical Physics)* (Oxford: Butterworth-Heinemann).

[23] Nehl C. L., Grady N. K., Goodrich G. P., Tam F., Halas N. J. and Hafner J. H. 2004. Scattering spectra of single gold nanoshells. *Nano Letters* **4**, 2355–2359.

[24] Rastinehad A. R., Anastos H., Wajswol E., Winoker J. S., Sfakianos J. P., Doppalapudi S. K., Carrick M. R., Knauer C. J., Taouli B., Lewis S. C., Tewari A. K., Schwartz J. A., Canfield S. E., George A. K., West J. L. and Halas N. J. 2019. Gold nanoshell-localized photothermal ablation of prostate tumors in a clinical pilot device study. *Proceedings of the National Academy of Sciences* **116**, 18590–18596.

[25] Vanderkooy A., Chen Y., Gonzaga F. and Brook M. A. 2011. Silica shell/gold core nanoparticles: Correlating shell thickness with the plasmonic red shift upon aggregation. *ACS Applied Materials & Interfaces* **3**, 3942–3947.

[26] Abadeer N. S., Brennan M. R., Wilson W. L. and Murphy C. J. 2014. Distance and plasmon wavelength dependent fluorescence of molecules bound to silica-coated gold nanorods. *ACS Nano* **8**, 8392–8406.

[27] Tzarouchis D. and Sihvola A. 2018. Light scattering by a dielectric sphere: Perspectives on the Mie resonances. *Applied Sciences* **8**, 184.

[28] Tzarouchis D. C. and Sihvola A. 2018. General scattering characteristics of resonant core–shell spheres. *IEEE Transactions on Antennas and Propagation* **66**, 323–330.

[29] Grady N. K., Halas N. J. and Nordlander P. 2004. Influence of dielectric function properties on the optical response of plasmon resonant metallic nanoparticles. *Chemical Physics Letters* **399**, 167–171.

[30] Rodríguez-Oliveros R., Paniagua-Domínguez R., Sánchez-Gil J. A. and Macías D. 2016. Plasmon spectroscopy: Theoretical and numerical calculations, and optimization techniques. *Nanospectroscopy* **1**, 67–96.

[31] Purcell E. M. and Pennypacker C. R. 1973. Scattering and absorption of light by nonspherical dielectric grains. *The Astrophysical Journal* **186**, 705.

[32] Draine B. and Flatau P. 1973. Discrete-dipole approximation for scattering calculations. *Journal of the Optical Society of America* **11**, 1491–1499.

[33] Zhang A.-Q., Qian D.-J. and Chen M. 2013. Simulated optical properties of noble metallic nanopolyhedra with different shapes and structures. *European Physical Journal D* **67**, 231.

[34] Discrete dipole approximation, Wikipedia, https://en.wikipedia.org/wiki/Discrete_dipole_approximation.

[35] Oubre C. and Nordlander P. 2004. Optical properties of metallodielectric nanostructures calculated using the finite difference time domain method. *Journal of Physical Chemistry B* **108**, 17740–17747.

[36] Lesina A. C., Vaccari A., Berini P. and Ramunno L. 2015. On the convergence and accuracy of the FDTD method for nanoplasmonics. *Optics Express* **23**, 10481–10497.

[37] Kristensen P. T. and Hughes S. 2014. Modes and mode volumes of leaky optical cavities and plasmonic nanoresonators. *ACS Photonics* **1**, 2–10.

[38] Sonnefraud Y., Leen Koh A., McComb D. W. and Maier S. A. 2012. Nanoplasmonics: Engineering and observation of localized plasmon modes. *Laser & Photonics Reviews* **6**, 277–295.

[39] Polman A., Kociak M. and García de Abajo F. J. 2019. Electron-beam spectroscopy for nanophotonics. *Nature Materials* **18**, 1158–1171.

[40] García De Abajo F. J. and Kociak M. 2008. Probing the photonic local density of states with electron energy loss spectroscopy. *Physical Review Letters* **100**, 106804–106807.

[41] Guiton B. S., Iberi V., Li S., Leonard D. N., Parish C. M., Kotula P. G., Varela M., Schatz G. C., Pennycook S. J. and Camden J. P. 2011. Correlated optical measurements and plasmon mapping of silver nanorods. *Nano Letters* **11**, 3482–3488.

[42] Prodan E., Radloff C., Halas N. J. and Nordlander P. 2003. A hybridization model for the plasmon response of complex nanostructures. *Science* **302**, 419–422.

[43] Nordlander P., Oubre C., Prodan E., Li K. and Stockman M. I. 2004. Plasmon hybridization in nanoparticle dimers. *Nano Letters* **4**, 899–903.
[44] Funston A. M., Novo C., Davis T. J. and Mulvaney P. 2009. Plasmon coupling of gold nanorods at short distances and in different geometries. *Nano Letters* **9**, 1651–1658.
[45] Baur S., Sanders S. and Manjavacas A. 2018. Hybridization of lattice resonances. *ACS Nano* **12**, 1618–1629.
[46] Gomez D. E., Teo Z. Q., Altissimo M., Davis T. J., Earl S. and Roberts A. 2013. The dark side of plasmonics. *Nano Letters* **13**, 3722–3728.
[47] Mohamed A. M. and El-Sayed M. A. 1999. Simulation of the optical absorption spectra of gold nanorods as a function of their aspect ratio and the effect of the medium dielectric constant. *Journal of Physical Chemistry B* **103**, 3073–3077.

Exercises

(1) **Gold nanorod for biosensing:** In a publication, Link and colleagues have calculated the extinction spectra of a nanorod in a dielectric medium with increasing dielectric permittivity. The *sensitivity factor* m was defined in Chapter 7 for nanosphere as $\Delta\lambda_{LSPR} = m \cdot \Delta n$. Use these data to calculate the *sensitivity factor* for these nanorods. Compare with the value of a sphere (see Chapter 7 and Exercise 5).

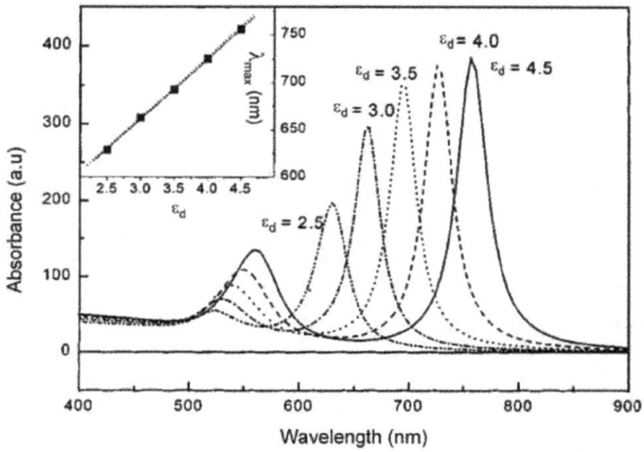

Simulated absorption spectra of gold nanorods of aspect ratio 3.3 in a medium with increasing dielectric permittivity. The inset shows a plot of the maximum of the longitudinal plasmon band determined from the calculated spectra.

Source: Reproduced with permission from Ref. [47]. Copyright © 1999 American Chemical Society.

(2) Consider a prolate ellipsoid made of gold with $a = 30\,\text{nm}$, $b = 10\,\text{nm}$, and $c = 10\,\text{nm}$.

(a) Express the Fröhlich condition on the polarizability that defines the plasmon resonance. Calculate the depolarization factors L_x, L_y, and L_z.

(b) Use the value of the complex dielectric permittivity ($\tilde{\varepsilon} = \varepsilon_1 + i\varepsilon_2$) of gold shown in the following figure to *calculate graphically* the value of the two LSPR modes of the ellipsoid.

(c) Same question with a larger a: $a = 50$ nm.

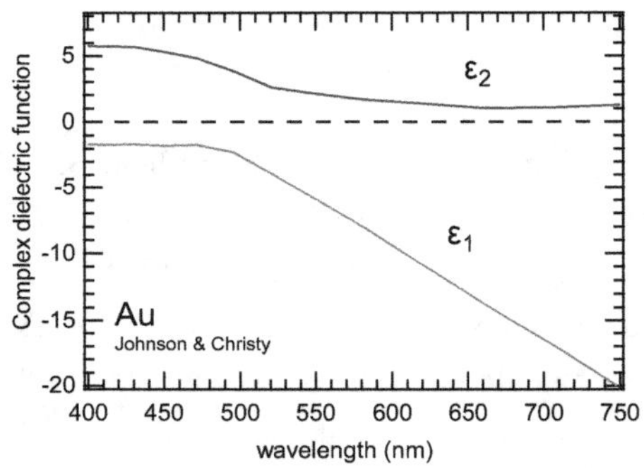

Complex dielectric functions of gold measured by Johnson and Christy (1972).
Source: https://refractiveindex.info/.

(3) In an article published in 2009, Mulvaney and his colleagues have prepared gold nanorods, as shown in the TEM image (a) in the following figure. They also measured their extinction spectra in water [44].

(a) Determine the average aspect ratio of these nanorods. Also measure as accurately as possible the two LSPR modes of the nanorods in water.

(b) We model these nanorods with ellipsoids of the same aspect ratio. Is this a prolate or oblate ellipsoid? Calculate the depolarization factors L_x, L_y, L_z.

(c) Use the results of the previous exercises to calculate the two LSPR modes of this ellipsoid. Compare these values with the experimental ones. Try to give some reasons for the poor accuracy of modeling nanorods with ellipsoids.

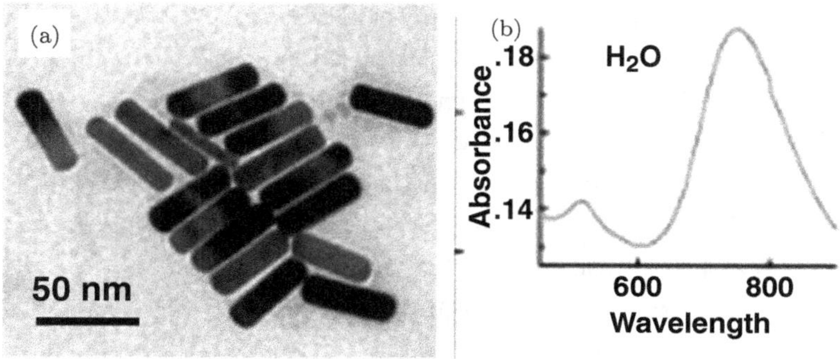

(a) TEM image of gold nanorods and (b) absorbance spectrum of these nanorods in water.
Source: Reproduced with permission from Ref. [44]. Copyright © 2009 American Chemical Society.

(4) Here, we consider spherical core–shell nanoparticles: with a metallic core and a dielectric shell placed in air. The metal is described by a simple Drude model: $\varepsilon = \varepsilon_\infty - \omega_p^2/\omega^2$. Demonstrate that the plasmon resonance frequency is

$$\omega_\pm = \frac{1}{2}\omega_p^2\left[1 \pm \frac{1}{3}\sqrt{1+8\eta^3}\right].$$

(5) [Exercise involves a calculation software] Consider a spherical Au@SiO_2 core–shell nanoparticle in water. The core diameter is 12 nm and the shell is 4 nm thick. Use the Equation (8.14) and calculate the polarizability of this nanoparticle. Use the complex refractive index from www.refractiveindex.info. Calculate the extinction cross section from α. Plot this value as a function of the wavelength.

(6) Use the online calculator available at https://bichromatics.com/calculator/.

  (a) Calculate the extinction cross section of the Au@SiO_2 core–shell nanoparticle of the previous exercise. Compare with the values of λ_{LSPR} obtained with the plot made in the previous exercise.

(b) Plot now the two core–shell spherical nanoparticles $Au@SiO_2$ and $SiO_2@Au$. In both cases, the inner diameter is 40 nm and the outer diameter is 60 nm. The nanoparticles are in water. Extract the values of λ_{LSPR}. Deduce and then explain the best strategy for the researchers should they wish to use these core–shell nanoparticles in the infrared for thermotherapy.

(7) **Near field in a core–shell nanoparticle:** Consider a spherical $Au@SiO_2$ nanoparticle with an inner diameter $D_1 = 60\,\text{nm}$ and an outer diameter $D_2 = 80\,\text{nm}$, as shown in the following figure. The nanoparticle is placed at the origin of a referential ($\boldsymbol{u}_x$, $\boldsymbol{u}_y$, $\boldsymbol{u}_z$). A plane wave $E_0\exp(i(kx - \omega t))\boldsymbol{u}_x$ excites this particle. The spherical coordinates are used.

(a) The electric potential scattered by the particle $V_{sca}(r, \theta)$ is given by Equation (8.11). Derive the electric field.

(b) The values for the dielectric permittivities at 550 nm are: $\varepsilon_1 = \varepsilon_{Au} = -6 + 2i$ and $\varepsilon_2 = \varepsilon_{SiO2} = 2.3$. Use the expression of the electric field and calculate the amplitude of the field at points A and B of the nanoparticle.

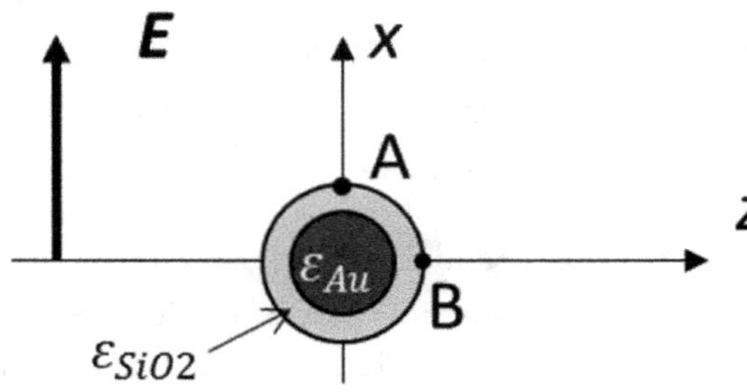

Sketch of core–shell $Au@SiO_2$ excited by a plane wave with the electric field polarized along the x-axis.

9

LSPR for Nano-optics and Biomedical Applications

This chapter mainly offers illustrative and selected examples of localized surface plasmon resonance (LSPR) for nano-optics and biomedical applications. The first part focuses on using LSPR to achieve direct biosensing. Then, we show the benefits of localized plasmons to enhance the sensitivity of some spectroscopy techniques, such as Raman spectroscopy.

9.1 Introduction

Localized surface plasmon resonance (LSPR), whose properties have been described in the previous chapters, can be used for biosensing. Indeed, the resonance of the localized plasmon, in particular, its intensity and wavelength position, is very dependent on the shape, composition, or size of the nanostructure. In addition to these parameters, the value of the refractive index of the surrounding medium also needs to be considered. As in the case of propagative surface plasmons, all these elements have led to the development of LSPR biosensors. One of the main challenges of this class of biosensors, especially LSPR sensors, is to achieve well-defined and uniform nanostructures and obtain a high reproducibility of the LSPR spectra. Therefore, the choice of the fabrication or synthesis method remains essential.

9.2 The Challenge of Nanoparticles Synthesis: Top-down vs. Bottom-up

Historically, the fabrication of nanostructures/nanoparticles has been categorized into two approaches: *top-down* and *bottom-up*. The first is based on the reduction of bulk material to the desired size, while the second is based on the growth of particles from primary nuclei (seeds of a few nm). Each approach has its own advantages: spatial arrangement and stability for the first one, crystallinity and tiny size for the second one. For a long time, there has been a strict frontier between these two approaches, but this has tended to blur nowadays. For example, nanosphere lithography combines both approaches.

This short section highlights three distinct ways to produce an LSPR biosensor. The first way is to use synthesized nanoparticles in a solution. The second way is to deposit these synthesized nanoparticles onto a surface. The last one directly produces an array of nanoparticles on a substrate with a "top-down" approach. In all three cases, we deal with LSPR biosensors since the principle of measurement is always based on the plasmonic properties of nanoparticles. However, some measurement techniques are used preferentially for one or the other configuration of nanoparticles, such as spectral measurement in solutions (detailed in Section 9.3.2).

9.3 Versatility of LSPR Biosensors

9.3.1 *LSPR shift: A refractive index measurement*

Several experimental methods can be used to detect the presence of biological elements via localized surface plasmons. The first approach is to measure the absorption or extinction spectrum and monitor the spectral shift of the maximum as a function of wavelength. Indeed, the presence of biological elements linked to the nanoparticle surface modifies the surrounding refractive index and induces a spectral shift. Then, we can define the wavelength shift as

$$\Delta\lambda = \lambda_{max,\,molecule} - \lambda_{max,ref}. \tag{9.1}$$

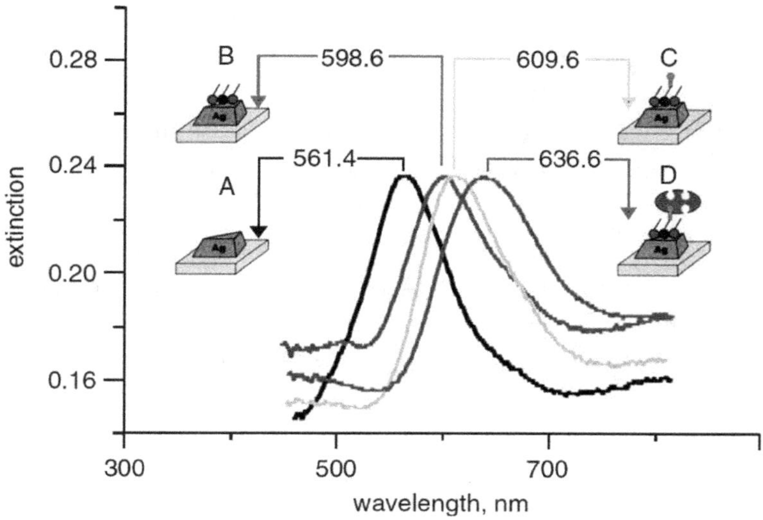

Fig. 9.1. LSPR spectra of the four steps to form a biotinylated Ag nanobiosensor from Ag nanoparticles: (A) unfunctionalized; (B) after immersion in 1 mM, 1:3 1-OT/11-MUA; (C) following absorption of 1 mM biotin; (D) after modification with 100 nM streptavidin. All extinction measurements were performed in a N2 environment.

Source: Reproduced with permission from Ref. [1]. Copyright © 2005 Institution of Electrical Engineers.

First, we present an experiment in which the extinction spectra of nanoparticles are measured with and without biomolecules.

For instance, Fig. 9.1 shows silver nanoparticles on an ITO substrate obtained by nanosphere lithography. The extinction spectra exhibit an LSPR peak $\lambda_{max,A}$ close to 561.4 nm. It corresponds to step A, which is the reference prior to molecule adsorption. This value is linked to the material (silver), size, and height of the nanostructures (~100 nm wide and 50 nm high) and their triangular shape. In step B, the nanoparticles are functionalized with a solution of 1 mM, 3:1 of 1-octanethiol/11-mercapto undecanoic acid (1-OT/11-MUA) in ethanolic solution for 24 h. This step corresponds to a surface coverage of 0.1 monolayer of carboxylate (~6,000 sites/nanoparticle). With this information, the authors could evaluate the number of biotin sites (~60–300 sites/particle) and calculate the resolution of their experiment. After this chemical functionalization in step B, the $\lambda_{max,B}$ is 598.6 nm, corresponding to a shift of 37.2 nm. Then, in step C, the substrate is exposed to a 1 mM solution of biotin, resulting in a shift of 11 nm ($\lambda_{max,C}$ = 609.6 nm). The biotin forms covalent links with the carboxyl group of 11-MUA.

At that stage, the Ag nanostructures are prepared to detect the target molecules, which is streptavidin here. This molecule is used at a concentration of 100 nM in step D. The new maximum $\lambda_{max,D}$ is at 636.6 nm, corresponding to a shift of 27 nm. This straightforward example shows how LSPR biosensors can be used to detect a concentration of streptavidin as low as 100 nM via a spectral shift. This measurement can be generalized to other proteins, antibodies, or molecules for many applications (health, military, food safety, etc.) by selecting the appropriate functionalization chemistry.

9.3.2 *Spectral measurement and color observed*

We have described a detection method that monitors the shift of the LSPR maximum in the extinction spectra. This spectral shift can be associated with a color change. Indeed, we know that nanoparticles will be seen with different colors depending on their nature, size, geometry, surrounding medium, and aggregation phase (see Chapters 8 and 10). Therefore, the LSPR biosensor can be considered as a color test. This type of sensor is essentially made of suspended nanoparticles. The word "colorimetric" can be found in scientific literature for this class of sensors. However, this term has different significations depending on the application field, and its use is often ambiguous in chemistry, physics, and other fields.

We know that the addition of molecules that eventually bind to the nanoparticles progressively shifts the maximum of the extinction spectrum. This variation in the wavelength is therefore associated with a color variation of the solution. When the shift is strong enough, the color variation may be perceptible to the naked eye, and different strategies are available to increase the detection threshold.

One of the first experiments in this field that revolutionized this approach was performed by Mirkin *et al.* in 1996. The idea was to bind nanoparticles to macroscopic materials by using DNA. They functionalized

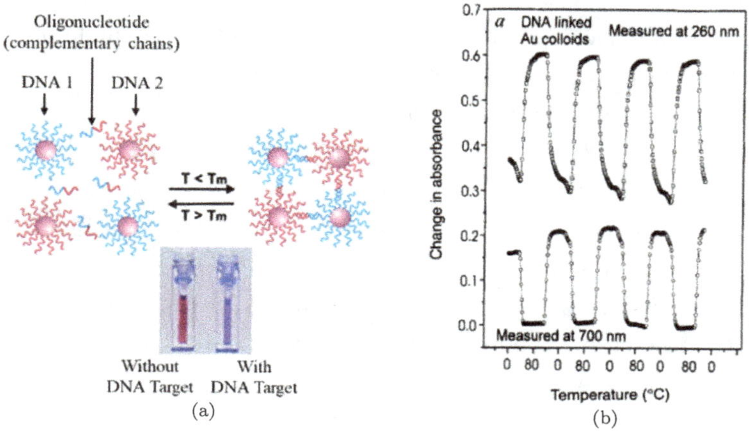

Fig. 9.2. (a) Schematic illustration of 13 nm gold nanoparticles functionalized with two different chains of DNA. When the target DNA (oligonucleotide with complementary chain) is injected, the color of the solution changes from red to purple. (b) Absorbance of the solution of DNA and particles as a function of temperature/time. The low temperature allows hybridization of DNA and aggregation of nanoparticles (purple color). The high temperature dehybridizes the complex, which causes it to return to its initial state (red color).

Source: (a) Reproduced with permission from Ref. [2]. Copyright © 2005 American Chemical Society. (b) Reproduced with permission from Ref. [3]. Copyright © 1996 Nature Publishing Group.

13 nm gold nanoparticles with two non-complementary DNA chains modified with a thiol group (DNA 1 and 2 in Fig. 9.2(a)). In a solution, due to the non-complementarity of each chain, nanoparticles remain separated. Then, by adding the oligonucleotide containing the complementary chain of each DNA chain, the nanoparticles aggregate. This agglomeration drastically changed the absorbance of the solution and therefore its color from red to purple. One of the outstanding features of this experiment was its reversibility by increasing the temperature (denaturation of DNA), as shown in Fig. 9.2(b). On increasing the temperature from 0 to 80°C, the nanoparticles become separated, thus modifying the absorbance and color of the solution.

Since this experiment in 1996, the observed color change based on LSPR has been advertised as the means to an easy, fast, and low-cost technique in the scientific community.

Here, we have seen the main strategy of inducing the aggregation of nanoparticles to cause a shift in the wavelength of the maximum of extinction. However, as we know, the color of the solution also depends on the size and shape of the nanoparticles (with sufficient variation to be visible to the naked eye being >10 nm). Thus, another strategy is based on the creation of reactions that will etch or expand nanoparticles and thereby induce a change in color. We have the following example, as shown in Fig. 9.3, of a multicolor glucose sensor.

In this work, the authors used a 3,3′,5,5′-tetramethylbenzidine(II) (TMB^{2+}) to etch Au nanorods (AuNRs). As a consequence, they obtained different shapes and sizes of nanorods corresponding to several colors of the solution.

To obtain TMB^{2+}, the product of the H_2O_2–HRP–TMB–HCl catalyzed oxidation system, the authors used several chemical reactions. First, glucose

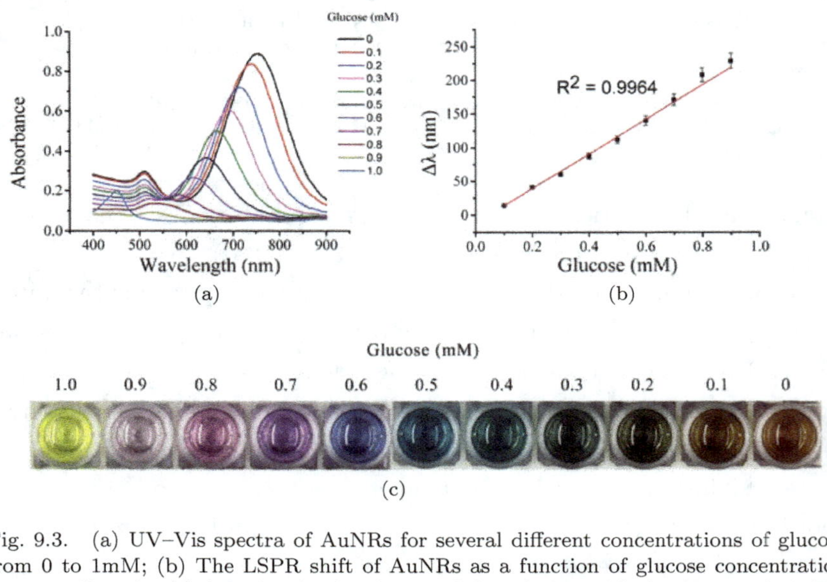

Fig. 9.3. (a) UV–Vis spectra of AuNRs for several different concentrations of glucose from 0 to 1mM; (b) The LSPR shift of AuNRs as a function of glucose concentration increases linearly; (c) Associated color change of the solution with the decrease in initial glucose concentration.

Source: Reproduced with permission from Ref. [4]. Copyright © 2016 Springer Nature.

reacts with dissolved oxygen in the presence of glucose oxidase (GOx) to produce H_2O_2. Next, the authors used H_2O_2 in the presence of horseradish peroxidase (HRP) to oxidize the TMB^0 to blue TMB^+. Then, the HRP is inactivated by adding hydrochloric acid (HCl) to the solution, and the TMB^{2+} is produced. Depending on the initial glucose concentration value, the authors obtained several shades of yellow. To enable naked eye visualization, a solution of AuNRs and cetyltrimethyl ammonium bromide (CTAB) is added to the TMB^{2+} solution.

The AuNRs react with TMB^{2+} to produce TMB^0, while Au oxidizes to Au(I) at the ends of the nanorods. Thus, different shapes and sizes of nanorods were obtained, which are directly linked to the initial concentration of glucose (Fig. 9.3(a)). The maximum of LSPR blueshifts from approximately 770 nm (0 mM of glucose) to 520 nm (0.9 mM of glucose). So, each concentration induces the oxidation etching process in differentiated stages, resulting in smaller nanorods. Based on this, the authors have reported several final solutions with a wide range of rich colors (Fig. 9.3(c)), from pink to green to red, which can be easily differentiated by the naked eye. When all the AuNRs are totally oxidized, only TMB^{2+} remains in the solution that induces the initial yellow color (see the case of 1 mM of glucose in Figs. 9.3(a) and (c)).

Under optimized conditions of concentrations of each catalyzer and inhibiter, the authors determined a linear correlation between the shift of the longitudinal mode of LSPR and the glucose concentration in the range 0.1–1.0 mM (Fig. 9.3(b)). Compared with the traditional colorimetric sensor, which is often monochromatic, this new multicolor glucose sensor provides several strong and diverse colors, easily distinguished by the naked eye. Moreover, this method has been successfully applied to detect glucose in serum samples with high specificity.

This second method highlights the strong correlation between the LSPR position and the particle geometry. By taking advantage of this dependence, the LSPR biosensors based on color variation are able to provide an easy way to detect the presence of molecules or elements in a solution.

9.4 Localized Plasmon and Electric Field Enhancement and Surface-Enhanced Raman Spectroscopy (SERS)

In each one of the previous cases, the working principle of the biosensors was based on the strong dependence of the LSPR on some properties of the nanostructures used as plasmonic transducer: their nature, morphology, or chemical environment. In addition to these properties, it should also be mentioned that localized plasmons are at the origin of the exaltation of local electric and magnetic fields. In the following, we focus more specifically on the Raman effect (inelastic scattering) and the key role of plasmon.

9.4.1 *Historical overview of SERS*

An excellent and detailed book written by Le Ru and Etchegoin describes the Raman effect's exaltation [5]. It covers all the physical mechanisms involved, the definitions and calculation methods of the exaltation factors. Here, we only mention the key points required to understand the mechanism and calculation methods. We also illustrate how the calculation of the enhancement factor is carried out through a practical example.

Experimentally observed in 1928 by Sir C.V. Raman, the Raman effect has not been widely used due to its very low yield: only one photon out of 10^6 contribute to the Raman signal. This being the case, for decades, infrared spectroscopy and fluorescence spectroscopies were used preferentially. However, in 1974, Fleischmann highlighted an increase in the intensity of the characteristic peaks of a molecule compared to the conventional measurement in solution by observing the Raman spectrum of pyridine on a rough silver surface. Even if he did not attribute this phenomenon to the presence of a rough metal film, it was the first experimental observation of the surface-enhanced Raman spectroscopy (SERS). In 1977, Professor R. Van Duyne and his colleague David L. Jeanmaire hypothesized that the increase in the signal was not due to the higher specific surface area (as supposed by Fleischmann) but linked to the intrinsic properties of the surface. Moskovits pointed out that the exaltation of the Raman signal was the result of two contributions: an electromagnetic effect linked to the substrate (shown by Van Duyne) and a chemical contribution (by charge transfer) highlighted by a second group in the same year (Creighton and Albrecht). The two contributions are in practice complementary and concomitant. Since the discovery of its higher intensity, SERS has become very attractive for a wide range of applications, including the detection of a single molecule.

9.4.2 *Brief description of Raman effect based on the classical approach*

Let's represent a diatomic molecule with a simple model made of a two-mass harmonic oscillator, and let's apply Newton's second law in a supposed Galilean reference frame. We consider that the only force applied to this oscillator is given by Hooke's law. We reach the following equation:

$$\mu\frac{d^2q}{dt^2} = -Kq, \tag{9.2}$$

where q corresponds to the total displacement, K is the spring constant, and μ the reduced mass $(m_1 \cdot m_2/m_1 + m_2)$. The solution to this equation is written as

$$q = q_0\cos(2\pi\upsilon_m t), \tag{9.3}$$

where q_0 is the oscillation amplitude and υ_m the frequency of molecular vibration defined as $\upsilon_m = \frac{1}{2\pi}\sqrt{\frac{K}{\mu}}$. This frequency depends on K and μ and is unique for each molecule. That is why the Raman measurement is used to identify molecules considered as pollutants, for instance.

Let us now consider that the molecule is exposed to monochromatic incident light with a frequency υ_0. The electric field is written as $\boldsymbol{E} = \boldsymbol{E}_0\cos(2\pi\upsilon_0 t)$. This induces a dipole moment denoted by $\boldsymbol{P}$ and expressed as $\boldsymbol{P} = \overline{\overline{\alpha}}\boldsymbol{E}$. $\overline{\overline{\alpha}}$ corresponds to the polarisability tensor associated with the molecule and describes the deformation of its electronic cloud. For small oscillations, i.e. small excitations, the first order of the linear approximation yields

$$\overline{\overline{\alpha}} \simeq \overline{\overline{\alpha_0}} + q\left(\frac{\partial\overline{\overline{\alpha}}}{\partial q}\right)_{q=0}. \tag{9.4}$$

After some calculations, we obtain

$$\vec{P} = \overline{\overline{\alpha_0}}\overrightarrow{E_0}\cos(2\pi \upsilon_0 t) + \frac{1}{2}\left(\frac{\partial\overline{\overline{\alpha}}}{\partial q}\right)_{q=0} q_0\overrightarrow{E_0}\left[\cos\left(2\pi\left(\upsilon_0 - \upsilon_m\right)t\right)\right.$$
$$\left. + \cos\left(2\pi\left(\upsilon_0 + \upsilon_m\right)t\right)\right]. \tag{9.5}$$

Expression (9.5) highlights three kinds of emissions. The first term only involves the incident frequency υ_0. There is no variation in frequency. It corresponds to an elastic diffusion called Rayleigh diffusion. The two other terms correspond to Raman scattering with a variation in the transmission frequency $\upsilon_0 \pm \upsilon_m$. The downshift of frequency $(\upsilon_0 - \upsilon_m)$ reflects a decrease in energy compared to the incident wave. It refers to the *Raman Stokes diffusion*. By analogy, the term of higher frequency corresponds to the *anti-Stokes Raman* scattering. The two phenomena are symmetrical on either side of Rayleigh scattering, and experimentally, only half of the spectrum is measured.

To explain the widening of Stokes and anti-Stokes rays and the difference in magnitude, it is useful to consider the problem from a quantum point of view. The energy levels of molecules are quantified and represented in a Jablonski diagram. Therefore, the Stokes and anti-Stokes scattering add to these discrete levels some supplemental peaks that are spectrally narrow and are a spectral signature of the molecule under study. The introduction of virtual vibrational levels helps to explain the experimental broadening observed in Stokes and anti-Stokes rays. In addition, the probability that the molecule is initially in an excited vibrational state is weaker than the probability that it is in its fundamental state. Thus, the transition corresponding to the Raman Stokes diffusion will be more intense (more likely) than the anti-Stokes one. It should be noted that temperature strongly influences this intensity ratio between the Stokes and anti-Stokes rays (explained by Boltzmann's law).

Now that the Raman effect is described (position of each peak and principle of experiments), we discuss the two contributions (electromagnetic and chemical) involved in the exaltation of the Raman signal from a molecule adsorbed on a surface. The discussion about these two contributions has given rise to intense debates in the spectroscopist community. Now, there is a general agreement on how to account for these two effects. These debates show that the research is continuously evolving and going through constant refinement. It is not impossible that better understanding, improved measurement techniques, well-designed and reproducible samples, and new analytical models modify the SERS theory.

9.4.2.1 *Electromagnetic contribution*

For ease of understanding, we separate into two steps the excitation and emission of the Raman dipole. In reality, the Raman scattering process is instantaneous, and both (excitation and emission) occur simultaneously. First, the interaction of the incident beam with the nanoparticle induces a local increase in the field strength (E_{loc}) near the molecule compared to the incident field (E_0). This increase is governed by the structure (tip enhancement) and is intensified when the excitation frequency comes closer to the resonance condition of the localized surface plasmon. The second process (emission) is induced by the presence of a molecule with an emission frequency (ω_R). Then, the final scattered intensity will be proportional to these two contributions:

$$|E_{scat}(\omega_0, \omega_R)|^2 \propto M_{loc}(\omega_0) M_{scat}(\omega_R) \, |E_0|^2 . \tag{9.6}$$

As a first approximation, $M_{loc} \simeq M_{scat}$. Moreover, both terms are proportional to the square of the ratio between the local field and the incident field. Therefore, a global factor G is defined as

$$G = M_{loc}(\omega_0) M_{scat}(\omega_R) \propto \frac{|E_{loc}(\omega_0)|^2}{|E_0(\omega_0)|^2} \times \frac{|E_{loc}(\omega_R)|^2}{|E_0\,(\omega_0)|^2} . \tag{9.7}$$

This expression is called the $|E|^4$ *approximation.*

If we assume that the Raman scattering frequency ω_R is close to the excitation frequency ω_0, i.e. $\omega_R \simeq \omega_0$, we can rewrite the expression as the following form:

$$G = \frac{|E_{loc}\,(\omega_0)|^4}{|E_0\,(\omega_0)|^4} . \tag{9.8}$$

9.4.2.2 *Chemical contribution*

This contribution suggested by Albrecht has been described by Moskovits as the modification of the Raman polarizability tensor of the molecule due to the formation of a molecule–substrate complex. This chemical contribution is characterized by a length in the order of ångström (in contrast to the electromagnetic one, which can occur from few nm until 100 nm). The modification of the polarizability is attributed to a charge transfer that results in a change in the electronic structure of the molecule and, consequently, its polarizability. Generally, the density functional theory is used to calculate this contribution, which remains quite low ($<10^2$).

Here, we consider a small difference between the Fermi level of the metal and one of the two molecular orbitals of interest (the highest occupied, or HOMO, and the lowest unoccupied, or LUMO, molecular orbital). When the incident beam has an energy close to the energy difference of Fermi-HOMO or Fermi-LUMO, it is possible that the electron initially occupying the HOMO is transferred to an unoccupied energy level close to the Fermi level. Similarly, an electron from an occupied energy level can be transferred to the LUMO. Therefore, the Fermi level of the metal is used as an intermediate level to facilitate these transitions. Thus, with a simple approach, we attribute the chemical enhancement to charge transfer and metal–molecular complex. However, we need to mention that this charge transfer approach does not involve the orientation of the molecule, whereas the SERS is a powerful technique to measure it. To take into account the orientation of the molecule, it is required to calculate the modification of the efficient Raman cross section of the molecule.

9.4.2.3 *SERS signal*

Considering the two contributions (electromagnetic and chemical), we can write the SERS intensity (I_{SERS}) in terms of the Raman intensity (I_{Raman}) as

$$I_{SERS} \simeq M_{loc}(\omega_0) M_{loc}(\omega_R) \frac{|\alpha_R|^2}{|\alpha_0|^2} I_{Raman}, \tag{9.9}$$

where α_R is the modified polarizability of the molecule at the Raman frequency. The M_{loc} product represents the electromagnetic contribution (10^4–10^7), whereas the ratio of the polarizabilities corresponds to the chemical contribution (10^2). This equation highlights the benefits of having a surface for Raman experiments. The Raman experiment shown in Fig. 9.4 is a good illustration of the Raman exaltation. With a detection limit as low as one single molecule, it demonstrates the true significance of SERS. For both experiments, the spectra of Rhodamine 6G were acquired after excitation by a laser with a power of 3 mW at 633 nm. For the non-SERS (or "classical" Raman experiment), a solution of 100 μM of RH6 G in water was used. The integration time was 400 s. The number of molecules (7.8×10^5) was estimated from the concentration and the scattering volume of the system ($13\,\mu m^3$). The bottom curve in Fig. 9.4, which was recorded in these conditions of *classical Raman* experiment, exhibits small peaks approximately

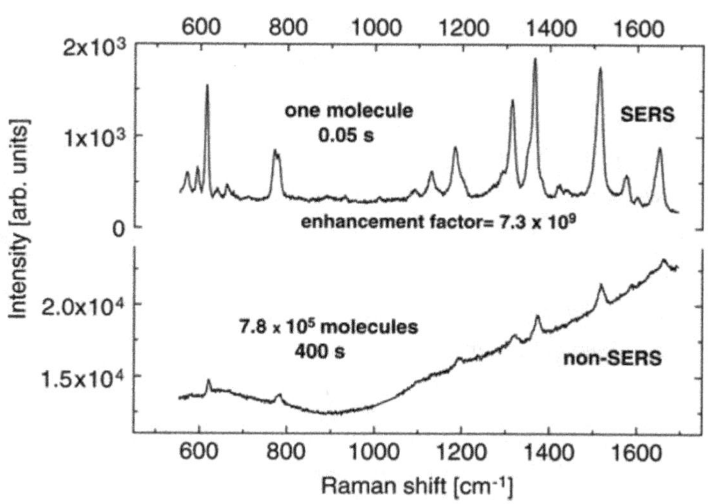

Fig. 9.4. Comparison of Raman (non-SERS, bottom) and SERS spectra (top) for rhodamine 6G molecules. For the Raman spectrum, the signal corresponds to 7.8×10^5 RH 6G molecules (100 M solution in a $13\,m^3$ scattering volume, the integration time is 400 s with a ×100 objective). Top: signal from a single RH 6G molecule under the same experimental conditions but with a 0.05 s integration time. To progress from the lower to the upper spectrum, an amplification of the Raman signal by an enhancement factor of 7.3×10^9 is required.

Source: Reproduced with permission from Ref. [5]. Copyright © 2009 Elsevier.

at 613, 775, 1184, 1312, 1364, 1512, and 1651 cm^{-1}. They are characteristic to Rhodamine 6G. In a second experiment with SERS conditions, only one molecule is measured, and the spectrum required an integration time of only 0.05 s. The spectrum displays stronger characteristic peaks. After normalization of both signals, this amplification corresponds to an enhancement factor of 7.3×10^9. This enormous amplification is attributed to the Ag colloidal solution that allows SERS. In addition, more characteristic peaks are available due to the stronger detection and the configuration of molecules on the surface of Ag nanoparticles.

Note that the spectra are given with the Raman shift in cm^{-1}. This value corresponds to the energy difference between the initial and final vibrational levels. To express the Raman shift, which is given in wave number (cm^{-1}), in terms of wavelength, the following formula with an appropriate coefficient is used:

$$\Delta\nu[cm^{-1}] = \frac{10^7}{\lambda_{ex}[nm]} - \frac{10^7}{\lambda[nm]}. \tag{9.10}$$

To compare different SERS experiments and identify the best sample (i.e. the best parameters of the nanostructured surface as well as the material), several enhancement factors (EFs) were defined during the past decade. Here, we give the two main EFs, following the definitions given in the book by Le Ru and Etchegoin [5].

The first enhancement factor is the analytical enhancement factor (AEF) and is the most intuitive. It relies on the variation in the intensity with the concentration of the analyte. The AEF is most suitable for solutions with nanoparticles, as opposed to "classical" SERS substrates, but remains a quick and reproducible factor. Its formula is:

$$AEF = \frac{I_{SERS}/c_{SERS}}{I_{Raman}/c_{Raman}},$$

where I_{SERS} and I_{Raman} are respectively the Raman intensities for the molecule adsorbed on the nanostructured surface (on the substrate) and the molecule in solution (Raman experiment used as reference). In the same way, c_{SERS} and c_{Raman} are the given concentration of the analyte for the two experiments (where the parameters, such as laser power, laser wavelength, and spectrometer settings, are kept identical).

In the case of a SERS substrate, it is better to use the substrate surface-enhanced factor (SSEF). The two intensities have the same definition as previously:

$$SSEF = \frac{I_{SERS}/N_{Surf}}{I_{Raman}/N_{Vol}} = \frac{I_{SERS}}{I_{Raman}} \times \frac{\mu_M \mu_S A_m}{c_{Raman}.H},$$

where N_{Surf} and N_{Vol} are the numbers of molecules on the sample surface and in the bulk (solution) illuminated by the laser, respectively. N_{Surf} is given by the average number of adsorbed molecules for a scattering volume V. $N_{Vol} = c_{Raman}V = c_{Raman}.H.A_{eff}$ corresponds to the average number of molecules in this volume. H is the height of the scattering volume, and A_{eff} is an intermediary parameter simplified in the equation $N_{Surf} = NA.\mu_M \mu_S A_m A_{eff}$, with μ_S the surface density of molecules (mol/m^2). A$_M$ is the metallic surface area of each nanostructure and μ_M (mol/m^2) the surface density of one nanostructure.

Although SSEF ignores the nonuniformity and orientation of the molecules on the sample (average), it is recognized as one of the best ways to characterize a serum substrate. The SSEF factor is valid only for a monolayer of molecules since the SERS is linked to the distance between the molecules and the active surface. The next layers will have less impact on the enhancement factor.

9.5 Strategies for Maximizing the SERS Signal and Perspectives

To optimize the SERS signal, several strategies can be employed [6]. We present two of them, on which there is a common agreement.

First, it is useful to select the laser excitation wavelength according to the LSPR resonance and the Raman vibration. Indeed, a phenomenological treatment predicts the maximal SERS response when $\lambda_{LSPR} = (\lambda_{Ex} + \lambda_{Raman})/2$ [7]. This equation is effective while the excitation is below the near-infrared range. Moreover, it has been verified for substrates with EBL or NSL nanostructures but not for clusters of randomly deposited particles where nanogaps occur.

In addition, depending on the size and concentration of the molecules, it is also important to measure the new position of extinction of the sample to adjust the excitation wavelength. Thus, for an array of uncoated Ag nanoparticles produced by nanosphere lithography, Van Duyne's group has measured the LSPR λ_{max} at 672 nm (Fig. 9.5). When incubated in 1 mM benzenethiol (for three hours) and then rinsed and dried, the sample provides a new maximum for $\lambda_{LSPR} = 729\,\text{nm}$ (blueshifted from 57 nm). In parallel, the measurement of the enhancement factor profile gives $\lambda_{ex,max} = 692$ nm. This profile is blueshifted with respect to the LSPR λ_{max} of the coated sample but redshifted in comparison to the LSPR λ_{max} of the uncoated nanoparticle array. This highlights that it is crucial to perform a characterization of the LSPR of a SERS substrate before and after

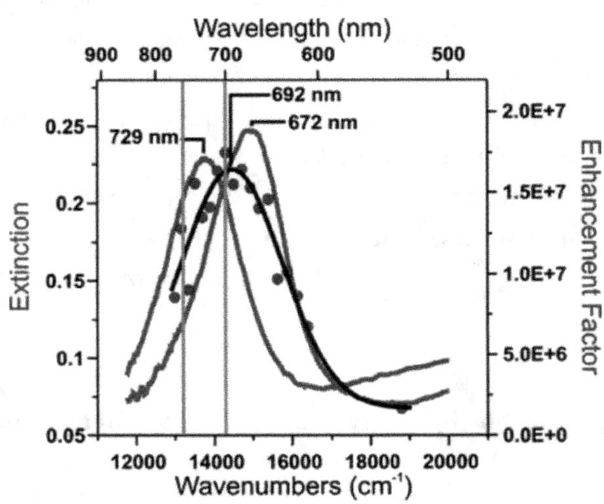

Fig. 9.5. LSPR spectra of Ag nanoparticles unfunctionalized (672 nm) and after functionalization in 1 mM of benzenethiol (729 nm), with the associated enhancement factors depending on the excitation wavelength in order to confirm the relation $\boldsymbol{\lambda_{LSPR} = (\lambda_{Ex} + \lambda_{Raman})/2}$.

Source: Reproduced with permission from Ref. [8]. Copyright © 2005 American Chemical Society.

adsorption of the analyte in order to select the appropriate laser excitation wavelength to maximize the EF or to establish conclusions [8]. In addition, we verified and confirmed that we have $(\lambda_{Ex}+\lambda_{Raman})/2 = (692+776)/2 = 734\,\text{nm}$ close to the $\lambda_{LSPR} = 729\,\text{nm}$. The λ_{Raman} has been determined for the characteristic peak at $1575\,\text{cm}^{-1}$.

Another key effect used to enhance the SERS signal is to bring very close several nanostructures in order to generate a "hot spot" of electromagnetic field. In this case, the LSP of each nanostructure can be coupled to create a strong near field in the space between them. The enhancement of the field is then far greater than the one obtained with individual nanostructures. Typically, the gap size to provide LSP coupling must be less than a few tens of nanometers (in regard to the exponential decrease in the LSP amplitude, the shape of nanoparticles, and the excitation wavelength). For instance, the team of Bachelot and Xiong has realized bowtie antenna with nanogaps [9]. The SERS spectra of their measurement are shown on Fig. 9.6. As the distance between the particles decreases, the SERS intensity increases (Fig. 9.6b). This effect is mainly visible for parallel polarization excitation

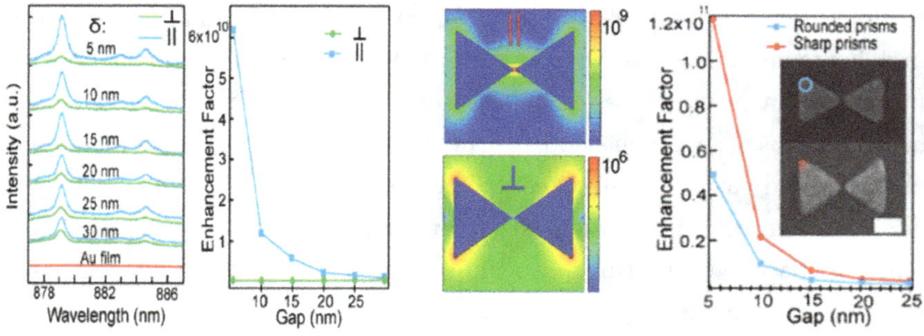

Fig. 9.6. (a) Raman spectra and (b) the enhancement factor of prism dimers with varying gaps between prisms for both polarizations. (c) Simulated electric field contour plot of both modes. (d) The enhancement factor of prism dimers with varying gaps between prisms for two different bowties. The inset shows the SEM image of round and sharp prism dimers. The scale bar is 100 nm.

Source: Reproduced with permission from Ref. [9]. Copyright © 2013 American Chemical Society.

(Fig. 9.6c), leading to a significant increase in the electric field between the tips (gap area) of the bowtie antenna. The value can reach 10^9 in the gap. As observed, the roughness and sharpness of the structures affect the value of the final enhancement factor.

9.6 Conclusion

The LSPR biosensors are very efficient and allow detection of the presence of molecules through an easy optical measurement or by color variation. With appropriate optimization, it is even possible to detect a single molecule (see SERS experiments). Nevertheless, many methods still need to be mastered, particularly in the calculation of LSPR resonances for complex geometries (nanostructures with metallic substrates, considering coupling with propagative surface plasmons).

This chapter has covered the important notion of the resonance of localized plasmon applied to biosensing. Mainly based on the shift of this resonance, the LSPR sensor can be optimized. We have shown the evolution of LSPR biosensors through the first historical experiment conducted by Mirkin *et al.* We have also provided an example for Raman spectroscopy and how the signal has been enhanced by several orders thanks to the presence of nanostructures. To characterize and compare the performance of SERS biosensors, a method to calculate their EF was given, including chemical and electromagnetic contributions. Then, we presented two optimization strategies using the position of the LSPR resonance as well as the nanogap between structures. An important point about LSPR resonance concerns the geometry of nanostructures that can considerably influence the final performance of the biosensors.

References

[1] Stuart D. A., Haes A. J., Yonzon C. R., Hicks E. M. and Van Duyne R. P. 2005. Biological applications of localised surface plasmonic phenomenae. *IEE Proceedings – Nanobiotechnology* **152**(1), 13.

[2] Rosi N. L. and Mirkin C. A. 2005. Nanostructures in biodiagnostics. *Chemical Reviews* **105**(4), 1547–1562.
[3] Mirkin C. A., Letsinger R. L., Mucic R. C. and Storhoff J. J. 1996. A DNA-based method for rationally assembling nanoparticles into macroscopic materials. *Nature* **382**(6592), 607–609.
[4] Lin, Y., Zhao, M., Guo, Y., Ma, X., Luo, F., Guo, L., Qiu, B., Chen, G. and Lin, Z. 2016. Multicolor colormetric biosensor for the determination of glucose based on the etching of gold nanorods. *Scientific Reports* **6**(1), 1–7.
[5] Principles of Surface-Enhanced Raman Spectroscopy. 2009. Elsevier.
[6] Guo L., Jackman J. A., Yang H.-H., Chen P., Cho N.-J. and Kim D.-H. 2015. Strategies for enhancing the sensitivity of plasmonic nanosensors. *Nano Today* **10**(2), 213–239.
[7] Félidj N., Aubard J., Lévi G., Krenn J. R., Hohenau A., Schider G. and Aussenegg F. R. 2003. Optimized surface-enhanced Raman scattering on gold nanoparticle arrays. *Applied Physics Letters* **82**, 3095–3097.
[8] McFarland A. D., Young M. A., Dieringer J. A. and Van Duyne R. P. 2005. Wavelength-scanned surface-enhanced Raman excitation spectroscopy. *The Journal of Physical Chemistry B* **109**(22), 11279–11285.
[9] Dodson S., Haggui M., Bachelot R., Plain J., Li S. and Xiong Q. 2013. Optimizing electromagnetic hotspots in plasmonic bowtie nanoantennae. *The Journal of Physical Chemistry Letters* **4**(3), 496–501.
[10] de la Rica R. 2017. One-Step fabrication of LSPR-tuneable reconfigurable assemblies of gold nanoparticles decorated with biotin-binding proteins. *Nanoscale*. Royal Society of Chemistry (RSC). **9**, 18855–18860.

Exercises

(1) **Experimental conditions:** The color test is widely used with LSPR sensors. Quick readout is one of their best assets. In both the following figures, we can observe a suspension of nanoparticles at 0.5 nM. Then, between each vial, the concentration of avidin increases from left to right and from 0 to 0.75 nM.

 (a) However, only the right-hand figure shows a color variation. Can you explain this result?
 (b) Which considerations arise from this result for the development of biosensors, especially in complex or uncontrolled media?

(2) **SERS experiments: preparation:** What are the main considerations for performing SERS experiments? Mention at least five elements.

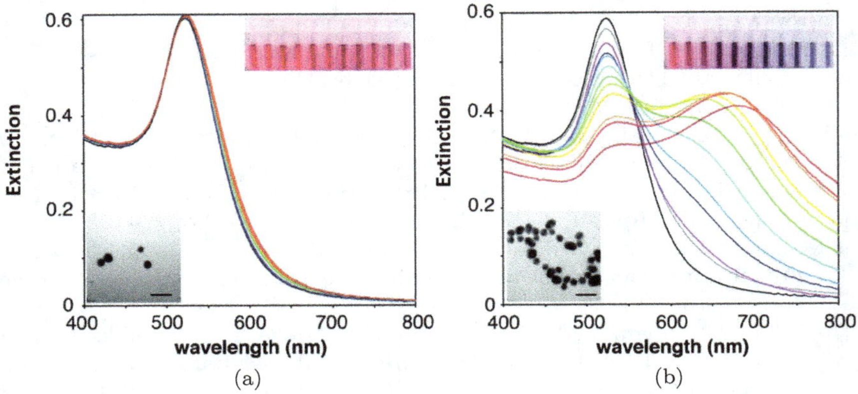

Extinction spectra of suspensions of gold nanoparticles (0.5 nM) upon addition of avidin with a final concentration of 0 (black), 0.04 (gray), 0.08 (purple), 0.15 (navy), 0.22 (dark blue), 0.30 (light blue), 0.38 (dark green), 0.45 (light green), 0.53 (yellow), 0.61 (orange), 0.68 (red), and 0.75 (maroon) nM in (a) 2.5 mM bicarbonate buffer of pH 11 and (b) 2.5 mM citrate buffer of pH 6. Corner inset in the graph on the right: photograph of the resulting supraparticle suspensions (avidin concentration decreases from left to right). Corner inset in the graph on the left: TEM image of nanoparticles upon addition of 0.22 nM avidin at each corresponding pH value. Scale bar: 100 nm.

Source: Reproduced with permission from Ref. [10]. Copyright © 2017 The Royal Society of Chemistry.

(3) **SERS experiments: optimization:** Let us consider a molecule with a characteristic peak at $1575\,\text{cm}^{-1}$. You have at your disposal a SERS device with three standard wavelengths (532, 660, and 785 nm). Your sample corresponds to a periodic array of nanostructures with one resonance at around 640 nm. Which one of these lasers will you use to maximize your detection? Justify your answer with a short calculation to support your choice.

(4) **SERS experiments: calculation of the enhancement factor:** The following figure corresponds to SERS experiments in a liquid solution. The scattering volume is 10 μm^3, and the concentration of molecules is 1 M. The SERS experiment was carried out at 0.1 mM. All the other parameters are the same (power, acquisition time, and excitation wavelength).

(a) Determine the exaltation factor with the peak at $1575\,\text{cm}^{-1}$.

(b) Consider now the area instead of the peak height. Is the value of the enhancement factor the same as in (a)? What is the impact on the measurement accuracy?

(c) By measuring the relative height of each characteristic peak, can we associate it with a specific measurement to confirm the E^4 approximation?

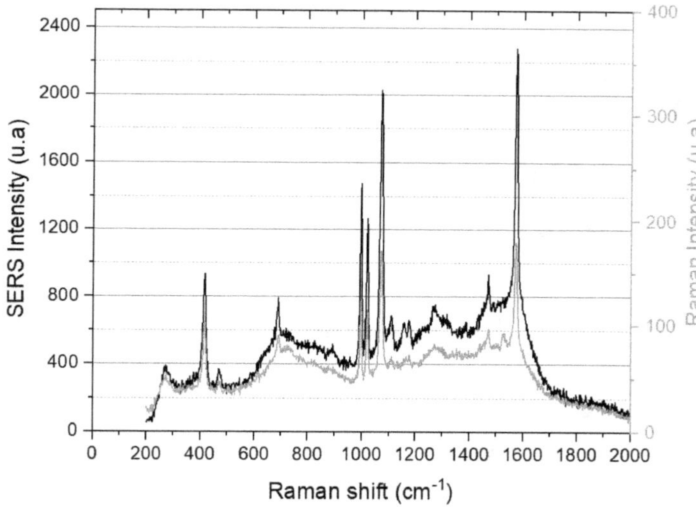

Raman and SERS spectra obtained with an excitation wavelength of 660 nm to detect thiophenol molecules. Copyright © J. F. Bryche.

10

Future Applications of Plasmonics

This chapter offers an overview of the most promising applications of plasmonics in the next decade. The topic discussed in each section is a field of research in itself, and just a brief perspective and "intellectual keys" are provided here. Plasmonics will open up new applications in **life sciences**, with ultrasensitive biosensors or advanced cancer therapies; it will give rise to new devices in **information technologies**, with the fabrication of plasmonic components in integrated circuits; it will also enable **new materials** with unexpected optical properties to be created, such as metamaterials or plasmonic colored paints.

10.1 Introduction

The interest in and the applications of plasmonics have been growing for the past 20 years. This trend will probably continue, as demonstrated by a publication in 2018 that offers a "Roadmap on plasmonics". This study is jointly authored by major contributors to plasmonics in the world [1]. Roadmaps were invented by the semiconductor industry to drive the efforts of all the partners in a rapidly evolving industry. The International Technology Roadmap for Semiconductors (ITRS) has been regularly published from 1993 until 2016. It has helped the major actors in this industry organize their efforts and maintain a very high pace in technological developments. In plasmonics, a similar roadmap stems from academic laboratories, and there is not yet an industrial push to generate a plasmonic industry. Three main areas benefit from the advances in plasmonics: life sciences, information technologies, and material sciences. These are reviewed in the following.

10.2 Advanced Biosensing: Converging Approaches

As indicated in Chapter 5, plasmonic biosensors need to overcome their current limitations. A clever way is to combine the properties of the different plasmonic modes. For example, nanodisks deposited on top of a metallic film lead to a coupling of SPP and LSPR [1]. This regime can be derived and analyzed with a coupled harmonic oscillator, as suggested by Novotny [2]. When the two wave vectors are equal, we can observe an anti-crossing phenomenon, as illustrated in Fig. 10.1, with the blue and purple curves giving

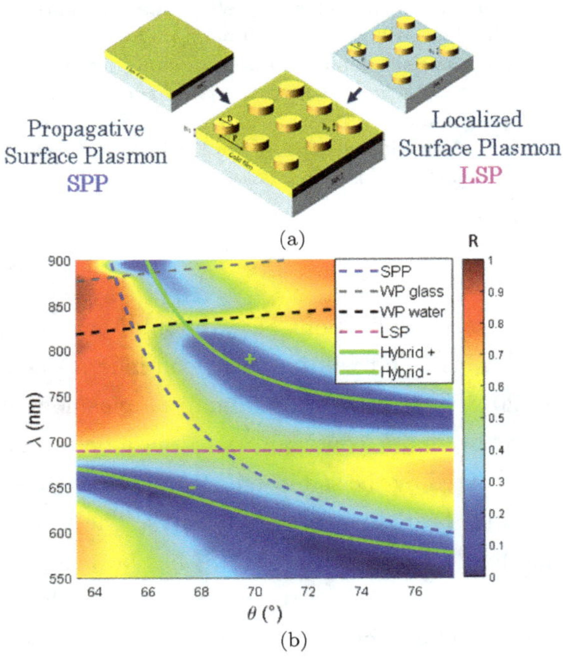

Fig. 10.1. (a) Biosensor chip where the SPP from a planar film is coupled with the LSPR from cylindric nanodisks. (b) 2D reflectance map showing the coupling between SPP (blue curve) and LSPR (purple curve). This coupling induces an anti-crossing visible in the spectro-angular SPR map (green curve). The Wood plasmon anomalies are also indicated (black and grey curves). Copyright © J. F. Bryche.

birth to the two green branches. It turns out that this anti-crossing enhances the performance (sensitivity and resolution) of the sensor by associating the exaltation effect produced by the nanoparticles and their localized plasmon resonance with the propagative surface plasmon. In addition, the so-called Wood plasmon anomalies are generated: They result from the modification of the SPP wave vector by the lattice vector ($\lambda_g = 2\pi/a$, with a being the distance between the nanodisks). They are represented by the grey and black curves.

A coupling between propagative and localized plasmons was used in SERS to increase the sensitivity of the biosensor to a test molecule, which was thiophenol in the example shown in Fig. 10.2 [3]. The presence of a metallic film leads to two beneficial effects: an increase in the collection efficiency due to the increase in light intensity collected by the microscope objective and an increase in the wavelength sensitivity factor. As a result, a gain of one order of magnitude was achieved (black curve) compared to the SERS signal obtained with the nano-pillars on a glass substrate (red curve). The green and blue curves are used as references in the absence of nanostructures, and the vibrational peaks of the molecule cannot be

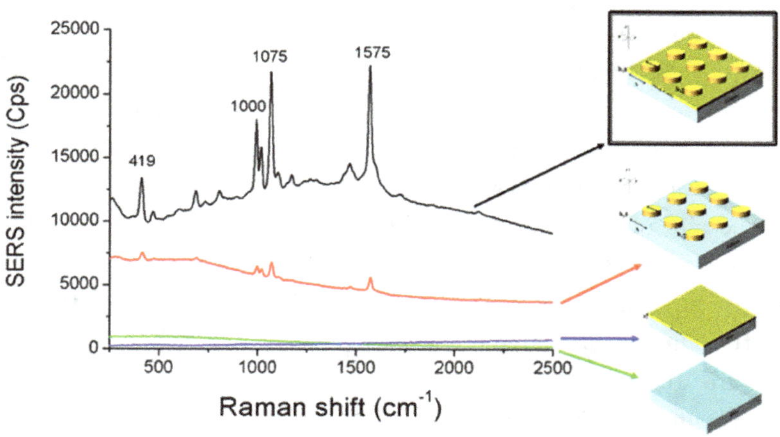

Fig. 10.2. SERS spectra of thiophenol for four types of samples: glass substrate (green curve), gold film (blue curve), gold nanostructures on glass substrate (red curve), and gold nanostructures on gold film (black curve). The array of 200 nm nanostructures is the same and correspond to a period of 400 nm.

Source: Reproduced from Ref. [3].

detected. For a given thiophenol concentration, we can obtain enhancement factors close to 9.7×10^6 for the 660 nm wavelength excitation of gold nanodisks on a gold film [3].

The plasmonic biosensors can thus be made more efficient by combining different modes. In addition to this fundamental approach, it is also an interesting possibility to take advantage of the distribution of the electric field and make a localized functionalization (see Section 10.8).

10.3 Nanoparticles for Enhancing Spatial Resolution in SPR Images

SPP has been also used for high-contrast optical microscopy. For example, living cells deposited on a planar gold surface can be observed under illumination where an SPP is excited. The presence of cells changes the optical index locally, which in turn modifies the SPP resonance condition and generates an optical contrast. This was demonstrated in 1999 and reported in Ref. [4]. However, these SPR images generally exhibit a blurring effect associated with the non-negligible propagation length of the plasmon wave. In this approach, plasmon imaging produces high contrast but degrades the spatial resolution. The propagative surface plasmons have a propagation length of several micrometers in the visible range (see Section 4.2.5), and this needs to be compared with the Rayleigh criterion that sets the limit for optical resolution at $\sim \lambda/2$, which is 0.35 μm. Nowadays, new strategies have emerged for improving the optical resolution of plasmon imaging, with the aim of approaching the limit of the Rayleigh criterion. This is achieved by adding nanostructures on the planar plasmonic surface. These nanostructures break down the propagation of SPP and improve the spatial resolution through better confinement [5].

Figure 10.3 shows the contrast in plasmon imaging obtained with different plasmonic substrates [1]. For a flat and uniform gold film whose thickness varies from 50 to 10 nm, the attenuation length is decreased from 12 μm to approximately 1 μm for the thinnest film (black curve in Fig. 10.3(a)). This shorter attenuation length leads to a higher lateral resolution. On the other hand, it also leads to a weaker contrast. When the

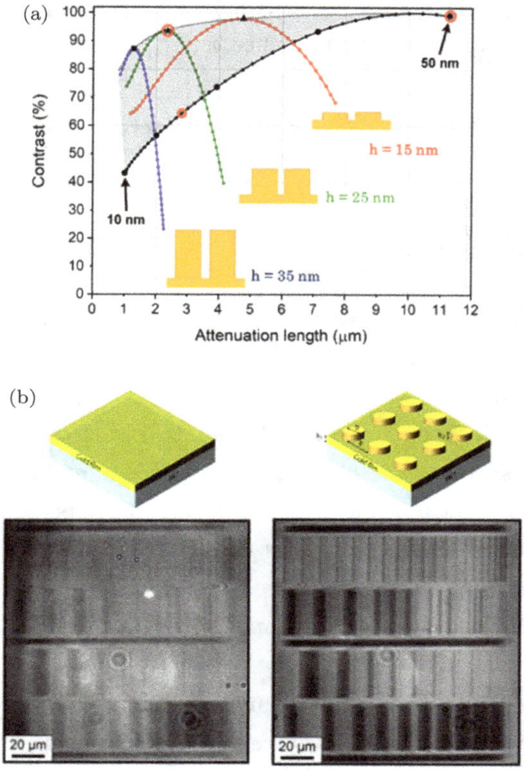

Fig. 10.3. (a) Calculated contrast in plasmonic imaging as a function of the plasmon propagation length. Various substrates are compared: planar film gold or gold nanodisks regularly organized on a planar gold film. (b) Comparison of SPR images with and without the gold nanodisks and the impact on lateral resolution.

Source: Reproduced from Ref. [1].

substrate includes nanostructures, the attenuation length of the propagative plasmon is also reduced, but the contrast may not be lost. It is thus possible to obtain a compromise between a sufficient resolution (low attenuation length) and a contrast higher than 90%, as shown with a 25 nm film and 25 nm nanostructures (green line in Fig. 10.3(a)). The black and white pictures in Fig. 10.3(b) illustrate the improvement obtained with the nanostructures on gold film by imaging a calibration slide. These research results can be further improved by tuning different geometrical parameters (period, diameter, film thickness, and height of the nanostructures), the experimental configuration (working wavelength), or the kind of materials used. In addition, plasmonic imaging also provides information on changes in optical indexes. This approach is being actively considered for research on living cells.

10.4 Cancer Therapies Based on Hyperthermia: Clinical Trials

Cancer is not an ordinary disease but a collection of diseases with more than 100 types. The common behavior in all types of cancer is that some of the body's cells begin to divide without stopping and spread to surrounding tissues. Cancer was responsible for 16% of human deaths in 2019. Nanomedicine, as well as the use of nanoparticles, is just one of many current strategies for fighting cancer. To date, there have been 120 clinical trials worldwide based on nanoparticles (for cancer or other diseases) [6, 7]. Out of these trials, the one that is especially of interest for plasmonics is the research initiated by Pr. Naomi Halas (Rice University, Houston). It resulted in the establishment of the company, Nanospectra Biosciences. They have developed a drug called AuroLase that went through the whole process of medical approval and is currently used for curing prostate cancers (see clinical trial NCT02680535) [7]. In this section, we focus on the plasmonic aspects of this research success story, which started around 2004 [8].

This therapy is based on core–shell nanoparticles of SiO_2 Au called AuroShells that have core and shell diameters of 120 and 150 nm, respectively. They are functionalized with PEG to make them biocompatible. These nanoparticles have a strong optical absorption in the near infrared at around 810 nm, as discussed in Chapter 8 [9]. This wavelength is crucial since it corresponds to the maximum penetration depth of light through human tissues. This is also the wavelength of conventional Ti–Sapphire lasers. When irradiated with a power of 1.06 W focused on a 5 mm spot for 200 s, the nanoparticles and their immediate surrounding undergo a temperature increase of 10°C [9]. This is sufficient to kill the cells where nanoshells were present (see Fig. 10.4). In the case of prostate cancer, nanoparticles accumulate in the tumor tissue via leaky tumor vasculature (known

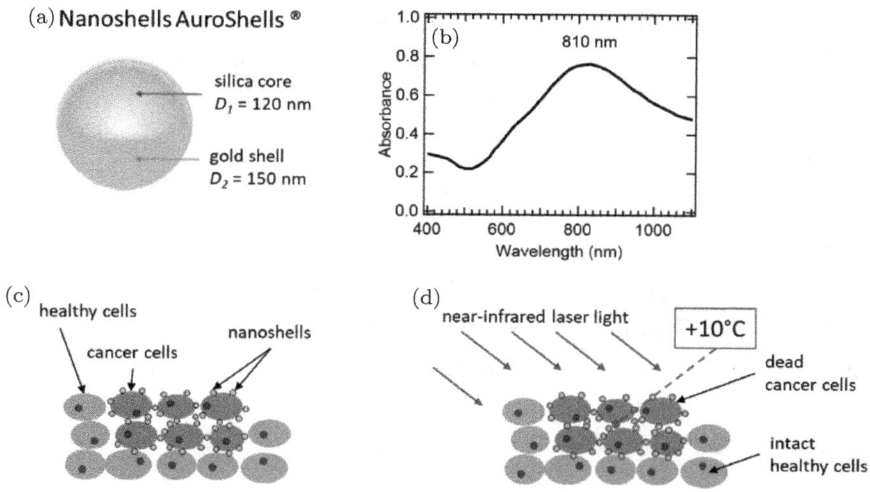

Fig. 10.4. Working principle of hyperthermia used for curing prostate cancer. The core–shell nanoparticles with a total diameter of 150 nm (a) exhibit a broadband absorption efficiency at 810 nm, (b) which results in localized heating when illuminated with a laser. These nanoshells are developed by the Nanospectra Biosciences company under the name AuroShell. (c) The nanoshells once injected into the tumoral regions, will concentrate around the cancer cells thanks to the EPR effect. (d) When the region is irradiated with a laser, a local temperature increase of 10° C kills the cancer cells, whereas the healthy cells remain alive.

as enhanced permeability and retention, or EPR, effect). Then, they are irradiated with a near-infrared laser. The tumor undergoes photothermal heating, resulting in selective hyperthermic cell death without heating the adjacent nontumorous tissues.

The AuroLase therapy based on hyperthermia is a very inspiring example of the use of plasmonics for medical applications. It demands a considerable effort and rigor and is a unique example.

10.5 Hot Carrier Physics

Most of the descriptions of light in this book considered light as an electromagnetic wave. However, a light beam can also be described as an assembly of photons, according to the theory of photoelectric effect theorized by Albert Einstein in 1905. Each photon carries a quantized energy E linked to its frequency ν by the Planck constant: $E = h\nu$. According to the photoelectric effect, if a photon has an energy higher than the work function of a metal, it will generate a photoelectron on hitting the metal. But if its energy is lower, an electron will not be ejected from the metal but remain as a *hot electron.* For most of the metals, the work function lies in the range 4.0–5.0 eV, which is slightly higher than the range for visible light (1.5–3.2 eV). Therefore, we expect visible light to generate such hot electrons, but these may not appear to be especially linked to plasmonics. The obstacles to generating hot electrons are twofold: The flux of photons is often too low to generate a substantial number of hot electrons and the lifetime of hot electrons is so short that they can hardly be detected or utilized. We shortly present an overview of how plasmonics and nanostructures bring new opportunities to hot electron physics. This investigation domain started in 2012 with some foundational review articles [10–12], and more recently, a Faraday Discussion gathered the most active researchers on that topic in London in 2019 [13].

10.5.1 *Hot spots, nano-antenna, and hot electrons*

The first step of hot electron generation is the absorption of a photon from incident light. The probability of photon absorption is proportional to the square of the local electric field inside the metal. Since the so-called *hot spots* in plasmonic nanostructures are nanometric regions where the electric field can be enhanced by up to 50 times, these regions are the sources of hot electrons. The nano-antenna effect of plasmonic nanostructures is therefore central to hot electron physics. For example, in Section 7.3, we calculated the absorption cross section of a 15 nm silver spherical nanoparticle and showed that the absorption efficacy was equal to 18.5, which means that silver has 18.5 times greater capacity to absorb photons when it is in the form of nanoparticles compared to a flat surface. In Chapter 8, we showed how electron energy loss spectroscopy (EELS) allowed us to map the local density of excited electrons on hot spots, such as the apex of a nanorod or the edges of two nano-triangles (see Section 8.4). This shows that the sources of hot electrons are of the order of magnitude of $10 \times 10 \times 10\,\mathrm{nm}^3$ or so in volume.

10.5.2 *Photoexcitation and hot-electron timescales*

The absorption of a photon by a metal may lead to the generation of an electron–hole pair if the photon energy fits within the conduction band, as shown in Fig. 10.5. It means that the conduction band should offer a filled state and a corresponding empty state, whose energies differ exactly by $h\nu$, as depicted in Fig. 10.5(b). This results in the creation of a *hot electron* and a *hot hole.*

Why are such electrons and holes called *hot*? This can be answered by comparing the Fermi–Dirac distribution, represented in Fig. 10.6, at three different temperatures. The Fermi–Dirac distribution $f(E)$ is the probability of finding an electron of energy E in a material of Fermi energy E_F at a temperature T. [14] $f(E)$ writes: $f(E) = 1/(1+\exp[(E-E_f)/k_BT])$. At room temperature (300 K), if we consider that a photon of energy 2.3 eV ($\lambda_0 = 540$ nm) is absorbed by a metal, it generates an electron–hole pair. Part of the energy (let's say 2.3 eV/2 = 1.15 eV) will be given to the electron and the remaining to the hole. According to the Fermi–Dirac distribution, the probability of generating electrons at 1.15 eV above the Fermi level is just 5×10^{-20}, and it is identical for generating a hot hole at 1.15 eV below E_F. This shows that at room temperature and at thermal equilibrium, there is no such hot carrier. This probability grows to 1.2% when temperature

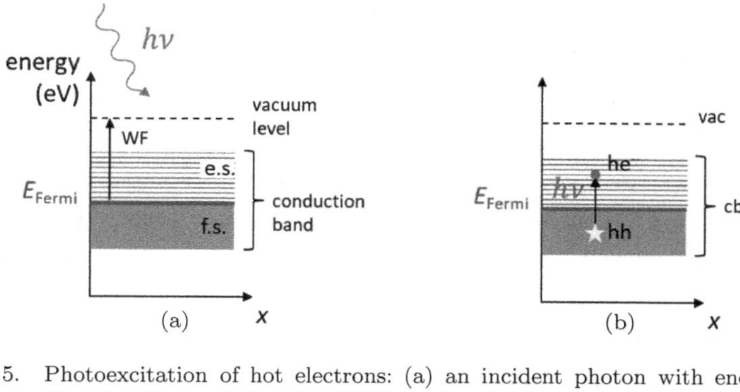

Fig. 10.5. Photoexcitation of hot electrons: (a) an incident photon with energy $h\nu$, which is smaller than the work function (WF), is absorbed by the conduction band (cb) of the metal. An electron is promoted from the filled states (f.s.) to an empty state (e.s.) of the cb. (b) This generates an electron–hole pair constituting a hot electron (he^-) and a hot hole (hh).

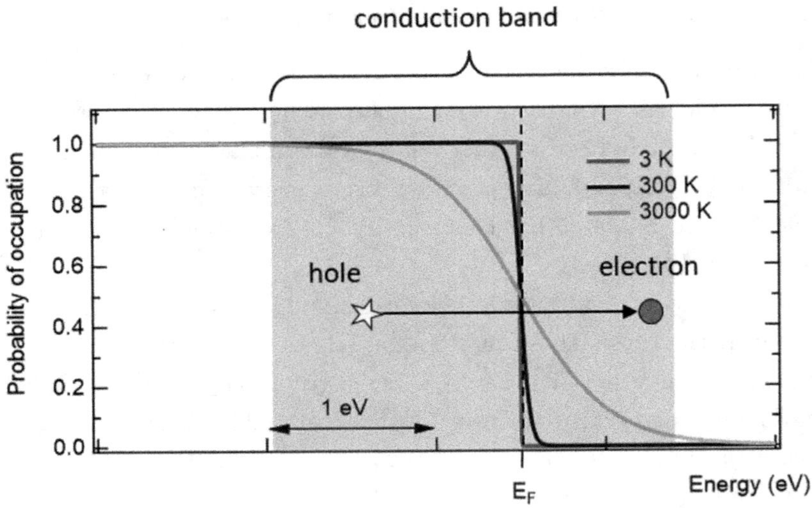

Fig. 10.6. Distribution of the electrons ruled by the Fermi–Dirac distribution $f(E)$, where E_F is the Fermi energy. The actual electron population is the product of $f(E)$ and the density of states of the level considered. In the current case, we consider a metal whose E_F lies within the conduction band. At low temperature (e.g. 3 K), the Fermi–Dirac distribution predicts an abrupt decay of the probabilities from 1 to 0 at E_F. When a photon is absorbed by the metal, it generates an excited electron, whose energy is identical to the energy of electrons of a very hot metal. This is why it is called a hot electron.

rises to 3000 K. Therefore, hot electrons can be defined as a population of electrons excited well above the thermal equilibrium defined by the Fermi–Dirac distribution. This out-of-equilibrium population is generated upon photon absorption.

After a photon has been absorbed, a hot electron–hole pair is created and a three-step process begins at three different timescales. (i) **Landau damping** over 1–100 fs: Landau damping is a pure quantum process that explains the energy transfer between the incoming electromagnetic wave and the charged particles (electrons) in a plasma. It results in the damping of the electromagnetic wave to the benefit of the electrons and was first theorized by Lev Landau in 1946 [15]. It gives rise to a set of nonlinear equations, the Vlasov–Poisson equations, that were ultimately solved by Mouhot and Villani in 2010 [16]. Note that Villani was awarded the Fields Medal

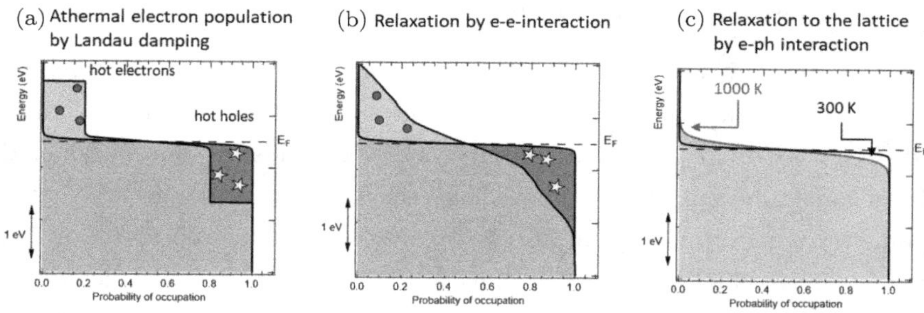

Fig. 10.7. Representation of the Fermi–Dirac distribution of electrons after they have been excited by an intense laser pulse. (a) First, the laser pulse transfers energy to the electrons via Landau damping and sets these electrons into a highly out-of-equilibrium state, which is measured by the electronic temperature. In step (b), this energy is redistributed among all the electrons within 1 ps or so. And finally, in (c), the electrons transfer their thermal energy to the atomic lattice within times of 100 ps–10 ns.

in 2010 for this. The Landau process gives rise to a population of electrons and holes, as depicted in Fig. 10.7(a). (ii) Then, the second process is the relaxation of this highly athermal population through **electron–electron scattering** from 100 fs to 1 ps, where the population of free electrons homogenize its temperature. The electronic temperature can be defined and reaches thousands of Kelvin. A representation is given in Fig. 10.7(b). (iii) Finally, this set of hot electrons undergoes **electron–phonon relaxation** that eventually results in a temperature elevation of the whole nanostructure due to thermal transfer from the electrons to the atomic lattice (100 ps–10 ns). Note that this temperature elevation is usually very limited and shown exaggerated in Fig. 10.7(c) since the population represented by the red line is at 1000 K (300 K for the black line), whereas the temperature increases are generally in the order of 10–100 K.

The sequence of processes described above explains the interaction of the optical wave with the electrons followed by their natural decay, which is accompanied by local heating. However, hot electron generation may also be driven so that this high energy electron–hole pair is split before their recombination, and their energy can be "harvested". In that case, it gives rise to new chemical reactions, such as H_2 dissociation on gold nanoparticles, desorption of chemical species, or enhanced photovoltaic efficiency.

10.5.3 *Thermoplasmonics and ultrafast spectroscopy*

As mentioned above, the first decay route of hot electrons is the *thermal route*. This is the field of thermoplasmonics, which is intrinsically linked to Joule losses in the nanostructure. The Joule losses were considered for a long time as a serious issue, but they have turned into a starting point for many applications. See, for example, hyperthermia described in Section 10.4. The field is experiencing now a strong growth with dedicated books [17]. For instance, Baffou has carried out various fundamental and experimental studies on nanoparticle arrays under a constant illumination (CW laser) [18, 19]. They generate microbubbles via thermal accumulation in a nanostructured network that concentrates the incident luminous energy via the thermoplasmonic effect. These microbubbles create local instabilities that trigger chemical reactions [20]. In the dynamic regime, several teams have focused on the transport of thermal energy at the nanometric scale to extract the constants inherent to the thermal mechanisms at the nanometer scale [21].

In order to rationalize the description of hot electron excitation by a laser pulse explained above in Section 10.5.2, it is necessary to introduce the so-called *two-temperature model.* Anisimov proposed this model in 1974 by considering that the metal should be divided into two systems: the electrons and the atomic lattice [22]. Each one has its own temperature, T_e and T_l, and its own heat capacity, C_e and C_l, for the electron and the lattice, respectively. When a light pulse hits the metal, the electrons absorb a part of the energy P_{abs} (power density absorbed per unit volume) and simultaneously transfer energy to the atomic lattice by collisions. The model consists in writing the energy balance for the two subsystems. If G represents the coupling constant between an electron and a phonon, it yields

$$\mathrm{C_e}\frac{\partial \mathrm{T_e}}{\partial \mathrm{t}} = P_{abs} - \mathrm{G}(\mathrm{T_e} - \mathrm{T_l}),$$

$$\mathrm{C_l}\frac{\partial \mathrm{T_l}}{\partial \mathrm{t}} = \mathrm{G}(\mathrm{T_e} - \mathrm{T_l}).$$

However, the two-temperature model is not accurate enough and has failed in determining the temperature dynamics accurately because it considers that the energy absorbed by the electrons is instantaneously thermalized. It also does not consider the thermal losses toward the surrounding medium and toward the substrate (corresponding essentially to conduction and labeled with the letter s to denote substrate). Therefore, Sun *et al.* have introduced the *three-temperature model* [23], which was recently completed by other groups [24, 25]. The idea of energy exchange as a function of time is shown in the diagram in Fig. 10.8(a). Here, N is the thermal energy density per unit volume contained in the non-thermalized electron gas. α and β are, respectively, the heating rates of the electron gas and the phonons by the hot electrons. One crucial point to consider in this model is the ballistic displacement of hot electrons in a very short time (<100 fs). An analogy between the displacement of these ballistic electrons is established with the dynamics of a fluid via a velocity field. This corresponds to the first line of the three-temperature model in the figure.

A second approach treats the thermalization of hot electrons with the Boltzmann transport equation. The Boltzman equation expresses the conservation of the number of heat carriers (electrons, phonons) over time. A good description of this approach can be found in Ref. [26].

Measuring these thermal effects that occur over very short timescales is extremely challenging. Interestingly, the temperature evolution can be monitored through a change in the optical reflectivity or the optical transmission. More precisely, the dielectric permittivity of materials depends on the temperature, and it is possible to measure them with very precise instruments. *Transient absorption spectroscopy* is a pump-probe spectroscopic technique aimed at measuring ultrafast dynamics of excited populations in a sample and the associated lifetimes of molecules, materials, and devices. One pulse (the *pump*) generates an excited population in the sample, and a white pulse (the *probe*) allows us to measure the sample's transmittance or reflectance spectrum. By varying the delay between the *pump* and the *probe*, the optical properties as a function of time can be recovered. Then, we obtain the transient spectra by comparing the transmittance T and reflectance R of the sample at rest with those of the excited sample, T^* and R^*:

$$\frac{\Delta R}{R} = \frac{R^* - R}{R}.$$

The pump-probe spectroscopy principle is illustrated in Fig. 10.8(b) for a gold film. Depending on the setup, the probe can be a white light pulse,

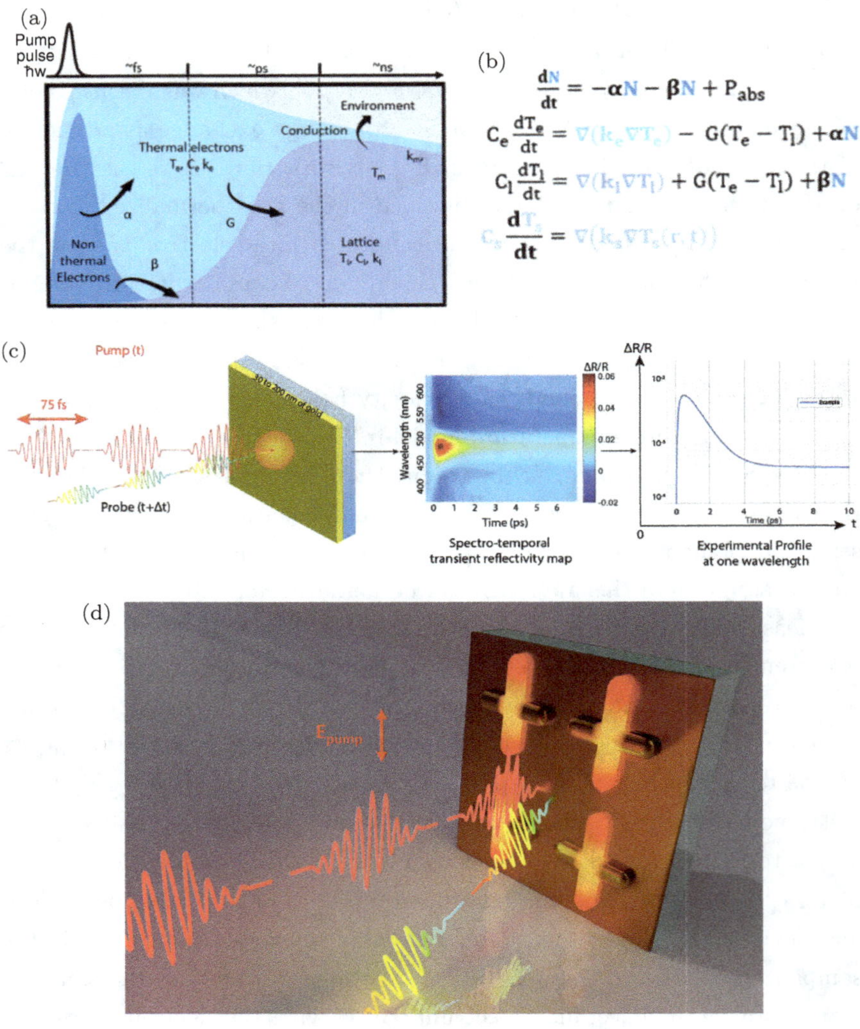

Fig. 10.8. Description of the three-temperature model: (a) Schematic representation provided by Vega of energy as a function of time within the different systems. (b) System of equations of the three-temperature model. (c) Illustration of pump-probe spectroscopy on gold film to obtain transient reflectivity map and an associated reflectivity profile. (d) Illustration of pump-probe spectroscopy on nanorod provided by Pace.

Source: Reproduced with permission from Ref. [25]. Copyright © 2020 American Physical Society.

which generates a transient reflectivity map in time and wavelength. Then, a profile $\Delta R/R$ can be extracted and compared with the variation determined with the 3T model [25]. Furthermore, the pump-probe spectroscopy can be performed on a nanostructured sample to explore the thermal repartition with nanoscale resolution, as conceptualized in Fig. 10.8(c) [24, 25].

10.6 Application of Hot Electrons to Photocatalysis

Heterogeneous catalysis typically deals with solid phase catalysts and gas phase reactants. A catalyst is a solid species that is not consumed by the chemical reaction and increases the chemical rate or orients the reaction (selectivity). One key property of catalysts is that they are not consumed by the chemical reaction and can act repeatedly although they are in minute quantities. In photocatalysis, the catalyst is activated by light through the generation of electron–hole pairs. This is where plasmonics enters the game to give rise to plasmonic photocatalysis [27]. Pioneering work in the area of hot electron photocatalysis includes liquid phase water splitting, H_2 production from alcohol, gas phase oxidation reactions, and hydrocarbon conversion. Here, we only illustrate the concept through the experiment of hydrogen dissociation [10].

A gold nanoparticle irradiated at the plasmon resonance generates electron–hole pairs, in particular a population of hot electrons extending slightly outside of the nanoparticle surface (see Fig. 10.9(a)). At the nanoparticle surface, hydrogen molecules are adsorbed, although this is a transient mild physisorption (and not a stable chemisorption). During this ultrashort stay of the molecule (femtoseconds), the hot electron in the vicinity of the surface can transfer into the lowest unoccupied molecular level (LUMO) of the adsorbed hydrogen, creating a transient negative ion, $H_2^{\delta-}$, depicted in Fig. 10.9(c). This additional charge in the molecule increases the Coulomb repulsion and results in the dissociation of molecular hydrogen into two atomic hydrogen. Calculations have shown that the H_2^- has a metastable state at 1.7 eV above the ground state of the neutral molecule that can be reached by hot electrons and that the dissociation energy of H_2^- is 2 eV lower than for isolated H_2 molecules. The process occurs at

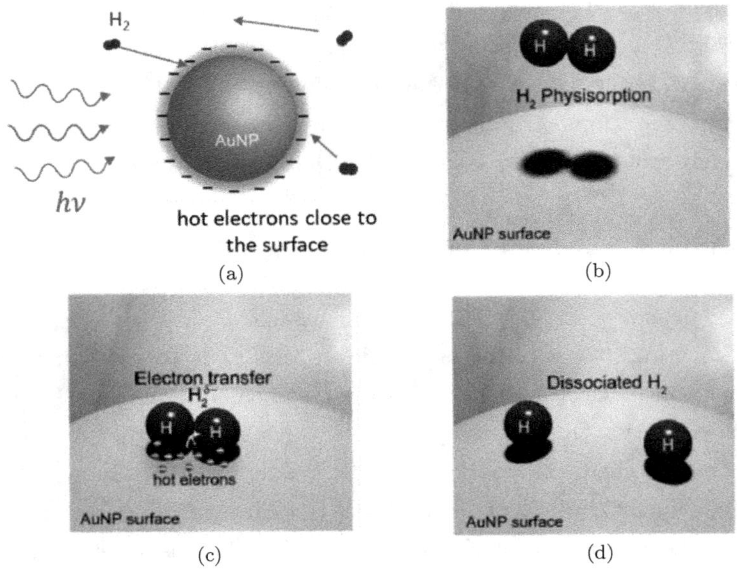

Fig. 10.9. Sequence of hydrogen dissociation triggered by hot electrons generated by a plasmonic wave. In (a), the incoming photons of energy hv generate hot electrons due to plasmonic enhancement in a gold nanoparticle. Simultaneously, in (b), a H_2 molecule passes close by the gold surfaces and captures one hot electron and is transformed into a metastable ionic state $H_2^{\delta -}$ in (c). This ionic hydrogen then quickly dissociates into atomic hydrogen in (d), which is released into the gaseous atmosphere.

Source: Reproduced with permission from Ref. [10]. Copyright © 2013 American Chemical Society.

room temperature with visible light at a wavelength that corresponds to the excitation of the localized surface plasmon of the nanoparticle.

Such photocatalytic reactions are presently at the heart of a fierce scientific debate concerning whether the enhanced catalytic activity is due to hot electron injection [27] or to local plasmonic heating that would accelerate the local chemical reaction thanks to the well-known Arrhenius equation, which expresses the chemical rate as $k = A\exp(-E_a/RT)$ [28, 29].

Nevertheless, a highly promising application of plasmonic photocatalysis is the water-splitting reaction, which is the decomposition of water molecules into hydrogen and oxygen through a redox reaction (oxidation of H_2O into O_2 and reduction of H_2O into H_2). This reaction is highly endothermic, and the harvesting of photon energy is an efficient way of making it possible. Recently, it has been shown that a clever combination of zinc–cadmium–sulfide nanowires with gold nanoparticles resulted in hydrogen production as high as 96 mmol/g · h under illumination [30].

10.7 Plasmonics for Photovoltaic Applications

Plasmonic structures pave the way for manipulating and controlling the electromagnetic field at a nanometer scale. Depending on the size and arrangement of the nanoparticles, it is therefore possible to modify resonance wavelengths, electromagnetic couplings, and induce effects beneficial for improving the efficiencies of photovoltaic solar cells [31, 32]. For example, there are at least three ways of using plasmonic nanoparticles for photovoltaic cells.

First, metallic nanoparticles can be used for scattering or trapping the incident radiation within an absorbing semiconductor thin film (Fig. 10.10(a)). The effective length of the optical path inside the cell can be increased by multiple scattering or high-angle scattering induced by the nanoparticles, and a larger amount of energy is converted into electricity. In the second approach, metallic nanoparticles can be used to enhance the coupling of the electromagnetic field with an active semiconductor. In this way, the nanoparticles operate as nano-antennas, and the effective absorption cross section is increased (Fig. 10.10(b)). The positioning of the particles close to the active surface also creates electron–hole pairs in the

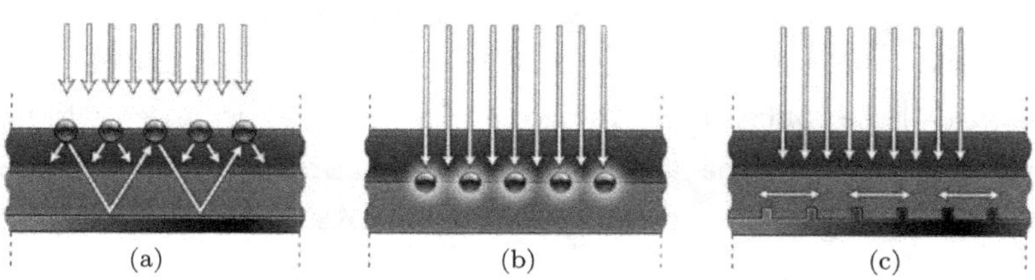

Fig. 10.10. Three uses of nanoparticles to improve photovoltaic cells: (a) nanoparticles on the top and used for diffusion and trapping sunlight; (b) nanoparticles in the vicinity of the semiconductor layer to improve the absorption cross section and generate electron–hole pairs; (c) nanoparticles in an array on the back surface to couple sunlight into SPP modes.

Source: Reproduced with permission Ref. [32]. Copyright © 2010 Springer Nature.

semiconductor. Third, a nanostructured surface on the back surface can couple the sunlight into SPP modes, after which this light is converted into photocarriers in the semiconductor, as shown in Fig. 10.10(c).

In photovoltaic cells, a recurrent challenge concerns the Shockley–Queisser limit of single bandgap devices. To overcome this limit, several approaches, such as thermal photon up-conversion, hot electron capture, light concentration, can be used. Particularly interesting among them is the tandem cells with multiple energy threshold devices. It consists of a series of varying bandgap materials, from the higher to the lower values. Positioning quantum dot structures inside each layer allows them to capture the electrons before they thermalize, resulting in an increase in the conversion efficiency. However, the generated photocurrent will be low and challenging to enhance due to the low conductance of this quantum dot assembly. On the contrary, surface plasmons can trap light in a wavelength-dependent manner. By adjusting the size of the nanoparticles incorporated into the layers, it is possible to cover the entire desired spectrum.

10.8 Plasmonic Nano-antenna for Advanced Chemistry

The strong confinement of the field around a nanoparticle and its dependence on polarization leads to very localized surface functionalization or chemical transformations [33]. The development of plasmonics for the control of surface chemistry has grown considerably in recent years and has spawned many applications, including analytical chemistry, detection, photocatalysis, quantum plasmonics, photovoltaics, and nanomedicine. It takes advantage of the confined electromagnetic fields, local heat generation, and hot carrier excitation that accompanies plasmon resonances for triggering photosensitive molecular functionalities or photoresists.

Bachelot's team has demonstrated the possibility to localize with high accuracy quantum dots next to gold nanoparticles (Fig. 10.11(a)) [34, 35]. They use the polarization-dependent field distribution to photopolymerize a resin containing the quantum dots in areas of strong fields. The two-photon-assisted photopolymerization shown in Fig. 10.11

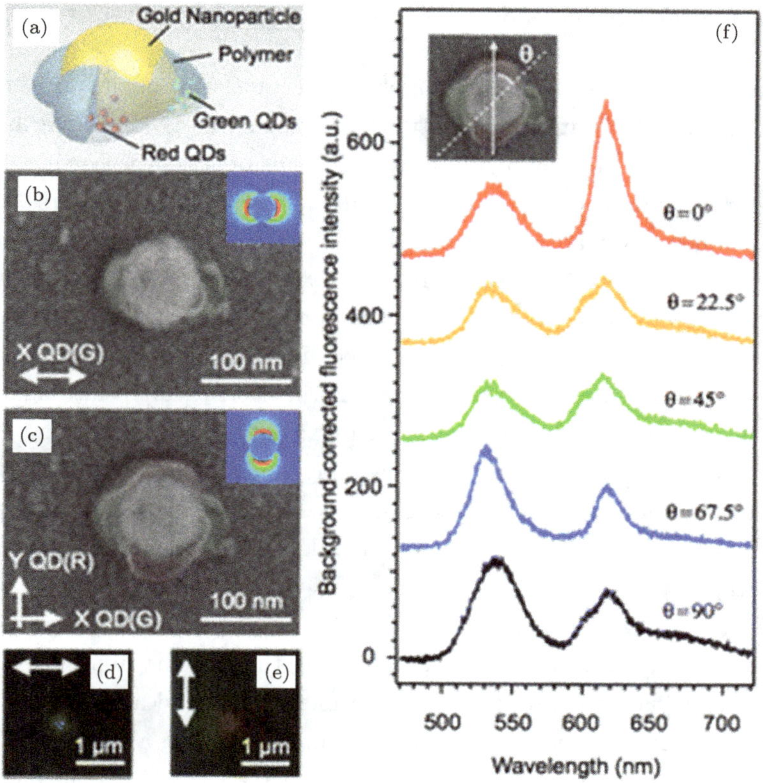

Fig. 10.11. (a) Schematic illustration of the localization of the QDs around the gold nanoparticle, (b) localization of green QD emitters along the x-axis, (c) localization of the red QD emitters along the y-axis, (d)–(e) corresponding far-field emission for the two directions of polarizations, (f) fluorescence intensity as a function of the incident polarization direction.

Source: Reproduced with permission from Ref. [35]. Copyright © 2015 American Chemical Society.

allows the spatial localization of two species of QD emitting in the red (Fig. 10.11(b)) and in the green (Fig. 10.11(c)). The positioning is confirmed by far-field fluorescence images (Fig. 10.11(d)–(e)). These nanoemitters can be tuned by rotating the polarization of the incident light to switch from green to red emission, as shown in Fig. 10.11(f), for different incident polarization directions.

10.9 Plasmonics and Nanoelectronics

Photonics is capable of providing information transfer through optical fibers at an unparalleled rate since the transportation speed is 1000 times higher than with electronic components. Yet, in terms of signal processing, there are almost no optical components. A computer is based on resistors, capacitors, diodes, and transistors, which have been hardly reproduced in optics, and when they have, their size is 1000 times larger than the electronic components. A good approach is to use each of the components at their best in terms of signal processing capacity, speed, and compacity. The crucial components will be transducers able to transform electric signals into light and vice versa. Here, we describe an intriguing solution for electron-to-photon transducers based on the tip of a scanning tunneling microscope (STM). The understanding of these nanoscale processes could probably offer some novel nanometric photonic sources capable of coupling inside plasmonic circuitry.

The working principle of the STM lies in monitoring an electric current based on quantum tunneling and going through the minuscule space ($d \approx 1\,\text{nm}$) that separates the apex of a metallic tip and a conductive surface. This nano-ampere current takes up upon applying a tip–surface bias of the order of $1\,\text{V}$, as shown in Fig. 10.12(a). The STM was invented in 1981, and it is well known for its ability to image surfaces at the atomic scale. The separation between the tip and the surface is a tunnel barrier that electrons can cross elastically, which means that the electrons ejected from the tip with a given energy reaches the surface with the same amount of energy, and this latter is dissipated later in the sample via non-radiative processes (Fig. 10.12(b)). A small number of electrons also undergo inelastic transition and release their energy partially that is converted into light (Fig. 10.12(c)). The interest in photonic emission was triggered by an STM tip developed in 2003 following the pioneering work of Ho on exciting the fluorescence of single molecules with an STM [36]. Several groups have since then contributed to the field, and we show here the connection with plasmonics [37, 38]. The presence of plasmonic nanostructures reinforce this electron-to-photon conversion, although the quantum efficiency do not

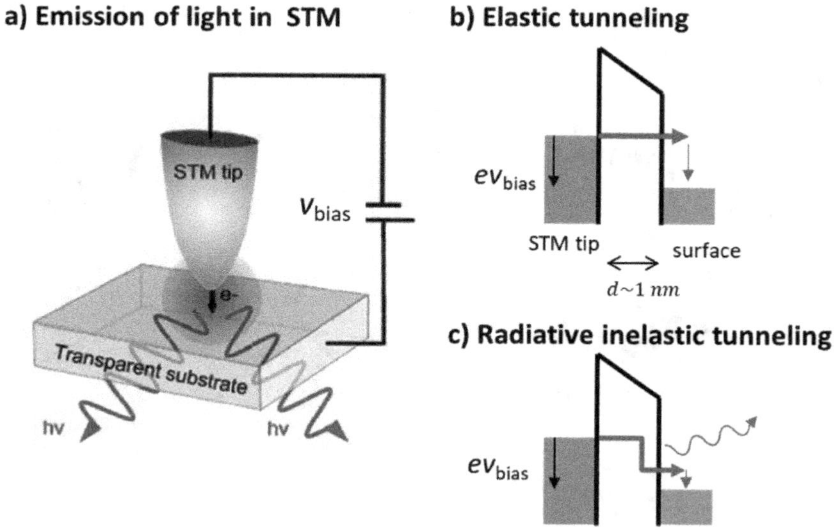

Fig. 10.12. (a) Example of a scanning tunneling microscope (STM) experiment carried out on a transparent conductive substrate, where the photons generated by inelastic tunneling are collected by an optical microscope objective (not shown). (b) In normal operation, the electrons tunnel elastically, and the STM is used for recording the surface topography. However, a few portion of electrons also tunnel inelastically, and part of their energy is converted into photons, as shown in (c).

overcome the value of 10^{-4} (1 photon generated for 10,000 electrons crossing the barrier). Therefore, this source of light is still very weak.

As an illustration, we show the case of a gold nanostructure of ~100 nm with a triangular shape deposited on indium tin oxide (ITO), which is conductive and transparent. Then, the STM tip is placed above the nanostructure and bias at $U_{\mathrm{b}} = 2.75\,\mathrm{V}$ while maintaining a set point current of $I_{\mathrm{t}} = 60\,\mathrm{pA}$, as shown in Fig. 10.13(a). The emitted light is collected by the objective and analyzed in several ways, which provides a wealth of information [39]. When the CCD camera is in focus on the nanostructure, the optical image demonstrates that the light is actually produced by the nanostructure. However, the spot size is diffraction limited and is equal to the point spread function (PSF) of the objective, which is around 500 nm. The STM tip, however, can be positioned with much more accuracy (~1 nm); therefore, it is possible to precisely correlate the intensity

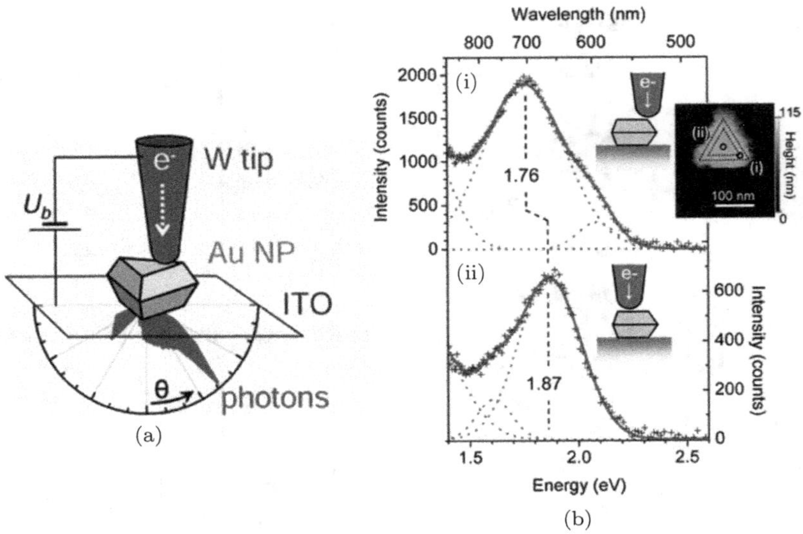

Fig. 10.13. Example of plasmonic emission from an STM tip. (a) The STM tip can be placed at different locations above a nanotriangle of size 128 nm placed on a transparent and conductive ITO substrate. The direction of photon emission can be detected. (b) The optical spectrum is recorded when the tip is biased at 2.75 V and shows distinctive characteristics depending on the exact location of electric excitation: on the edge of the triangle (i) or in its center (ii).

Source: Reproduced with permission from Ref. [39]. Copyright © 2013 American Chemical Society.

of the light with the exact excitation point. This is used for mapping the *hot spots* of the plasmonic nanostructure. Another piece of information can be obtained by placing the CCD on the Fourier plane of the objective. In this configuration, the angular distribution of light can be recorded, as sketched in Fig. 10.13(a). This is extremely interesting because this gives access to the direction of wave vector $\boldsymbol{k}$ of the photons. Finally, it is also possible to record the optical spectrum, as shown in Fig. 10.13(b). Note that since the process is inelastic, the spectrum between 1.4 and 2.6 eV is obtained by polarizing the STM tip at 2.75 eV and with no voltage sweeping. Figure 10.13(b) shows that when the nanotriangle is excited on the edge marked (i), it exhibits a plasmonic emission at 1.76 eV (705 nm) and a weaker emission at 1.87 eV (665 nm) when excited at its center marked (ii).

This new approach for exciting local plasmonic excitations is very promising since it offers the nanometric resolution that conventional optics (focusing with optical objectives) cannot reach. It will probably allow testing of various geometries for fabricating ultrasmall light sources, paving the way for plasmonic circuitry.

10.10 Quantum Plasmonics

10.10.1 *From classical to quantum plasmonics*

Quantum plasmonics deals with using quantum physics to describe the plasmon wave. Let's first state that most of the results in this book were established within classical electrodynamics, where light is described by electromagnetic waves and matter by the dielectric permittivity built up from the initial ideas of the Drude model. Two levels of quantization could be considered, which bring new insights into plasmonics: the semi-classical approach, where quantization is applied to the electron cloud in the metal while the electromagnetic field remains classical. This first approach was already mentioned earlier in this book when dealing with the computational approach to plasmonic nanostructures (see Chapter 8 and, for example, the DDA and FDTD methods). This first approach is necessary for accurate computing of LSPR resonance when nanostructures are smaller than 1000 electrons ($D < 4$ nm for a one-electron metal) or with non-spherical shapes. The second approach, which will be shortly described here, is the full quantization, where the electromagnetic field is also quantized. This is the so-called second quantization. Here, we briefly explain why quantum plasmonics is relevant in certain situations, such as closely packed small plasmonic nanoparticles.

10.10.2 *Plasmon and quasiparticles*

Light can be described to be quantized in terms of particles with a given energy and momentum. These particles are known as photons with $E = h\nu$ and $\boldsymbol{p} = \hbar\boldsymbol{k}$. Within quantum electrodynamics (QED), the assembly of photons is produced by the deexcitation of atoms. The process is described by a mathematical formalism where the photon population is controlled by mathematical operators known as *creation* and *annihilation* operators.

Quasiparticles are a way of simplifying complex situations with many-body interactions that make the behavior too complex to be solved. Just three particles exist in solids: electron, proton, and neutrons. But when considering the interaction of an electron inside a crystalline semiconductor, it becomes incredibly complex because that electron interacts with many other electrons and ionic cores. However, it is proven that its trajectory (energy, momentum) can be accurately modeled by a quasiparticle that is affected by its effective mass m_e^* and traveling in vacuum with no interaction with electrons and ions. This electron with effective mass m_e^* is a quasiparticle, and this is an effective way to simplify the many-body problem in quantum mechanics [14, 40]. It has no specific name, other than *electron*. Quasiparticles exist only inside a solid. Other quasiparticles are **holes**, **polarons**, and **excitons**. They are fermions (see Table 10.1).

On the other hand, there are also collective oscillations, such as **phonons** in an elastic crystal. A phonon is associated with the vibrations of atoms. It results from the quantization of sound waves in solids, and each phonon is described by its energy E linked to the oscillation eigenfrequencies ω_n by $E = \hbar\omega_n$ and by its wave vector (or momentum) $\boldsymbol{k}_n$. The link between E and $\boldsymbol{k}_n$ is provided by the relation of dispersion. A **magnon** is a collective excitation associated with the electrons' spin structure in a

Table 10.1. Some quasiparticles and collective excitations in the quantum description of matter and light.

Quasiparticles (fermions)		Collective excitations (bosons) in a crystal	
Electron quasiparticle	Modification of the electron behavior inside a solid	Photon	Light oscillation in free space
Hole	Movement of the electrons in the valence band of semiconductors	Phonon	Lattice mechanical oscillation
Exciton	Electron and hole bound together	Polariton	Light oscillation in materials, close to a resonance
Polaron	Movement of electron partially screened by ionic displacement	Plasmon	Concerted electron oscillations in a plasma
		Magnon	Oscillation of the spin of electrons

crystal lattice. It is a quantum of a spin wave. In that perspective, a **plasmon** is also a collective excitation associated with the eigenfrequencies of plasma oscillations, with this plasma being either a "real plasma" at high temperature or the electron clouds in solids, which is our focus presently (see Table 10.1).

Rigorously, the term *quasiparticle* is used for real particles whose properties are amended due to the collective effects inside a solid (holes, electrons, polarons, and excitons). They are also called the *dressed-particle* by Feynman. On the other hand, a *collective excitation* is an ensemble behavior (such as vibrations) of a system with no real particle at its origin. This distinction is not fully acknowledged though.

10.10.3 *Quantum optics and the second quantization*

The principle of the second quantization is a process that starts from classical results with the existence of a resonant phenomenon. In the case of a sphere, let's denote ω_p^ξ the unique dipolar plasmon mode oriented in the $\boldsymbol{\xi}$ direction (vector decomposed on $\boldsymbol{u}_x$, $\boldsymbol{u}_y$, $\boldsymbol{u}_z$ basis). The Hamiltonian can be written as [41]

$$\hat{H}_{p,0} = \sum_{\xi} \omega_p^\xi \hat{b}_\xi^\dagger \hat{b}_\xi. \tag{10.1}$$

In this relation, $\hat{b}_\xi$ is the annihilation operator that generates a supplemental plasmon, and $\hat{b}_\xi^\dagger$ is the creation operator that destroys a plasmonic quantum. The creation and annihilation operators are central to the second quantization theory and find its root in the full quantum formulation of the harmonic oscillator theory. If the creation operator on a state $|A_\xi\rangle$ that has already A_ξ plasmons in mode $\boldsymbol{\xi}$, a new quantum is created according to [41]:

$$\hat{b}_\xi^\dagger \left|A_\xi\right\rangle = \sqrt{A_\xi + 1} \left|A_\xi + 1\right\rangle, \tag{10.2}$$

$$\hat{b}_\xi \left|A_\xi\right\rangle = \sqrt{A_\xi} \left|A_\xi - 1\right\rangle. \tag{10.3}$$

A step-by-step explanation of the quantization procedure is provided by the review article of Tame [42]. This description becomes highly relevant when we want to study the coupling between several oscillators, such as two interacting plasmonic particles, or a fluorescent molecule and a plasmonic particle [42, 43].

10.10.4 *The charge transfer plasmon mode*

In this section, we illustrate the success of quantum plasmonics in describing a plasmonic dimer constituting two closely interacting nanoparticles. We reproduce the calculations setup by Aizpurua within the quantum corrected model (QCM) [44]. Considering two close spheres that are slowly brought closer at a distance d, as depicted in Fig. 10.14(a), electrons of the spheres start exchanging when $d < 0.5$ nm. A local classical description of the response of metallic nanoparticles does not properly account for the correlated motion of the conduction electrons and their spill out from the nanoparticle surfaces (see Fig. 10.14(b)). Therefore, the tunneling effect should be accounted for. This gives rise to the charge transfer plasmon (CTP). With the classical approach, the energy of the plasmon resonance starts decreasing (redshift) for dimer separation in the 2–5 nm range. In this range, all models coincide. However, the classical model exhibits a dramatic divergence near the contact point ($d = 0$) and a much too pronounced redshift. The QCM, however, also predicts a redshift but also provides a smooth transition between interacting dimer and the connected nanoparticles. This model accurately predicts the CTP that is active when $d < 0.3$ nm.

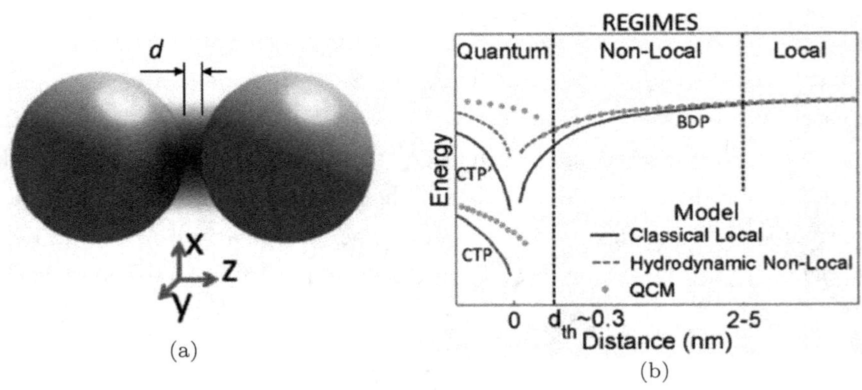

Fig. 10.14. (a) Case of two plasmonic oscillators in close interaction so that electrons can tunnel from one particle to the other; when the inter-distance is very small. (b) Comparison of the predicted energy for the plasmon resonance with three models: classical local, hydrodynamic non-local, and the quantum corrected model.

Source: Reproduced with permission from Ref. [44]. Copyright © 2015 The Royal Society of Chemistry.

10.11 Plasmon for Generating Colors

10.11.1 *Brief reminder about color perception by the human eye*

What is a color? It is far too naïve to restrict one color to just one wavelength in the electromagnetic spectrum. A color is a combination of several wavelengths with a set of relative weightings. The human eye perceives colors with the *cone cells* located in the retina, and normal eyes have three types of cones called S, M, and L (for short, medium, and long wavelengths), whose maximum sensitivities are respectively at 445, 545, and 565 nm. The signals from the S, M, and L cones are combined together in the retina and in the brain so that the human perception is described with the so-called "tristimulus model". The three sensitivity response curves are given in Fig. 10.15(a). If an object is illuminated with a light source with spectral energy $I(\lambda)$ (called the illuminant) and has a spectral reflective efficiency $R(\lambda)$, the tristimulus values X, Y, and Z are calculated the following way:

$$
\begin{aligned}
X &= \frac{1}{k}\int \bar{x}(\lambda)\cdot I(\lambda)\cdot R(\lambda)\,d\lambda,\\
Y &= \frac{1}{k}\int \bar{y}(\lambda)\cdot I(\lambda)\cdot R(\lambda)\,d\lambda,\\
Z &= \frac{1}{k}\int \bar{z}(\lambda)\cdot I(\lambda)\cdot R(\lambda)\,d\lambda.
\end{aligned}
\tag{10.4}
$$

In the case where the material is transparent, the transmitted color is calculated by replacing $R(\lambda)$ by the transmission coefficient $T(\lambda)$. This is known as the *Commission Internationale de l'Eclairage* (CIE) 1931 color space and is acknowledged as the standard in industrial colorimetry [45, 46]. In this system, Y corresponds to the luminance, and the chromaticity is described by x and y, which are obtained from the normalization of the X, Y, and Z values:

$$
x = \frac{X}{X+Y+Z},\quad y = \frac{Y}{X+Y+Z},\quad z = \frac{Z}{X+Y+Z} = 1 - x - y.
\tag{10.5}
$$

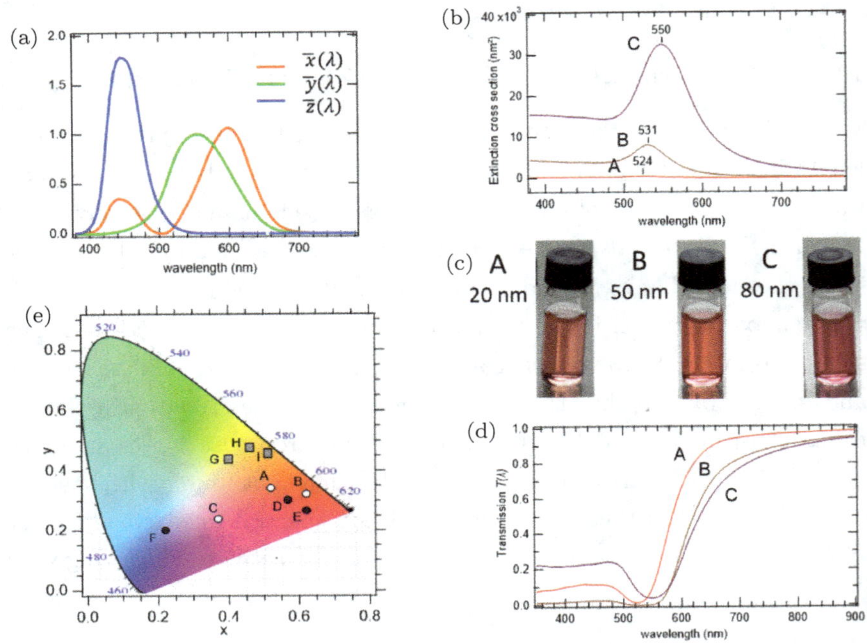

Fig. 10.15. (a) The CIE established in 1931 the CIE color space. The human eye sensitivity is described by the $\bar{x}\bar{y}$, and $\bar{z}$ standard observer color matching functions, as shown in the graph. These functions are linked to the sensitivity of the eye cones to the red, green, and blue lights. (b) Calculated extinction cross sections spectra for three suspensions of plasmonic nanoparticles with diameters of (A) 20 nm, (B) 50 nm, and (C) 80 nm in water. (c) Photographs of actual nanoparticle suspensions of the same size. (d) Calculated transmission spectra of the three type of nanoparticles with parameters indicated in Table 10.2. (e) The CIE chromaticity diagram with the transmitted colors corresponding to the various cases of spherical nanoparticles discussed here: gold in water (A, B, C), gold in glass (D, E, F), and silver in water (G, H, I). Colors are represented with the normalized tristimulus values x and y. See Tables 10.2 and 10.3 for details.

In this CIE XYZ color space, a color is described by the two parameters x and y whose values are between 0 and 1, and the luminance is given by the value of Y. The diagram in Fig. 10.15(e) represents the chromaticity diagram with all the colors that the human eye can see, called the *gamut* of the human eye. The point $x = 0.3$, $y = 0.3$ corresponds to white. The more a color strays away from this point, the more vivid it becomes. The values indicated on the edges of the visible area correspond to the colors of monochromatic lights of wavelengths from 440 nm (blue) to 650 nm (red).

Finally, a color is decomposed into its *hue* and its *saturation*. Saturated colors are located closer to the gamut edge. For example, a brick and a tomato are both red (*hue* is red), but usually, the red color of the tomato is more statured than the color of the brick. We will now apply these concepts to the plasmonic colors.

10.11.2 *Colors obtained by absorption*

The absorption spectra of various types of nanoparticles can be measured or calculated, and therefore, the resulting color can be predicted with relations (10.4) and (10.5). As shown in Chapter 7, for spherical nanoparticles, the transmission spectra are easily obtained (see Section 7.3). For example, three sizes of nanoparticles, 20, 50, and 80 nm, are considered, and the corresponding extinction cross sections are plotted in Fig. 10.15(b) (spectra A, B, and C, respectively) in the case where the nanoparticles are in water. Photographs of the nanoparticle suspensions are also presented in Fig. 10.15(c). From the cross sections, the transmitted intensity through a 100 μm film is calculated and shown in Fig. 10.15(d). The corresponding transmission coefficients $T(\lambda)$ are then used in relation (10.4) to calculate the normalized chromaticity coordinates, which are represented in the CIE diagram in Fig. 10.15(e). It shows that the transmitted colors of spherical gold nanoparticles are always in the red region of the visible spectrum.

Additional examples are also presented, where these nanoparticles are placed in glass instead of water. They are the points denoted by D, E, and F, and this pushes the colors further toward the blue region. Finally, using silver nanoparticles instead of gold, it sets the colors seen in transmission in the yellow region, as shown by the points denoted by G, H, and I. The detailed results are given in Tables 10.2 and 10.3. The colors depend only lightly on the nanoparticle diameters but more strongly on the optical index of the surrounding medium (water vs. glass in the current case). It also strongly depends on the nanoparticle concentration or, equivalently, the material thickness. These properties have opened up the approach to plasmonic colors where plasmonic particles are treated as pigments that are used for designing special inks.

Table 10.2. Sets of transmitted colors calculated for six different materials containing spherical gold nanoparticles. Three nanoparticle diameters are considered: 20, 50, and 80 nm. Nanoparticles are placed in either water (A, B, C) or in glass (D, E, F). The material thickness is 100 μm and the nanoparticle density is indicated in the second column. The resulting color is given with the normalized tristimulus values (x, y) and as colored squares.

Gold NPs		In water ($n = 1.33$)				In glass ($n = 1.51$)			
D (nm)	NP/μm^3		x	y			x	y	
20	100	A	0.52	0.34		D	0.57	0.30	
50	10	B	0.62	0.32		E	0.62	0.27	
80	1	C	0.37	0.24		F	0.22	0.20	

Table 10.3. Sets of transmitted colors calculated for six different materials containing spherical silver nanoparticles.

Silver NPs		In water ($n = 1.33$)				In glass ($n = 1.51$)			
D (nm)	NP/μm^3		x	y			x	y	
20	500	G	0.40	0.44		J	0.48	0.48	
50	10	H	0.46	0.48		K	0.54	0.45	
80	1	I	0.51	0.46		L	0.55	0.37	

10.11.3 *Example of an image obtained with plasmonic colors*

Plasmonic nanostructures and colloidal solutions have been considered as good candidates for coloring materials for several years and can be summarized under two approaches: (1) using lithographic process to create nanostructures on rigid supports (see review article in Ref. [47]) and (2) developing liquid pigments made of metallic nanoparticles and colloids, which are later included in a binder and printed with conventional printers (inkjet, offset, flexography, 3D printing, etc.). A nice example is the printing of the Lena image with plasmonic colors [48]. The image of the Swedish model Lena Forsén has been used since 1973 as a standard for checking image-processing algorithms. In this case, the image was coded into a $50 \times 50\ \mu m^2$ image of 200×200 plasmonic pixels. Each pixel was made of posts of lithography resin named HSQ, which are surmounted by silver (15 nm) plasmonic nanodisks covered with a thin gold protective layer (5 nm). The color of the pixel is controlled by the diameter of the nanodisks and their relative distance. A pixel is made of four interacting nanodisks, as shown in Fig. 10.16(a).

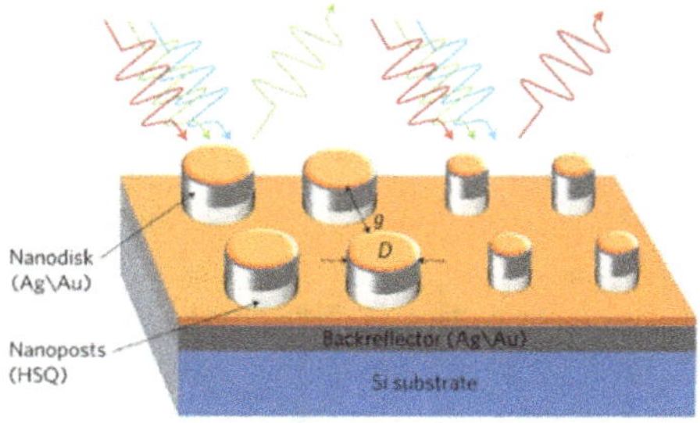

(a)

(b)

Fig. 10.16. (a) Plasmonic pixels made by an array of nanodisks of silver protected by a thin gold layer. Each colored pixel is made of four nanodisks that are placed on top of cylindrical posts made of the HSQ resin. The color is controlled by diameter D of the disks and their relative distance g. (b) An example of the famous image of Lena printed with plasmonic colors. Pixels are made of the nanodisk arrays prepared by lithography. The image is 200×200 pixels and $50 \times 50\ \mu m^2$ wide.

Source: Reproduced from Ref. [48].

10.11.4 *Colors obtained by scattering*

Plasmonic colors, however, allow unique properties since the color of a material can also be controlled by the scattering cross sections and not just by the extinction cross section as with any other pigment. In order to clarify this property, let's consider again the three types of spherical gold nanoparticles with diameters of 20, 50, and 80 nm dispersed in water. Their absorption, scattering, and extinction cross sections are readily calculated using the Mie theory (see Chapters 7 and 8) and are shown in Fig. 10.17. For 20 nm nanoparticles, the scattering cross section is completely negligible compared to the absorption cross section (110 times smaller). As a consequence, the color is fully controlled by the absorption spectra even for a non-transparent object. In this case, light bounces back from the object after having interacted within the very surface of the object where light was absorbed mostly at the peak absorption. This happens with molecular pigments and with small plasmonic nanoparticles. However, when the nanoparticle diameter grows, the scattering cross section grows more rapidly than the absorption cross section (see Section 7.2.4). For 80 nm gold nanoparticles, the absorption is 19,000 nm^2 whereas that of scattering is 14,800 nm^2. These quantities are now of the same order of magnitude so that the scattering phenomenon is not negligible anymore. Subsequently, the object will be seen with two colors: one color in transmission set by the extinction

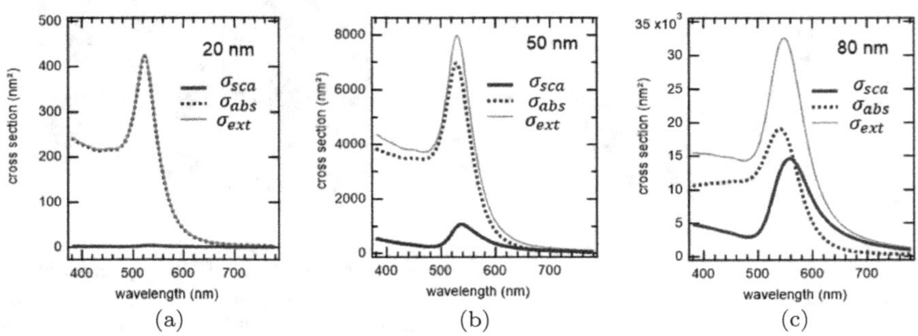

Fig. 10.17. Calculated cross sections of gold nanoparticles in water with diameters of (a) 20 nm, (b) 50 nm, and (c) 80 nm.

spectrum (which is more or less the complementary color of the peak extinction spectrum) and the other color by the diffused light. This effect is known as bichromatism or dichroism [49].

10.11.5 *Bichromatic effect*

The bichromatic effect is illustrated by the photographs in Fig. 10.18. A vial containing 80 nm gold spherical nanoparticles is illuminated from one side: The shadow shows the transmitted color, which is purple, whereas the diffused light is brown-orange. These nanoparticles were incorporated into epoxy, and they exhibit a clear bichromatic appearance: blue in transparency and orange when placed on a black fabric that ensures that only scattered light beams are perceived by the observer. This bichromatic effect is responsible for the surprising colors of the Lycurgus cup (see Chapter 7). This certainly offers a nice coloring possibility for objects but has not been used so far.

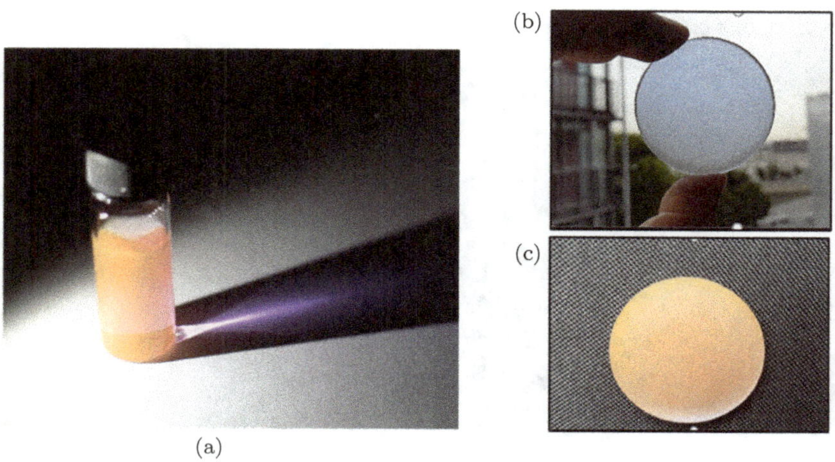

Fig. 10.18. (a) Vial containing 3 mL of gold spherical nanoparticles of 80 nm diameter. They exhibit a clear bichromatic effect, with purple color in transmission and orange in diffusion. (b) The same nanoparticles inserted in an epoxy polymer creating the bichromatic effect for a solid pellet observed in transmission (blue color) or (c) in reflection (orange color). Copyright © O. Pluchery.

10.12 Metamaterials

Metamaterials are artificial materials with exceptional and unique properties created by the tailored arrangement of individual nanostructures [50, 51]. These composite materials therefore have properties that are different from natural materials whose properties are primarily determined by their chemical constituents and bonds.

Generally, the electric permittivity ε and magnetic permeability μ can be used to classify materials. Figure 10.19 shows the four possible regions. The first one corresponds to $\varepsilon > 0$ and $\mu > 0$, where most of the dielectric materials can be found. The second quadrant covers metals, ferroelectric materials, and doped semiconductors that may exhibit negative permittivity values at certain frequencies (below the plasma frequency). Region IV is composed of some ferrite materials with negative permittivity. However, their magnetic responses fade rapidly over microwave frequencies. The area of greatest interest is quadrant III, in which both permittivity and permeability are simultaneously negative. Such materials do not exist in nature, so they are classified as *metamaterials*.

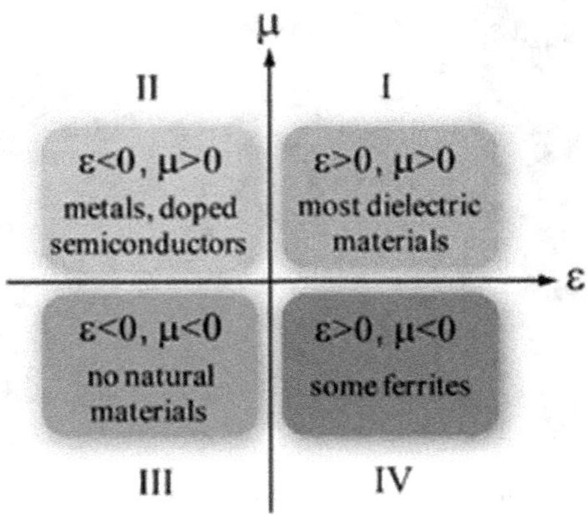

Fig. 10.19. The four areas describing through the electric permittivity (ε) and magnetic permeability (μ) the behavior of materials. The III area corresponds to the metamaterials.
Source: Reproduced with permission from Ref. [50]. Copyright © 2011 The Royal Society of Chemistry.

The fields of metamaterials and plasmonics have been progressing in parallel for a long time without meeting each other. Finally, by approaching the visible spectrum, where most of the plasmonic metal resonances are observed, the two fields have led together to an effervescence of interaction and applications unachievable before [52].

Many examples are available in the literature, including the use of negative refractive index, the development of sub-wavelength resolution imaging, invisibility coatings, data processing, hypersensitive biosensing, as well as nano-optic tunable devices and system generations.

Figure 10.20(a) corresponds to a plasmonic metasurface, i.e. a metamaterial structured in two dimensions [51]. Therefore, it is a particular case

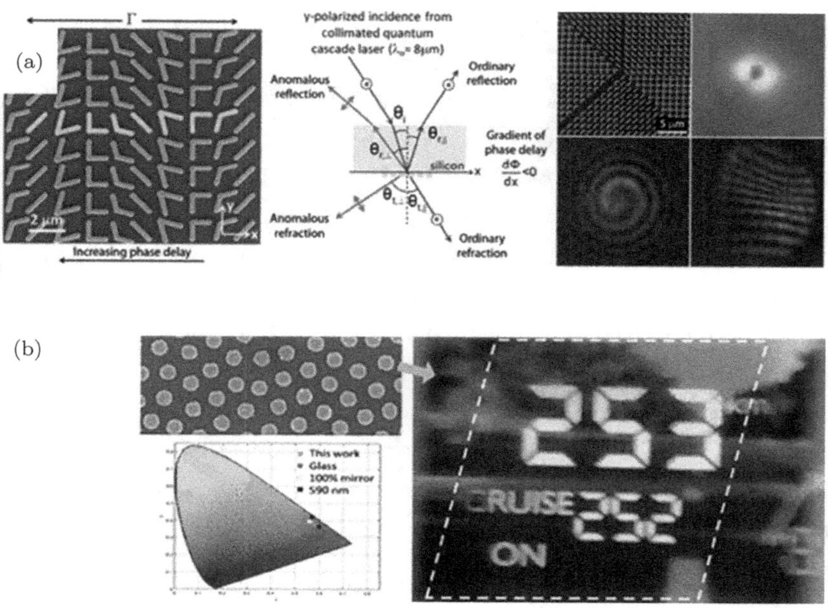

Fig. 10.20. Illustration of plasmonic metamaterials organized on transparent surface: (a) The dephasing caused by the plasmonic nanostructure induces an anomalous reflection and an anomalous refraction. (b) A set of disordered plasmonic nanostructures removes the usual diffraction patterns and increases the readability of the head-up display for automotives.

Source: (a) Reproduced with permission from Ref. [54]. Copyright © 2011 Science. (b) Reproduced with permission from Ref. [53]. Copyright © 2018 American Chemical Society.

where the phase of the device is modulated by nano-structuration. Such modulation is achieved from a plasmonic grating with sub-wavelength sizes and periods. Figure 10.20(b) shows another system with correlated disordered arrangements of gold nanostructures [53]. Such disposition offers a control on reflectance and transmittance, removes diffraction effects induced by a classic array, and can be used, for instance, for head-up display in cars.

10.13 Plasmonics Without Gold: Alternative Materials

10.13.1 *What makes a material plasmonic?*

The fundamental property of a plasmonic material is that its structure supports free electrons. Their concerted oscillation coupled with an electromagnetic wave is the essence of plasmonics, as much for SPP as for LSPR. Free electrons can be found in metals and in doped semiconductors. Therefore, are all metals plasmonics? And what about semiconductors? Why are the majority of plasmonics applications illustrated with gold and silver?

The optical behavior of materials is captured by the complex dielectric function $\varepsilon = \varepsilon' + i\varepsilon''$ (see Chapters 1 and 2). Here, we assume non-magnetic materials ($\mu = 1$). Materials are plasmonics if the real part of the dielectric permittivity is negative: $\varepsilon' < 0$ (see Section 4.2.4 in Chapter 4 for SPP and Chapter 7 for LSPR). The imaginary part ε'' of the dielectric permittivity reflects the losses in the metal. The higher the ε'', the stronger the losses. In the case of the SPP, the propagation length of the plasmon wave at the metal–dielectric interface is inversely proportional to ε'', as expressed by relation (4.29) in Chapter 4. Therefore, the conditions for finding which material is plasmonic at a given wavelength λ are

$$\varepsilon' < 0 \text{ and } \varepsilon' \text{ the smallest as possible.} \tag{10.6}$$

If we use the simple Drude model for the dielectric function of the metal, we know that

$$\varepsilon' = 1 - \frac{\omega_p^2}{\omega^2 + \gamma^2} \text{ and } \varepsilon'' = \frac{\omega_p^2 \gamma}{(\omega^2 + \gamma^2)\omega}.$$

Therefore, $\varepsilon' < 0$ only if $\omega < \omega_p$, which means that the corresponding metals have a chance to be plasmonic only for $\omega < \omega_p$ (see Table 2.1

for values of ω_p). Most of the metals have plasma frequencies in the UV, and this means that most of metals are plasmonics in the visible and IR range. Regarding losses, the ideal plasmonic metal would have $\varepsilon' = 0$. However, this is impossible to have simultaneously a negative permittivity and zero losses over the entire optical spectrum because of the causality principle [55]. For Drude metals, it is clear that $\varepsilon' \neq 0$ since $\gamma \neq 0$. Khurgin and Sun have explored theoretically some special cases where the band structure of a material would allow having zero losses for a small range of wavelengths. This would be the elusive lossless metal. However, they conclude that none of the existing metals and semiconductors exhibit this special band structure [56, 57].

Plasmon waves are surface waves (SPP) or surface oscillations (LSPR), and the properties of the dielectric permittivity discussed above should apply to the metal surface. This sets two other conditions for having a plasmonic material. First condition is its chemical stability. For example, copper is plasmonic, but its surface readily oxidizes in air, and the Cu_2O and CuO oxide layers are not conductive and kill the surface plasmon waves. Similarly, silver is not very stable either. Aluminum also gets covered with 2–3 nm thin Al_2O_3 oxide layer. The second aspect to be taken into account is that the dielectric permittivity of the surface of a metal differs from that of the bulk. Very often, the surface roughness induces higher values of γ and an increase in losses. These two conditions favor the chemically stable metals, unless working in vacuum environments.

10.13.2 *Comparison of the plasmonic metals*

From the discussion above, six metals are considered as good or acceptable plasmonic metals. First, there are the noble metals: gold (Au), silver (Ag), and copper (Cu). However, copper oxidizes in air and completely loses its plasmonic behavior, and this poor chemical stability makes it a tricky plasmonic to handle. Silver also oxidizes but more gently, and the plasmonic properties are partially maintained. Alkali metals have also been investigated and exhibit very low losses [58]. This is the case of sodium (Na) and potassium (K), but they are very reactive in ambient conditions and

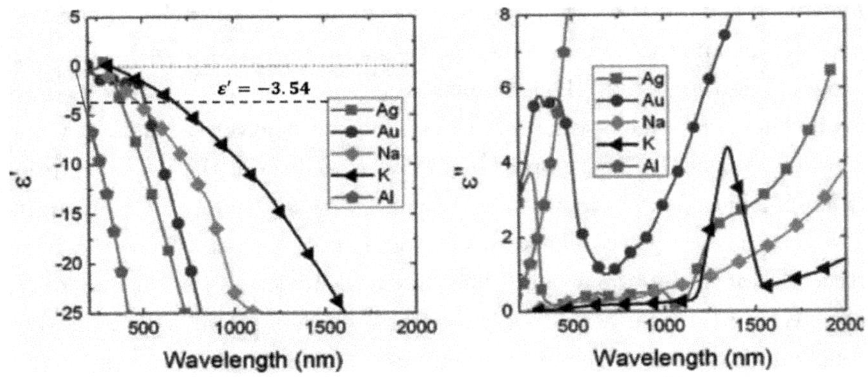

Fig. 10.21. The values of the complex dielectric function for five plasmonic materials: real part in (a) and imaginary part in (b). A good plasmonic metal should exhibit a strong negative value for ε' and small values for ε'' in order to reduce losses. In (a), the horizontal line at -3.54 corresponds to the Fröhlich resonance condition for exciting an LSPR in water.

Source: Reproduced with permission from Ref. [59].

therefore cannot be integrated into any device. Finally, aluminum (Al) is also a plasmonic metal in the UV. A practical way of comparing the plasmonic properties of metals is to plot the ε' and ε'' and seek the wavelength range where simultaneously ε' is strongly negative and ε'' is the smallest. Such a data compilation was carried out by Shalaev and Boltasseva in Ref. [59] and is reproduced in Fig. 10.21. An efficient database for retrieving the values of ε' and ε'' for most of the metals is available at https://refractiveindex.info/ [60].

Let's be more specific now. The efficiency of a plasmonic metal can be illustrated by computing the propagation length L_{SPP} of the surface plasmon polariton at the metal–air interface. The formula was derived in Chapter 4 and is

$$L_{SPP} = \frac{1}{2}\frac{\lambda_1}{2\pi}\sqrt{\frac{\varepsilon_1' + \varepsilon_2}{\varepsilon_1'\varepsilon_2}}\frac{2\varepsilon_1'(\varepsilon_1' + \varepsilon_2)}{\varepsilon''_1\varepsilon_2}. \tag{10.7}$$

The propagation length L_{SPP} is computed with the ε_1 values plotted in Fig. 10.21 for the metal and $\varepsilon_2 = 1$ (since air is taken as the dielectric). Here, we consider the ideal situation where the metals are not covered with any oxide layer and do not exhibit any surface roughness. Results are summarized in Table 10.4. Silver is clearly the most effective plasmonic metal that combines the highest efficiency with the lowest losses. L_{SPP} reaches almost 1 mm if the surface oxidation would not alter these performances. Gold however shows more modest performances with $L_{SPP} = 25$ μm, but in this case, these values were also experimentally obtained thanks to its chemical stability. Aluminum could be used in the UV region with a propagation length of just 2.2 μm. But this metal is cheap and currently used in microelectronics, which is a strong practical advantage.

The plasmonic quality can also be tested for the localized plasmon. The condition for an LSPR being excited by an optical wave in a nanosphere placed in a dielectric medium ε_{diel} is given by the Fröhlich condition (7.15) of Chapter 7, which states that the LSPR occurs at the wavelength where $\varepsilon' + 2\varepsilon_{diel} = 0$. For a nanosphere in water ($\varepsilon_{water} = 1.77$), the condition is simply $\varepsilon' = -3.54$. The resulting plasmon wavelengths are shown in Table 10.4 and compared to more accurate computations from Blaber and Cortie [58].

The five metals in Table 10.4 exhibit localized plasmon resonances. However, due to the losses represented by ε'', these resonances might be flattened and eventually hardly detected.

Table 10.4. Comparison of the plasmonic performances of five metals. The values of the complex dielectric function ($\varepsilon' + i\varepsilon''$) are given at the wavelength λ_1 taken from Ref. [60]. L_{SPP} is the propagation length in intensity of the SPP. In case of the LSPR, the plasmon resonance is calculated either using the Fröhlich condition or more accurately from the Mie calculation.

	SPP (metal–air interface)			LSPR in water	
Metal	λ_1	$\varepsilon' + i\varepsilon''$ @ λ_1	L_{SPP} @ λ_1	λ_{LSPR} (Fröhlich)	λ_{LSPR} Ref. [58]
Ag	1100 nm	$-62.3 + i0.71$	934 μm	383 nm	361 nm
Au	700 nm	$-16.5 + i1.06$	25 μm	515 nm	516 nm
Na	700 nm	$-8.75 + i0.29$	39 μm	470 nm	387 nm
K	700 nm	$-3.97 + i0.19$	6.0 μm	673 nm	555 nm
Al	250 nm	$-8.55 + i1.08$	2.2 μm	172 nm	143 nm

Source: Reproduced from Ref. [58].

10.13.3 *Plasmonics with semiconductor and transparent conductive oxides (TCO)*

A semiconductor behaves like metals when the carrier concentration is greater than 10^{21} cm^{-3}. For a metal like gold, $n = 5.9 \times 10^{22}$ cm^{-3}. Reaching $n > 10^{21}$ cm^{-3} requires a very heavy doping, and this is close to the maximum value in the case of silicon. Nevertheless, the possibility of adjusting the electron concentrations in the material by doping opens up an interesting method for controlling the plasmon frequency. This is definitely a strong motivation for investing efforts in semiconductor plasmonics. Moreover, the targeted plasmonic frequencies should be smaller than the semiconductor bandgap Eg. If $\hbar\omega > E_g$, then the photons induce interband transitions and heavy losses. From these considerations, we see that semiconductors are difficult candidates for supporting plasmon oscillations, although some publications have demonstrated SPP generation at the silicon–air interface [61] or InP–air interface [62]. A more in-depth discussion can be found in the review article by Milliron, where the authors also briefly discuss applications such as infrared detectors, smart windows, two-photon upconversion, enhanced luminescence, and infrared metasurfaces [63].

Transparent conductive oxides (TCOs) are a class of materials, such as zinc oxide (ZnO), cadmium oxide, and indium tin oxide (ITO), that can be heavily doped so that they become conductive. They are essential for the manufacture of displays, and the screens of the mobile phones are made with ITO.

10.13.4 *Plasmonics with graphene*

The Nobel Prize in physics was awarded in 2010 to Geim and Novoselov "for their groundbreaking experiments regarding the two-dimensional material graphene". Graphene is a two-dimensional lattice of carbon atoms that is characterized by its unique electronic properties: for example, as a conductor of electricity, it performs as well as copper. Moreover, graphene is a zero-gap semiconductor with a linear dispersion relation near the Fermi level. These linear dispersion curves cross at the Dirac point, which can be modulated by an external electrostatic bias. Therefore, the nature and

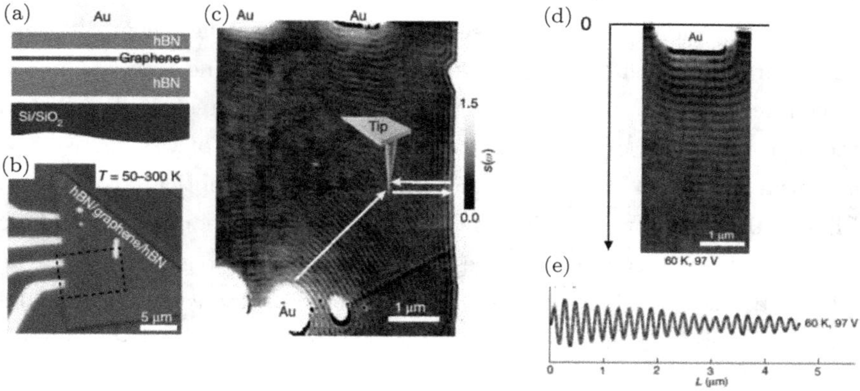

Fig. 10.22. (a) Sketch of a heterostructure that includes a graphene sheet that is able to propagate an SPP wave. (b) Optical microscopy image of the device with gold nanostructures that are used to launch the SPP wave. (c) Nanoscale infrared image obtained with an AFM tip that transforms the SPP wave into a far-field wave that is detected. Images are taken at $\lambda = 11.28$ μm. The ripples are plasmonic interferences recorded at low temperature (60 K) and $V_g = 97$ V. (d) Nanoscale infrared image of the normalized scattering amplitude of the plasmon wave. The gold microstructure on the top works as an antenna that emits graphene plasmons. (e) Intensity profile of the plasmonic fringes that shows that the plasmon wave propagate over ~5 μm.

Source: Reproduced with permission from Ref. [64]. Copyright © 2008 Springer Nature.

concentration of the charge carriers can be intentionally selected: conductivity can be achieved by electrons or holes.

These unique properties are appealing for application in plasmonics, which is based on the existence of a free electron population. Actually, plasmonic nanodevices based on graphene have been fabricated. We show one example of the detection of the SPP wave on a device made of a single graphene sheet embedded in two hexagonal boron nitride (hBN) layers [64]. This assembly is positioned on a silicon surface that plays the role of a back gate and serves to adjust the carrier density in the graphene sheet, as shown in Fig. 10.22(a). An SPP wave is launched in the graphene layer by shining an infrared laser at a wavelength of 11.28 μm. The gold electrodes that can be seen in Fig. 10.22(b) play the role of a nano-antenna and transform this far-field wave into a near-field one. The group of Basov has also developed an AFM coupled with a very sensitive photonic detector. The tip of the AFM is also a nano-antenna that plays two roles: First, it generates a punctual plasmonic source, and second, it transforms the plasmonic wave into a far-field signal that can be detected. This allows the detection of the wave intensity with an accuracy of 10 nm. This lateral resolution relies on the high resolution of the AFM and allows a mapping of the intensity at a spatial resolution that is 10 times smaller than the optical wavelength. Figure 10.22(c) shows the ripples due to standing plasmonic waves. These waves result from the interferences of the plasmon wave generated by the AFM tip (sketched in Fig. 10.22(c)) and the wave launched by the gold microstructure. In the right part of the images, the fringes are produced by the reflection of the plasmonic wave from the tip of the graphene edges. The authors have measured the SPP wavelength at $\lambda_{SPP} = 170\,\text{nm}$. They have also studied the effect of the back-gate voltage. The higher the voltage, the more abundant the electrons in the graphene sheet. They have demonstrated a decrease in the plasmonic losses when the carrier density is increased (high back-gate voltage). In Fig. 10.22(d), the plasmonic interference fringes are visible and propagates over 5 μm as shown in Fig. 10.22(e) with $V_g = 97$ V. This distance is zero when V_g is smaller than 50 V.

This experimental demonstration of graphene plasmon waves shows that it is possible to launch an SPP wave in complex architectures and manipulate light well below the wavelength. The interesting result is their ability to turn on and off the plasmonic propagation by adjusting the carrier concentration.

10.14 Conclusion

Plasmonics has been experiencing strong interest for 20 years and is still a field with intense patenting activity. Since 2015, 500 patents per year have been filed in Europe. During the period 2011–2016, this represented 3,200 patents. The industrial promises of plasmonics lie in the domain of healthcare, diagnostics, telecommunications, and high-speed data processing. Some applications have found their place on the market and are well established, such as medical diagnostics, where the SPR sensors have generated revenues of USD 628 million in 2015 [65]. This strong dynamic is an echo of the fascination generated by a field that lies at the frontiers between optics, nanoscience, and condensed matter physics and is also able to produce visible and intriguing phenomena that could inspire eager artists to introduce new coloring effects in their artworks. Maybe in 10 years, we will be able buy plasmonic colors, use a mobile phone whose display is made of plasmonic pixels, and our high-speed internet would rely on optical fibers controlled by plasmonic switches.

References

[1] Banville F. A., Moreau J., Sarkar M., Besbes M., Canva M. and Charette P. G. 2018. Spatial resolution versus contrast trade-off enhancement in high-resolution surface plasmon resonance imaging (SPRI) by metal surface nanostructure design. *Optics Express* **26**, 10616–10630.

[2] Novotny L. and Hecht B. 2012. *Principles of Nano-Optics (2nd Ed.)* (Cambridge: Cambridge University Press).

[3] Bryche J.-F., Gillibert R., Barbillon G., Gogol P., Moreau J., Lamy de la Chapelle M., Bartenlian B. and Canva M. 2016. Plasmonic enhancement by a continuous gold underlayer: Application to SERS sensing. *Plasmonics* **11**, 601–608.

[4] Giebel K. F., Bechinger C., Herminghaus S., Riedel M., Leiderer P., Weiland U. and Bastmeyer M. 1999. Imaging of cell/substrate contacts of living cells with surface plasmon resonance microscopy. *Biophysical Journal* **76**, 509–516.

[5] Kim D. J. and Kim D. 2010. Subwavelength grating-based nanoplasmonic modulation for surface plasmon resonance imaging with enhanced resolution. *Journal of the Optical Society of America B* **27**, 1252–1259.

[6] Anselmo A. C. and Mitragotri S. 2019. Nanoparticles in the clinic: An update. *Bioengineering & Translational Medicine* **4**, e10143.

[7] The National Library of Medicine 2020. Clinical Trials of privately and publicly funded clinical studies conducted around the world. https://clinicaltrials.gov/.

[8] Rastinehad A. R., Anastos H., Wajswol E., Winoker J. S., Sfakianos J. P., Doppalapudi S. K., Carrick M. R., Knauer C. J., Taouli B., Lewis S. C., Tewari A. K., Schwartz J. A., Canfield S. E., George A. K., West J. L. and Halas N. J. 2019. Gold nanoshell-localized photothermal ablation of prostate tumors in a clinical pilot device study. *Proceedings of the National Academy of Sciences* **116**, 18590–18596.

[9] MacLellan C. J., Fuentes D., Elliott A. M., Schwartz J., Hazle J. D. and Stafford R. J. 2014. Estimating nanoparticle optical absorption with magnetic resonance temperature imaging and bioheat transfer simulation. *International Journal of Hyperthermia* **30**, 47–55.

[10] Mukherjee S., Libisch F., Large N., Neumann O., Brown L. V., Cheng J., Lassiter J. B., Carter E. A., Nordlander P. and Halas N. J. 2013. Hot electrons do the impossible: Plasmon-induced dissociation of H-2 on Au. *Nano Letters* **13**, 240–247.

[11] Clavero C. 2014. Plasmon-induced hot-electron generation at nanoparticle/metal-oxide interfaces for photovoltaic and photocatalytic devices. *Nature Photonics* **8**, 95–103.

[12] Brongersma M. L., Halas N. J. and Nordlander P. 2015. Plasmon-induced hot carrier science and technology. *Nature Nanotechnology* **10**, 25.

[13] Weatherby S., Seddon J., Vallance, C. 2019. Hot-electron science and microscopic processes in plasmonics and catalysis. *Faraday Discussions* **214**, 517–519.

[14] Ashcroft N. and Mermin D. 1976. *Solid State Physics* (Brooks Cole: Pacific Grove, CA).

[15] Landau L. D. 1946. On the vibrations of the electronic plasma. *Journal of Physics* **10**, 25–34.

[16] Mouhot C. and Villani C. 2011. On Landau damping. *Acta Mathematica* **207**, 29–201.

[17] Baffou G. 2017. *Thermoplasmonics: Heating Metal Nanoparticles Using Light* (Cambridge: Cambridge University Press).

[18] Baffou G., Quidant R. and Girard C. 2009. Heat generation in plasmonic nanostructures: Influence of morphology. *Applied Physics Letters* **94**, 153109.

[19] Baffou G., Girard C. and Quidant R. 2010. Mapping heat origin in plasmonic structures. *Physical Review Letters* **104**, 136805.

[20] Baffou G. and Quidant R. 2014. Nanoplasmonics for chemistry. *Chemical Society Reviews* **43**, 3898–3907.

[21] Lozan O., Sundararaman R., Ea-Kim B., Rampnoux J.-M., Narang P., Dilhaire S. and Lalanne P. 2017. Increased rise time of electron temperature during adiabatic plasmon focusing. *Nature Communications* **8**, 1656.

[22] Anisimov S. I., Kapeliovich B. L. and Perelman T. L. 1974. Electron emission from metal surfaces exposed to ultrashort laser pulses. *Journal of Experimental and Theoretical Physics* **39**(2), 375–377.

[23] Sun C. K., Vallée F., Acioli L., Ippen E. P. and Fujimoto J. G. 1993. Femtosecond investigation of electron thermalization in gold. *Physical Review B* **48**, 12365–12368.

[24] Schirato A., Maiuri M., Toma A., Fugattini S., Proietti Zaccaria R., Laporta P., Nordlander P., Cerullo G., Alabastri A. and Della Valle G. 2020. Transient optical symmetry breaking for ultrafast broadband dichroism in plasmonic metasurfaces. *Nature Photonics* **14**, 723–727.

[25] Bresson P., Bryche J. F., Besbes M., Moreau J., Karsenti P. L., Charette P. G., Morris D. and Canva M. 2020. Improved two-temperature modeling of ultrafast thermal and optical phenomena in continuous and nanostructured metal films. *Physical Review B* **102**, 155127.

[26] Sun C. K., Vallée F., Acioli L. H., Ippen E. P. and Fujimoto J. G. 1994. Femtosecond-tunable measurement of electron thermalization in gold. *Physical Review B* **50**, 15337–15348.

[27] Halas N. J. 2019. Spiers Memorial Lecture Introductory lecture: Hot-electron science and microscopic processes in plasmonics and catalysis. *Faraday Discussions* **214**, 13–33.

[28] Atkins P. W. and De Paula J. 2006. *Atkins' Physical Chemistry* (Oxford; New York: Oxford University Press).

[29] Baffou G., Bordacchini I., Baldi A. and Quidant R. 2020. Simple experimental procedures to distinguish photothermal from hot-carrier processes in plasmonics. *Light-Science & Applications* **9**, 108–123.

[30] Chen Y.-C., Huang Y.-S., Huang H., Su P.-J., Perng T.-P. and Chen L.-J. 2020. Photocatalytic enhancement of hydrogen production in water splitting under simulated solar light by band gap engineering and localized surface plasmon resonance of ZnxCd1-xS nanowires decorated by Au nanoparticles. *Nano Energy* **67**, 104225.

[31] Pillai S. and Green M. A. 2010. Plasmonics for photovoltaic applications. *Solar Energy Materials and Solar Cells* **94**, 1481–1486.

[32] Atwater H. A. and Polman A. 2010. Plasmonics for improved photovoltaic devices. *Nature Materials* **9**, 205–213.

[33] Kherbouche I., Luo Y., Félidj N. and Mangeney C. 2020. Plasmon-mediated surface functionalization: New horizons for the control of surface chemistry on the nanoscale. *Chemistry of Materials* **32**, 5442–5454.

[34] Ge D., Marguet S., Issa A., Jradi S., Nguyen T. H., Nahra M., Béal J., Deturche R., Chen H., Blaize S., Plain J., Fiorini C., Douillard L., Soppera O., Dinh X. Q., Dang C., Yang X., Xu T., Wei B., Sun X. W., Couteau C. and Bachelot R. 2020. Hybrid plasmonic nano-emitters with controlled single quantum emitter positioning on the local excitation field. *Nature Communications* **11**, 3414.

[35] Zhou X., Wenger J., Viscomi F. N., Le Cunff L., Béal J., Kochtcheev S., Yang X., Wiederrecht G. P., Colas des Francs G., Bisht A. S., Jradi S., Caputo R., Demir H. V., Schaller R. D., Plain J., Vial A., Sun X. W. and Bachelot R. 2015. Two-color single hybrid plasmonic nanoemitters with real time switchable dominant emission wavelength. *Nano Letters* **15**, 7458–7466.

[36] Qiu X. H., Nazin G. V. and Ho W. 2003. Vibrationally resolved fluorescence excited with submolecular precision. *Science* **299**, 542–546.

[37] Kuhnke K., Große C., Merino P. and Kern K. 2017. Atomic-scale imaging and spectroscopy of electroluminescence at molecular interfaces. *Chemical Reviews* **117**, 5174–5222.

[38] Le Moal E., Dujardin G. and Boer-Duchemin E. *Gold Nanoparticles for Physics, Chemistry and Biology*, pp. 365–391.

[39] Le Moal E., Marguet S., Rogez B., Mukherjee S., Dos Santos P., Boer-Duchemin E., Comtet G. and Dujardin G. 2013. An electrically excited nanoscale light source with active angular control of the emitted light. *Nano Letters* **13**, 4198–4205.

[40] Kittel C. 2004. *Introduction to Solid State Physics* (New York: John Wiley & Sons, Inc.).

[41] Nascimento D. R. and III A. E. D. 2015. Modeling molecule-plasmon interactions using quantized radiation fields within time-dependent electronic structure theory. *The Journal of Chemical Physics* **143**, 214104.

[42] Tame M. S., McEnery K. R., Özdemir Ş. K., Lee J., Maier S. A. and Kim M. S. 2013. Quantum plasmonics. *Nature Physics* **9**, 329–340.

[43] Lee C. and Tame M. 2017. *Gold Nanoparticles for Physics, Chemistry and Biology* (London: World Scientific Publishing), pp. 131–164.

[44] Esteban R., Zugarramurdi A., Zhang P., Nordlander P., García-Vidal F. J., Borisov A. G. and Aizpurua J. 2015. A classical treatment of optical tunneling in plasmonic gaps: Extending the quantum corrected model to practical situations. *Faraday Discussions* **178**, 151–183.

[45] Hunt R. W. G. and Pointer M. R. 2011. *Measuring Colour, 4th Edition* (Hoboken, New Jersey: Wiley).

[46] Sève R. 2009. *Science de la couleur* (Marseille: Chalagam Edition).

[47] Song M., Wang D., Peana S., Choudhury S., Nyga P., Kudyshev Z. A., Yu H., Boltasseva A., Shalaev V. M. and Kildishev A. V. 2019. Colors with plasmonic nanostructures: A full-spectrum review. *Applied Physics Reviews* **6**, 041308.

[48] Kumar K., Duan H., Hegde R. S., Koh S. C. W., Wei J. N. and Yang J. K. W. 2012. Printing colour at the optical diffraction limit. *Nature Nanotechnology* **7**, 557–561.

[49] Pluchery O., Schaming D. and Remita H. 2017. Patent: Packaging with two-color visual effect for decoration or identification. ed INPI (Europe: WO2017013373A1).

[50] Liu Y. and Zhang X. 2011. Metamaterials: A new frontier of science and technology. *Chemical Society Reviews* **40**, 2494–2507.

[51] Yao K. and Liu Y. 2014. Plasmonic metamaterials. *Nanotechnology Reviews* **3**, 177–210.

[52] Sikdar D. and Zhu W. 2019. Editorial: A special issue on plasmonic metamaterials. *Journal of Physics: Condensed Matter* **31**, 310401.

[53] Bertin H., Brûlé Y., Magno G., Lopez T., Gogol P., Pradere L., Gralak B., Barat D., Demésy G. and Dagens B. 2018. Correlated disordered plasmonic nanostructures arrays for augmented reality. *ACS Photonics* **5**, 2661–2668.

[54] Yu N., Genevet P., Kats M. A., Aieta F., Tetienne J.-P., Capasso F. and Gaburro Z. 2011. Light propagation with phase discontinuities: Generalized laws of reflection and refraction. *Science* **334**, 333–337.

[55] Dressel M. and Grüner G. 2002. *Electrodynamics of Solids: Optical Properties of Electrons in Matter.*

[56] Khurgin J. B. and Sun G. 2010. In search of the elusive lossless metal. *Applied Physics Letters* **96**, 181102.

[57] Maradudin A., Sambles J. R. and Barnes W. L. 2014. *Modern Plasmonics* (Amsterdam: Elsevier Science).

[58] Blaber M. G., Arnold M. D., Harris N., Ford M. J. and Cortie M. B. 2007. Plasmon absorption in nanospheres: A comparison of sodium, potassium, aluminium, silver and gold. *Physica B: Condensed Matter* **394**, 184–187.

[59] West P. R., Ishii S., Naik G. V., Emani N. K., Shalaev V. M. and Boltasseva A. 2010. Searching for better plasmonic materials. *Laser & Photonics Reviews* **4**, 795–808.

[60] Polyanskiy M. N. 2008. Refractive index database.
[61] Ginn J. C., Jr. R. L. J., Shaner E. A. and Davids P. S. 2011. Infrared plasmons on heavily-doped silicon. *Journal of Applied Physics* **110**, 043110.
[62] Panah M. E. A., Takayama O., Morozov S. V., Kudryavtsev K. E., Semenova E. S. and Lavrinenko A. V. 2016. Highly doped InP as a low loss plasmonic material for mid-IR region. *Optics Express* **24**, 29077–29088.
[63] Agrawal A., Cho S. H., Zandi O., Ghosh S., Johns R. W. and Milliron D. J. 2018. Localized surface plasmon resonance in semiconductor nanocrystals. *Chemical Reviews* **118**, 3121–3207.
[64] Ni G. X., McLeod A. S., Sun Z., Wang L., Xiong L., Post K. W., Sunku S. S., Jiang B. Y., Hone J., Dean C. R., Fogler M. M. and Basov D. N. 2018. Fundamental limits to graphene plasmonics. *Nature* **557**, 530–533.
[65] Viste P. 2018. Brevets en plasmonique: les tendances depuis 20 ans. *Les Photoniques* **90**, 19–20.

Index

T

V

W

Y

www.ingramcontent.com/pod-product-compliance
Lightning Source LLC
LaVergne TN
LVHW052032200926
848481LV00005B/1651

* 9 7 8 1 8 0 0 6 1 9 7 8 4 *